Wastewater Purification

Aerobic Granulation
in
Sequencing
Batch Reactors

Wastewater Purification

Aerobic Granulation in Sequencing Batch Reactors

Edited by
Yu Liu

CRC Press
Taylor & Francis Group
Boca Raton London New York

CRC Press is an imprint of the
Taylor & Francis Group, an **informa** business

CRC Press
Taylor & Francis Group
6000 Broken Sound Parkway NW, Suite 300
Boca Raton, FL 33487-2742

© 2008 by Taylor & Francis Group, LLC
CRC Press is an imprint of Taylor & Francis Group, an Informa business

First issued in paperback 2019

No claim to original U.S. Government works

ISBN-13: 978-0-367-45283-4 (pbk)
ISBN-13: 978-1-4200-5367-8 (hbk)

Visit the Taylor & Francis Web site at
http://www.taylorandfrancis.com

and the CRC Press Web site at
http://www.crcpress.com

Library of Congress Cataloging-in-Publication Data

Wastewater purification : aerobic granulation in sequencing batch reactors / editor, Yu Liu.
    p. cm.
Includes bibliographical references and index.
ISBN 978-1-4200-5367-8 (hardback : alk. paper)
1. Sewage--Purification--Microbial granulation process. 2. Sewage--Purification--Sequencing batch reactor process. I. Liu, Yu.

TD765.65.W37 2006
628.3'2--dc22
                                        2007030300

Contents

Preface

Biogranulation is a process of microbial self-immobilization, and it can be divided into two general groups, that is, anaerobic and aerobic granulation. Anaerobic granulation has been studied extensively for decades, whereas the interest in aerobic granulation was started just a few years ago. Aerobic granulation is an environmental biotechnology developed for the purpose of high-efficiency wastewater treatment. The distinguishing characteristics of aerobic granules attribute superiority to this technology in comparison with the conventional activated sludge processes. Thus far, intensive research has been conducted to understand the mechanism of aerobic granulation in sequencing batch reactor (SBR) and its application in treating a wide variety of municipal and industrial wastewater. Obviously, the basic research of aerobic granulation has promoted this technology from laboratory study to the present pilot- and full-scale applications. This book aims to discuss the up-to-date research and application of this environmental biotechnology tailored for enhanced wastewater purification.

First, chapter 1 presents experimental evidence showing that aerobic granulation in SBR is indeed insensitive to the substrate type and its concentration applied, although the carbon source seems to influence the physical properties and microbial diversity of mature aerobic granules. It appears from this chapter that aerobic granulation technology is applicable to the purification of a wide spectrum of wastewater. Hydrodynamic shear force resulting from intensive aeration in SBR plays an essential role in aerobic granulation. Chapter 2 elaborates on how hydrodynamic shear force would influence aerobic granulation, with special focus on shear force-associated changes in microbial activity, cell surface property, and production of extracellular polysaccharides. Hitherto, almost all successful aerobic granulations are achieved in SBR that is featured by its cyclic operation. Chapter 3 further looks into the role of SBR cycle time in aerobic granulation.

Chapter 4 focuses on understanding the role of settling time in aerobic granulation, which is a unique operating parameter of SBR as compared to conventional activated sludge reactors. Settling time is shown as an essential driving force of aerobic granulation. Aerobic granulation would fail if settling time is not properly controlled. Aerobic granulation seems to be an effective defensive or protective strategy of the microbial community against external selection pressure. Chapter 5 identifies the volume exchange ratio and discharge time of SBR as two other possible driving forces of aerobic granulation in SBR. Further, chapter 6 shows that all the major selection pressures identified so far can be unified to an easy concept of the minimal settling velocity that ultimately determines aerobic granulation in SBR. This selection pressure theory offers useful guides for up-scaling, manipulating, and optimizing aerobic granular sludge SBR.

Aerobic granulation is a gradual process that can be quantitatively described as change in granule size in the course of SBR operation. In this regard, some kinetic models have been developed and presented in chapter 7. Because of the large size of the aerobic granule, mass diffusion limitations exist in the aerobic granule. Chapter 8 looks into the diffusion behaviors of substrate and dissolved oxygen in aerobic granules and presents a comprehensive modeling system, which describes the dynamic diffusion of substrate and oxygen in various-sized aerobic granules. This model system can provide an effective and useful tool for predicting and optimizing the performance of aerobic granular sludge SBR.

It is believed that cell-to-cell self-aggregation initiates aerobic granulation. Cell surface hydrophobicity serves as an essential affinity force that initiates the first contact of cell to cell. Existing evidence shows that a number of culture conditions can induce cell surface hydrophobicity. Chapter 9 discusses the factors known to influence cell surface hydrophobicity. Furthermore, a thermodynamic interpretation of the role of cell surface hydrophobicity in aerobic granulation is also given. The enrichment culture of highly hydrophobic bacteria thus appears to greatly facilitate aerobic granulation. Chapter 10 discusses the essential roles of extracellular polysaccharides in the formation and maintenance of structural stability of aerobic granules. It appears that both the quantity and the quality of extracellular polysaccharides determine the matrix structure and integrity of aerobic granules.

Chapter 11 reveals that the internal structure of the aerobic granule experiences a shift from homogenous to heterogeneous as the aerobic granule grows to a big size due to mass diffusion limitation. Uneven distributions of granule biomass, extracellular polysaccharides, and cell surface hydrophobicity are also discussed in chapter 11. Chapter 12 mainly focuses on biodegradability of extracellular polysaccharides produced by aerobic granules. Only nonbiodegradable extracellular polysaccharides can play a crucial protective role in the granule integrity stability, while biodegradable extracellular polysaccharides accumulated at the central part of the aerobic granule can serve as an additional energy reservoir when an external carbon source is no longer available for microbial growth. Chapter 13 provides a plausible explanation for the observed high calcium accumulation in acetate-fed aerobic granules from both experimental and theoretical aspects. It is shown that the calcium ion may not be an essential element required for successful aerobic granulation.

Unlike the continuous activated sludge process, a substrate periodic starvation exists in aerobic granular sludge SBR due to its cyclic operation. Chapter 14 discusses different, even controversial, views with regard to the role of such a periodic starvation in aerobic granulation. As filamentous growth has been frequently observed in aerobic granules, chapter 15 looks into causes and control of filamentous growth in aerobic granular sludge SBRs. In view of its industrial application, long-term stability of aerobic granular sludge SBR remains a main concern. For this purpose, chapter 16 sheds light on the possible operation strategy that can help improve the stability of aerobic granules, including the selection of slow-growing bacteria and control of granule age. After nearly ten years of laboratory research, aerobic granulation technology has achieved pilot- and full-scale applications. Chapter 17 shows that successful aerobic granulation can be achieved in pilot-scale SBR using fresh or stored aerobic granules as seeds.

This book presents readers all aspects of aerobic granulation in SBR. The successful test of this technology in pilot-scale study foresees its promising application in practical wastewater treatment. I sincerely hope that the publication of this book will provide a platform for the further development of this technology and promote its quick application in the wastewater treatment industry.

Yu Liu

Contributors

Yu Liu, Ph.D.
School of Civil and
  Environmental Engineering
Nanyang Technological University
Singapore

Zhi-Wu Wang, Ph.D.
School of Civil and
  Environmental Engineering
Nanyang Technological University
Singapore

Qi-Shan Liu, Ph.D.
Singapore Polytechnic
Singapore

Lei Qin, Ph.D.
School of Chemical and
  Environmental Engineering
Shanghai University
People's Republic of China

Yong Li, M.Eng.
School of Civil and
  Environmental Engineering
Nanyang Technological University
Singapore

1 Aerobic Granulation at Different Carbon Sources and Concentrations

Qi-Shan Liu and Yu Liu

CONTENTS

1.1 INTRODUCTION

Granulation is a process in which microorganisms aggregate to form a spherical, dense biomass. Granules have been grown successfully in either anaerobic or aerobic environments (Lettinga et al. 1984; Morgenroth et al. 1997; Beun et al. 1999; J. H. Tay, Liu, and Liu 2001; Su and Yu 2005). The characteristics of the substrate have been considered to influence the formation and structure of anaerobic granules (Wu 1991; Chen and Lun 1993). Filamentous anaerobic granules developed on volatile fatty acids (VFAs) tend to be mechanically fragile and larger in size, whereas more robust, rod-type anaerobic granules were grown on sugar beet or potato processing wastewater (Adebowale and Kiff 1988). However, the formation of aerobic granules seems to be independent of the characteristics of the organic substrate (J. H. Tay, Liu, and Liu 2001).

Another important parameter that affects the anaerobic granulation process and the characteristics of anaerobic granules is the substrate concentration (Hulshoff Pol, Heijnekamp, and Lettinga 1988; Campos and Anderson 1992). An appropriate substrate concentration is critical to the microbial granulation in anaerobic systems. Morvai et al. (1990) found that anaerobic granulation developed well in upflow anaerobic sludge blanket (UASB) reactors fed with influent chemical oxygen demand (COD) concentrations of 1000 to 3000 mg L^{-1}, but not in a reactor with influent concentration of 500 mg L^{-1}. The substrate concentration also has direct impact on the biofilm structure where high surface loading rate leads to the increase of the average biofilm thickness (van Loosdrecht et al. 1995; Tijhuis et al. 1996; Kwok et al. 1998). This chapter discusses the effect of substrate carbon source and its concentration on the formation and characteristics of aerobic granules.

1.2   AEROBIC GRANULATION WITH ACETATE AND GLUCOSE

1.2.1   Microscopic Observation of Aerobic Granulation

J. H. Tay, Liu, and Liu (2001) investigated the evolution process of aerobic granulation in two sequencing batch reactors (SBRs) that were fed with glucose and acetate, respectively, and monitored by means of optical microscope, image analysis (IA) technique, and scanning electronic microscope (SEM), and found that aerobic granulation is a gradual process from seed sludge to aggregates and finally to compact mature granules.

1.2.1.1   Seed Sludge

Microscopic examination of seed sludge taken from a sewage treatment plant showed a typical morphology of conventional activated sludge, in which filaments were observed (figure 1.1). A SEM micrograph further revealed that seed sludge had a very loose and irregular three-dimensional structure (figure 1.1C). The average floc size of the seed sludge was about 70 μm, with a sludge volume index (SVI) value of 280 mL g^{-1}, which suggests filamentous bacteria were predominant in the seed sludge due to its high SVI value (Crites and Tchobanoglous 1998).

1.2.1.2   Formation of Compact Aggregates after Operation for One Week

One week after the reactor startup, filamentous bacteria gradually disappeared in the acetate-fed SBR, but still prevailed in the glucose-fed SBR. Figure 1.2A shows the

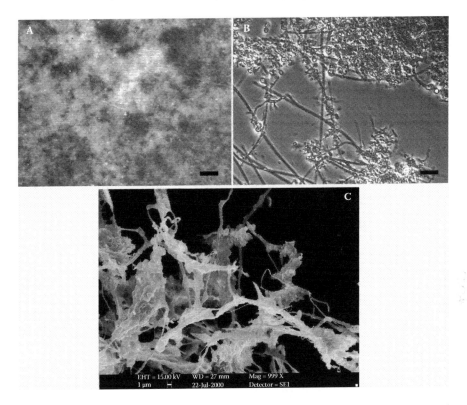

FIGURE 1.1 Morphology of seed sludge used for cultivation of aerobic granules. (A) Viewed by image analysis (scale bar: 2 mm); (B) viewed by optical microscope (scale bar: 5 μm); (C) viewed by SEM. (From Liu, Q. S. 2003. Ph.D. thesis, Nanyang Technological University, Singapore. With permission.)

morphologies of 1-week-old sludge in the glucose-fed SBR observed by imagine analysis. The compact and dense sludge aggregates can be seen and at this stage, the sludge aggregates exhibited much more compact and denser structure than the seed sludge.

1.2.1.3 Formation of Granular Sludge after Operation for Two Weeks

Figure 1.2B shows the sludge morphology after operation for 2 weeks. It is clear that granular sludge with a clear round outer shape was formed. Filamentous bacteria were still predominant in the reactor fed with glucose, while filaments completely disappeared in the reactor fed with acetate after operation for 2 weeks. It is known that a high-carbohydrate substrate composed of glucose or maltose supports the growth of filamentous bacteria (Chudoba 1985). This might be the reason for the filaments-dominant situation in the glucose-fed sludge. As can be seen in figures 1.2A and 1.2B, the major differences between microbial aggregates and granular sludge can be attributed to their sizes, compactness, and outer shapes. It should be realized that the evolution of sludge in both the glucose- and acetate-fed SBRs indeed followed a similar evolution pattern in the course of operation. These indicate that the carbon source has an insignificant influence on the formation of aerobic granules in SBR.

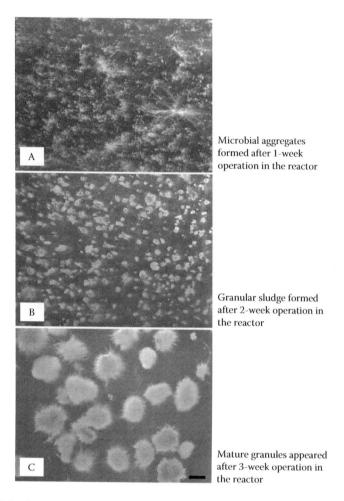

Microbial aggregates formed after 1-week operation in the reactor

Granular sludge formed after 2-week operation in the reactor

Mature granules appeared after 3-week operation in the reactor

FIGURE 1.2   Image analysis of the sludge morphology at different operation times in the sequencing batch reactors fed with glucose as substrate. Scale bar: 2 mm. (From Tay, J. H., Liu, Q. S., and Liu, Y. 2001. *J Appl Microbiol* 91: 168–175. With permission.)

1.2.1.4   Appearance of Mature Granules after Operation for Three Weeks

Mature aerobic granules were obtained after 3 weeks of operation (figure 1.2C). Aerobic granules had an average roundness of 0.79 in terms of the aspect ratio, defined as the ratio between the minor axis and the major axis of the ellipse equivalent to the granule. Mature granules had a much more regular, homogeneous and clearer outer morphology than the granular sludge observed after operation for 2 weeks. Figure 1.2 clearly exhibits the visual evolution track of the aerobic granulation process. The SEM micrograph further shows the detailed microstructures of glucose- and acetate-fed aerobic granules (figure 1.3). Glucose-fed granules had a filaments-dominant outer surface, whereas the acetate-fed granules showed a very compact bacterial structure, in which rod bacteria, tightly linked cell to cell, were found to be predominant. Such a tight cellular structure was not found in the seed sludge.

It can be seen from figures 1.1 and 1.2 that the formation of aerobic granules is a gradual process from seed sludge to dense aggregates, then to granular sludge, and finally to mature granules. Microscopic observations clearly revealed that microbial structure could be significantly strengthened, and further shaped, that is, they became more and more regular and dense, as the granulation process proceeded. In fact, the sludge-settling property could be improved significantly after granulation. Seed sludge for the reactor startup had a SVI value of 280 mL g^{-1} with many filamentous bacteria present (figure 1.1). However, an average SVI of 50 to 85 mL g^{-1} was achieved for granules formed from both substrates, which is almost three times higher than the original seed sludge. It is clear that granulation leads to a significant improvement in the sludge settleability. The granulation process could take 1 to 2 weeks or even a few more weeks depending on the substrate and the condition of operation. The process will normally take longer for slow-growing bacteria, for example nitrifying bacteria, and for toxic wastewater (Tsuneda et al. 2003; S. T. L. Tay, Zhuang, and Tay 2005; Yi et al. 2006). Aerobic granules can form with different carbon sources. It seems that the formation of aerobic granules is a process independent of or insensitive to the characteristics of the substrate (J. H. Tay, Liu, and Liu 2001). However, the substrate component has a profound impact on the microbial structure and the diversity of mature granules, as discussed above. In fact, the microstructure of anaerobic granules formed in UASB reactors is also strongly associated with the substrates (Wu 1991).

1.2.2 CHARACTERISTICS OF GLUCOSE- AND ACETATE-FED AEROBIC GRANULES

The physical characteristics of aerobic granules were more compact compared with the sludge flocs, while the microbial activity was comparable or somewhat lower compared with sludge flocs, depending on the size and structure of the granules. The characteristics of granules cultivated from glucose and acetate substrate are compared in the following section.

1.2.2.1 Morphology

The photographs by image analysis exhibited that mature granules formed from both glucose and acetate substrates had a regular round-shaped structure with an average roundness of 0.79 in terms of aspect ratio for glucose-fed granules, and 0.73 for acetate-fed granules (table 1.1). The glucose-fed granules had a mean diameter of 2.4 mm, whereas the granules grown on acetate had a mean diameter of 1.1 mm. The glucose-fed granules had filamentous bacteria extruding out from the surface (figure 1.3C and D). However, the acetate-fed granules had a smooth surface with a very compact bacterial structure and few filaments were observed (figure 1.3A and B).

1.2.2.2 Sludge Settleability

The sludge-settling property is a key operation factor that determines the efficiency of solid–liquid separation, which is essential for the proper functioning of a wastewater treatment system. The settleability of aerobic granules was much better than the sludge flocs of a conventional activated sludge process. The sludge volume

TABLE 1.1

Characteristics of Glucose- and Acetate-Fed Mature Aerobic Granules

Items	Glucose-Fed Granules	Acetate-Fed Granules
Average diameter (mm)	2.4 ($\pm$ 0.71)	1.1 ($\pm$ 0.43)
Aspect ratio	0.79 ($\pm$ 0.06)	0.73 ($\pm$ 0.04)
Sludge volume index (mL g^{-1})	51–85	50–80
Settling velocity (m h^{-1})	35 ($\pm$ 8.5)	30 ($\pm$ 7.1)
Granule strength (%)	98 ($\pm$ 0.9)	97 ($\pm$ 1.2)
Biomass density (g L^{-1})	41.1 ($\pm$ 6.9)	32.2 ($\pm$ 9.1)
Hydrophobicity (%)	68 ($\pm$ 3.9)	73 ($\pm$ 5.3)
Specific oxygen uptake rate (mg O$_2$ g^{-1} h^{-1})	69.4 ($\pm$ 8.8)	55.9 ($\pm$ 7.1)

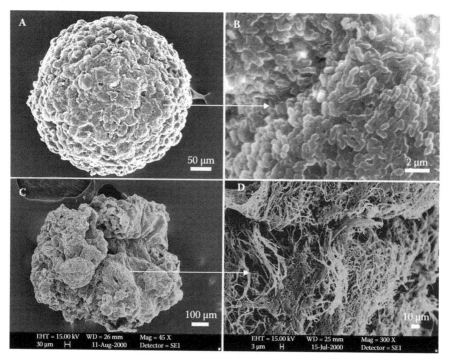

FIGURE 1.3 Scanning electron micrographs of aerobic granule cultivated from acetate substrate (A) and its surface microstructure (B), and granule cultivated from glucose substrate (C) and its surface microstructure (D). (From Liu, Q. S. 2003. Ph.D. thesis, Nanyang Technological University, Singapore. With permission.)

index (SVI) of the mature granules was 51 to 85 mL g^{-1} for glucose-fed granules and 50 to 80 mL g^{-1} for acetate-fed granules (table 1.1). The low SVI values indicated the high compactness of the granules. Compared with the seed sludge of SVI 280 mL g^{-1}, it is obvious that the settleability of sludge had improved significantly for aerobic granules. The average settling velocity of glucose-fed granules was 35 m h^{-1}, and 30 m h^{-1} for acetate-fed granules. Such settling velocities of aerobic granules are comparable with that of anaerobic granules cultivated in UASB (Hulshoff Pol et al. 1986; Beeftink 1987), and at least three times higher than those of activated sludge flocs having a settling velocity of less than 10 m h^{-1}. In fact, high settling velocity of 72 m h^{-1} was also reported for aerobic granules (Etterer and Wilderer 2001). It can be understood that the settling velocity will be influenced by the size and compactness of the aerobic granules.

1.2.2.3   Granule Physical Strength and Biomass Density

The physical strength of aerobic granules, expressed as the integrity coefficient (%), which is defined as the ratio of residual granules to the total weight of the granular sludge after 5 min of shaking at 200 rpm on a platform shaker (Ghangrekar et al. 1996), was 98% for glucose-fed granules and 97% for acetate-fed granules. The higher the integrity coefficient, the higher is the physical strength of granules. A high integrity coefficient represents the granule's ability to withstand high abrasion and shear. Aerobic granules cultivated in both substrates had a high strength. Meanwhile, the mature granules had a dry biomass density of 41.1 g L^{-1} for glucose-fed granules, as determined by the method of Beun et al. (1999), while it was 32.2 g L^{-1} for acetate-fed granules. The higher biomass density of aerobic granules reflects a denser microbial structure. The better settling ability of aerobic granules is consistent with higher biomass density, which is the result of a denser microbial structure.

1.2.2.4   Cell Surface Hydrophobicity

The seed sludge flocs had a cell surface hydrophobicity of 39% measured by the hydrocarbon partitioning method of Rosenberg, Gutnick, and Rosenberg (1980). After the formation of aerobic granules, the respective hydrophobicity of the cell surface increased to 68% for glucose-fed granules and 73% for acetate-fed granules. The hydrophobicity of aerobic granules was nearly twice higher than that of the seed sludge. High cell surface hydrophobicity favors cell attachment and then the aggregation of the sludge. Cell surface hydrophobicity is considered an important affinity force in cell attachment and self-immobilization (Del Re et al. 2000; Y. Liu et al. 2003).

1.2.2.5   Microbial Activity

The glucose-fed granules had a microbial activity expressed by specific oxygen uptake rate (SOUR) at 69.4 mg O$_2$ g^{-1} MLVSS h^{-1}, and 55.9 mg O$_2$ g^{-1} MLVSS h^{-1} for acetate-fed granules. The microbial activity of the granules would be strongly associated with the granule size and structure, which influence the oxygen and substrate transfer. The most beneficial aspect of aerobic granules is their excellent physical characteristics, which could lead to a high biomass concentration in the reactor, and subsequently smaller footprint for the reactor system.

1.2.2.6   Storage Stability of Aerobic Granules

Similar to anaerobic granules, aerobic granules have good storage stability (J. H. Tay, Liu, and Liu 2002; Zhu and Wilderer 2003). J. H. Tay, Liu, and Liu (2002) found that

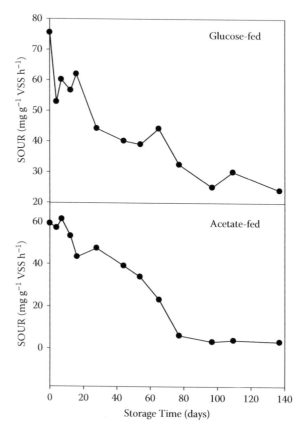

FIGURE 1.4 Effect of storage time on the microbial activity of aerobic granules cultivated from glucose and acetate, respectively. (Data from Tay, J. H., Liu, Q. S., and Liu, Y. 2002. *Environ Technol* 23: 931–936.)

aerobic granules cultivated from both glucose and acetate substrates had little reduction in microbial activity after 10 days of storage at 4°C in a refrigerator (figure 1.4). The SOUR then gradually decreased for both types of granules, from the initial value of 57 mg O_2 g^{-1} biomass h^{-1} to 6 mg O_2 g^{-1} biomass h^{-1} for acetate-fed granules at day 77, while the SOUR only decreased to 24 mg O_2 g^{-1} biomass h^{-1} for glucose-fed granules after 137 days, that is, a 60% reduction after more than 4 months of storage. Glucose-fed granules could be stored longer than those acetate-fed granules with less SOUR reduction. The loss in microbial activity would be associated with the length of storage time, the type of feed carbon, and the culture history. Both types of granules were visually in good granular shape after being stored for 4 months, and no disintegration was observed. Compared with the fresh granules, the strength of the stored granules decreased from 98% to 91% for glucose-fed granules, and from 97% to 89% for acetate-fed granules. Therefore, aerobic granules could still maintain good physical strength. Similar to the anaerobic granules, stored aerobic granules could be used as the seeding material for reactor startup because of its long storage ability. In fact, Zhu and Wilderer (2003) found that after 7 weeks of storage of aerobic granules in ambient environment, these aerobic granules could regain their microbial activity in less than a week.

1.3   AEROBIC GRANULATION ON OTHER CARBON SOURCES

Aerobic granules can be formed with different organic carbon sources of acetate and glucose as discussed above, while nitrifying granules can be formed with various N/COD ratios, as discussed above. In fact, aerobic granulation has been demonstrated in a variety of substrates, including organic/inorganic carbon, toxic wastewater, real municipal and industrial wastewater, for example, sucrose (Zheng et al. 2006), ethanol (Beun et al. 1999), phenol (Jiang, Tay, and Tay 2002; J. H. Tay, Jiang, and Tay 2004; Jiang et al. 2006), pentachlorophenol (Lan et al. 2005), and *tert*-butyl alcohol (S. T. L. Tay, Zhuang, and Tay 2005; Zhuang et al. 2005); particulate organic matter-rich wastewater (Schwarzenbeck, Borges, and Wilderer 2004), domestic sewage (de Kreuk and van Loosdrecht 2006), and industrial wastewater (Arrojo et al. 2004; Inizan et al. 2005; Schwarzenbeck, Erley, and Wilderer 2005; Su and Yu 2005; Wang et al. 2007). Phosphorus-accumulating granules have also been developed in the SBR (Y. Liu, Lin, and Tay 2005).

It was reported that nitrifying aerobic granules cultivated possess excellent nitrification ability (J. H. Tay, Yang, and Liu 2002), while aerobic granules grown on phenol can enhance the ability of bacteria to tolerate the toxic effect of phenol (Jiang, Tay, and Tay 2002). It is believed that cell immobilization is a useful strategy for bacteria to overcome the substrate inhibition associated with high-strength phenolic wastewater. It was shown that the kinetic behaviors of the phenol-degrading granules are subject to the Haldane model, indicating that the phenol-degrading aerobic granules could counteract the adverse effects of phenol inhibition (Jiang, Tay, and Tay 2002). The aggregation of microbial cells into compact granules may serve as an effective protection against the high phenol concentration. Aerobic granules would be powerful bioagents for the removal of inhibitory or toxic organic compounds present in industrial wastewater.

A novel strategy to add the benign co-substrate to toxic substrate to accelerate the granulation process and the performance of the granular sludge was proposed (Yi et al. 2006). It was found that with the addition of glucose to a toxic substrate, *p*-nitrophenol (PNP), the PNP metabolic activity could be enhanced through the formation of granules. The improvement of metabolic activity is most likely due to the retention of specific PNP-degrading microorganisms in the granules. This could be the result of syntrophic interactions between the community members in the granules, while the metabolic enhancement could result from the increase of specific degradation activity through the exchange of genetic material among the bacteria in granules. Use of a co-substrate strategy in the granulation process could improve the biodegradation of toxic and recalcitrant organic compounds. This shows another beneficial aspect of applying aerobic granulation technology in wastewater treatment.

1.4   AEROBIC GRANULATION AT DIFFERENT
COD CONCENTRATIONS

Aerobic granules can be formed in a relatively wide range of COD concentrations (Q. S. Liu, Tay and Liu 2003; J. H. Tay et al. 2004). Q. S. Liu, Tay and Liu (2003) studied the effect of substrate concentration on the formation, structure, and characteristics of aerobic granules and found that granules can be successfully formed with COD concentrations from 500 to 3000 mg L^{-1}, corresponding to an organic loading rate of 1.5 to 9.0 kg COD m^{-3} d^{-1}.

The formation of aerobic granules at different COD concentrations was a gradual process from seed sludge to the mature granules, with the same process as discussed earlier. It was found that a fast increase in sludge particle size was observed in the reactors supplied with high COD concentrations, for example, the size increased to 0.7 mm at day 10 of the operation in the reactor fed with influent COD of 500 mg L^{-1}, while it was 1.0, 1.1, and 1.4 mm for reactors fed with influent COD of 1000, 2000, and 3000 mg L^{-1}, respectively. It appeared that high substrate concentration favored a fast increase in granule size. This would be due to the fact that a high substrate concentration can sustain fast microbial growth.

The sludge particle size and SVI variation with the operation time in the reactor supplied with the highest substrate COD concentration of 3000 mg L^{-1} is shown in figure 1.5. It can be seen that the size of sludge particles gradually increased from 0.09 mm to a stable value of 1.9 mm after 20 days of operation. With an increase in size, the sludge SVI decreased from 208 mL g^{-1} to about 35 mL g^{-1} accordingly. These results clearly indicated that the settleability of aerobic granules was much better than that of seeding sludge bioflocs. Figure 1.6 compares the size distributions of seed sludge and the granular sludge cultivated, indicating that over 90% of the seed sludge particles had a size less than 0.2 mm, while more than 98% of the granular sludge fell in the size range of 0.4 mm to 3.2 mm. It is clear that the formation of aerobic granules seems to be independent of substrate concentration in the range from 500 to 3000 mg COD L^{-1} as discussed above. In fact, other researchers also suggest that aerobic granules can be formed in a wide COD range (Moy et al. 2002; J. H. Tay et al. 2004). This is probably due to the nature of aerobic bacteria. Compared to the biofilm process, aerobic granulation is a phenomenon of cell-to-cell self-immobilization instead of cell attachment to a solid surface. Similar to biofilm, it seems that substrate concentration is not a governing factor for the formation of aerobic granules.

1.4.1 EFFECT OF COD CONCENTRATION ON THE PROPERTIES OF AEROBIC GRANULES

A comparison of the average size of aerobic granules formed at different substrate concentrations found that the granule size slightly increased with the increase of substrate concentration (figure 1.7). The granule size was 1.57 mm at 500 mg COD L^{-1}, while it increased to 1.79 mm at 1000 and 2000 mg COD L^{-1}, and further increased to 1.89 mm at 3000 mg COD L^{-1}. Moy et al (2002) also found the size of aerobic granules increased when the loading rate was increased. A similar phenomenon was observed in anaerobic granulation (Grotenhuis et al. 1991). This can be easily understood because high substrate concentration would lead to the fast biomass production, and finally to a large size. It can be expected that the granules either of aerobic or anaerobic would have similar growth pattern in relation to the substrate concentration.

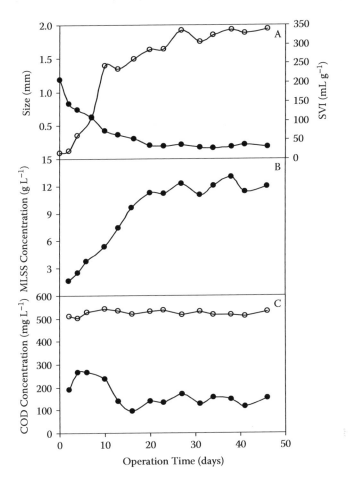

FIGURE 1.5   Size and SVI variation: (A, ○: size; ●: SVI), biomass (B) and COD concentration (C, ○: influent (×6); ●: effluent) in the operation time in a reactor fed with influent COD of 3000 mg L^{-1}. (From Liu, Q. S. 2003. Ph.D. thesis, Nanyang Technological University, Singapore. With permission.)

The relationship between substrate concentration and granule morphology expressed by the roundness and aspect ratio found that granules become more irregular under high substrate concentration, as shown in figure 1.8. The average roundness of granules was 0.69 at substrate concentration of 500 mg COD L^{-1}, while it was 0.66, 0.67, and 0.64 at substrate concentrations of 1000, 2000, and 3000 mg COD L^{-1}. The irregularity of the granule surface at high substrate concentration could be due to the fast growth rate. In fact, a heterogeneous and porous biofilm structure with extrusion was observed under high loading rate condition (van Loosdrecht et al. 1995; Kwok et al. 1998).

As shown in figure 1.9, the granule strength decreased with the increase of substrate concentration. The granule strength, expressed as integrity coefficient, was 97% at 500 and 1000 mg COD L^{-1}, but dropped to 95% at 2000 mg COD L^{-1}, and

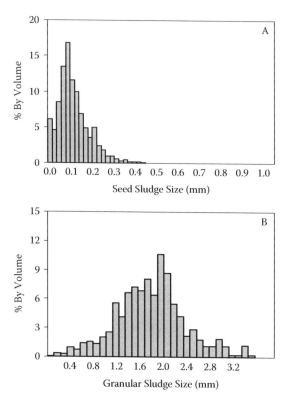

FIGURE 1.6 Comparison of size distribution by volume between seed sludge (A) and granular sludge (B) cultivated in a reactor fed with influent COD of 3000 mg L^{-1}. (From Liu, Q. S., Tay, J. H., and Liu, Y. 2003. *Environ Technol* 24: 1235–1242. With permission.)

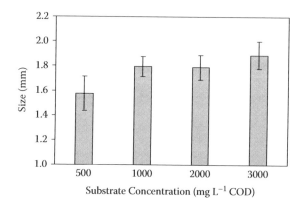

FIGURE 1.7 Effect of COD concentration on the size of mature granules. (From Liu, Q. S. 2003. Ph.D. thesis, Nanyang Technological University, Singapore. With permission.).

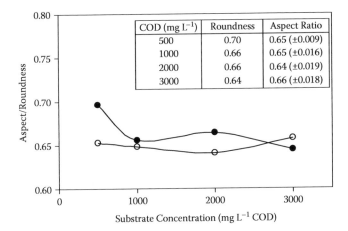

FIGURE 1.8 Effect of substrate concentration on granule morphology. ○: aspect; ●: roundness. (From Liu, Q. S. 2003. Ph.D. thesis, Nanyang Technological University, Singapore. With permission.)

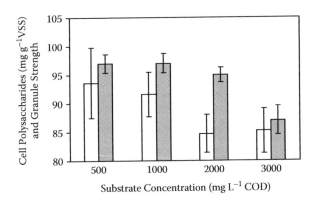

FIGURE 1.9 Effect of substrate concentration on the granule polysaccharides contents (white bar) and granule strength (dark bar) (From Liu, Q. S. 2003. Ph.D. thesis, Nanyang Technological University, Singapore. With permission.)

further to 87% at 3000 mg COD L^{-1}. The granules formed at the highest substrate concentration have a very loose structure. The lower strength of aerobic granules at high substrate concentrations could also possibly be due to the high biomass production rate. This has been reported in anaerobic granules where a high substrate concentration/loading rate resulted in a reduced strength of anaerobic granules, that is, the granules would easily lose their structural integrity, and disintegration would occur (Quarmby and Forster 1995). Morvai, Mihaltz, and Czake (1992) also found that increased loading rate raises the biomass growth rate, and high growth rate of anaerobic microorganisms would reduce the strength of the three-dimensional structure of a microbial community.

The specific gravity and density of sludge reflect the compactness of a microbial community. The specific gravity of granular sludge formed in various substrate COD

TABLE 1.2

Characteristics of Aerobic Granules Cultivated at Different Substrate Concentrations

Parameters	Seed Sludge	Influent Substrate COD Concentration (mg L^{-1})			
		500	1000	2000	3000
Size (mm)	0.09	1.57 ($\pm$ 0.14)	1.79 ($\pm$ 0.08)	1.79 ($\pm$ 0.10)	1.89 ($\pm$ 0.11)
SVI (mL g^{-1})	208	41 ($\pm$ 4.6)	43 ($\pm$ 4.3)	36 ($\pm$ 4.6)	34 ($\pm$ 3.1)
Biomass density (g L^{-1})	~	54.3 ($\pm$ 6.3)	54.7 ($\pm$ 8.4)	54.6 ($\pm$ 5.5)	56.1 ($\pm$ 7.6)
Specific gravity of sludge (kg L^{-1})	1.001	1.01 ($\pm$ 0.001)	1.01 ($\pm$ 0.001)	1.012 ($\pm$ 0.003)	1.012 ($\pm$ 0.002)
Biomass concentration in reactor (g L^{-1})	~	8.4 ($\pm$ 1.0)	9.5 ($\pm$ 1.5)	11.2 ($\pm$ 1.3)	12.3 ($\pm$ 1.6)
Cell surface hydrophobicity (%)	49.4	81.1 ($\pm$ 4.1)	84.2 ($\pm$ 5.5)	77.7 ($\pm$ 3.8)	79.1 ($\pm$ 4.2)
Effluent COD concentration (mg L^{-1})	~	27 ($\pm$ 9.4)	48 ($\pm$ 12.3)	68 ($\pm$ 15.1)	156 ($\pm$ 35.7)
COD removal efficiency (%)	~	95 ($\pm$ 1.7)	95 ($\pm$ 1.2)	97 ($\pm$ 1.1)	95 ($\pm$ 1.2)

concentrations of 500 to 3000 mg L^{-1} was around 1.010 kg L^{-1}, which is a significant increase compared with the seed sludge flocs of 1.001 kg L^{-1} (table 1.2). The SVI of 30 and 40 mL g^{-1} for granular sludge also improved significantly as compared with that of seed sludge of 208 mL g^{-1}. The granule density was around 55 g L^{-1} at different substrate concentrations. However, it seems that substrate concentration has an insignificant effect on the settleability of the granules.

The effect of substrate concentration on cell polysaccharides contents of aerobic granules is shown in figure 1.9. The cell polysaccharides content of aerobic granules formed at 500 mg COD L^{-1} was 93.6 mg g^{-1} VSS, while it was 91.6, 84.7, and 85.2 mg g^{-1} VSS at 1000, 2000, and 3000 mg COD L^{-1}, respectively. There is a slight decrease in polysaccharide content with the increase of substrate concentration. However, compared with the cell polysaccharide content of seed sludge of 60.9 mg g^{-1} VSS, a very significant increase was observed between the seed sludge and granular sludge. It should be pointed out that the lower physical strength of aerobic granules observed at high COD concentration could be reasonably attributed to its lower polysaccharides content. It has been generally agreed that cell polysaccharides are important in maintaining the structural integrity of a cell-immobilized community. It has also been reported that the production of cell polysaccharides was closely associated with hydrodynamic shear force, as discussed in chapter 2. It is not surprising that there is no significant change in cell polysaccharide content with varied substrate concentration because of its similar shear condition in the reactors.

Cell surface hydrophobicity of granular sludge seems to have no relationship to the substrate concentration, as listed in table 1.2. The cell surface hydrophobicity of aerobic granules was 81% at 500 mg COD L^{-1}, and 84%, 78%, and 79% at 1000, 2000, and 3000 mg COD L^{-1}, respectively. However, cell surface hydrophobicity of aerobic granules was much higher than that of seed sludge. Cell surface hydrophobicity has been generally considered to be an important affinity force in the self-immobilization and attachment of cells, for example biofilm and anaerobic granules (Del Re et al. 2000; J. H. Tay, Xu, and Teo 2000). The role of cell surface hydrophobicity in aerobic granulation will be discussed in detail in chapter 9. It can be considered that hydrophobicity might be the force for the initiation of aerobic sludge granulation. However, the substrate concentration did not seem to affect the cell surface hydrophobicity.

1.4.2 Effect of COD Concentration on the Reactor Performance

The reactor performance, in terms of the effluent COD concentration, exhibited an increased trend with the substrate concentration. An average effluent COD of 27 mg L^{-1} was obtained in the reactor fed with 500 mg L^{-1} influent COD, while it was 48, 68, and 156 mg L^{-1} for reactors fed with 1000, 2000, and 3000 mg L^{-1}, respectively. However, COD removal efficiency showed a comparable level in all reactors, which was above 95%. It should be mentioned that a high biomass concentration of 8.4 to 12.3 g L^{-1} was achieved among different reactors at steady state after granulation. Aerobic granulation could lead to more biomass being retained in the reactor due to good settling property of granular sludge. The high biomass concentration in reactors favored the performance and stability of biological reactors.

1.5 AEROBIC GRANULATION AT DIFFERENT SUBSTRATE N/COD RATIOS

Yang, Tay, and Liu (2005) investigated the effect of substrate N/COD ratio on the formation and physical characteristics of aerobic granules, and found that aerobic granules can be formed successfully in a relatively wide range of N/COD ratios, from 5/100 to 30/100. Figure 1.10 further illustrates the size change of microbial aggregates in the course of SBR operation. It can be seen that the mean size of microbial aggregates cultivated at different substrate N/COD ratios gradually increased and stabilized over the culture time, while small aerobic granules were obtained at the high substrate N/COD ratio. Figure 1.10 clearly shows that aerobic granular sludge reactors can be started up within about four weeks. Figure 1.11 further displays a correlation of SVI of mature aerobic granules with substrate N/COD ratio, indicating that SVI tended to decrease with the increase of the substrate N/COD ratio and the lowest SVI of 51 mL g^{-1} was found at the highest substrate N/COD ratio of 30/100. It is reasonable to consider that the substrate N/COD ratio would have a significant effect on the structure of microbial granules, that is, a more compact microbial structure could be expected at a higher substrate N/COD ratio.

It seems certain that the use of aerobic granules for upgrading the existing wastewater treatment plants towards simultaneous organics removal and nitrification would be feasible and beneficial, meanwhile such granules can also be used to bioaugment municipal wastewater treatment plant in which washout of nitrifying biomass is encountered.

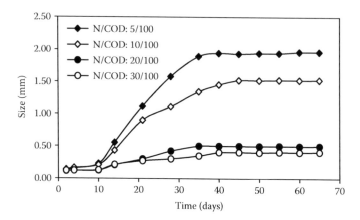

FIGURE 1.10 Change in aggregate size versus time in reactors operated at different substrate N/COD ratios. (From Yang, S. F., Tay, J. H., and Liu, Y. 2005. *J Environ Eng* 131: 86–92.)

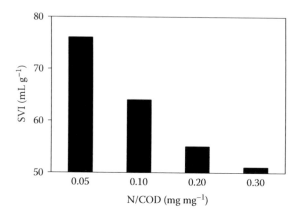

FIGURE 1.11 Sludge volume index of aerobic granules developed at different substrate N/COD ratios. (From Yang, S. F. 2005. Ph.D. thesis, Nanyang Technological University, Singapore. With permission.)

1.5.1 Effect of N/COD Ratio on the Properties of Aerobic Granules

The changes of SVI with the time of SBR operation at different N/COD ratios are shown in figure 1.12. It clearly indicates that after the formation of aerobic granules on day 20 onwards, the SVI was found to be as low as 50 mL g^{-1}. Moreover, it appears from figures 1.10 and 1.12 that smaller and denser aerobic granules would be formed at higher substrate N/COD ratios. This seems to imply that one may expect to manipulate the size of aerobic granules by controlling substrate N/COD ratio. Figure 1.13 reveals that the specific gravity of aerobic granules seems to be positively related to the applied substrate N/COD ratio, that is, aerobic granules developed at higher substrate N/COD ratios would have a much more compact and denser structure. In fact, this observation is consistent with the SVI trend as shown in figure 1.11.

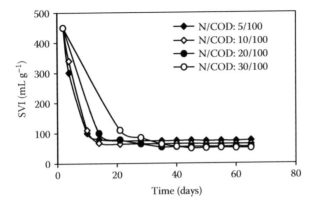

FIGURE 1.12   Change of sludge volume index versus operation time in reactors operated at different substrate N/COD ratios. (From Yang, S. F. 2005. Ph.D. thesis, Nanyang Technological University, Singapore. With permission.)

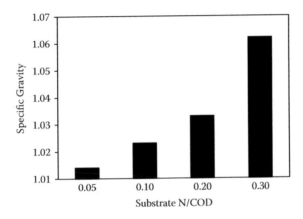

FIGURE 1.13   The specific gravity of aerobic granules developed at different substrate N/COD ratios. (From Yang, S. F. 2005. Ph.D. thesis, Nanyang Technological University, Singapore. With permission.)

Y. Liu et al. (2005) proposed the following equation to describe the settling velocity of bioparticles:

$$V_s = \alpha \frac{d_p^2}{SVI} e^{-\beta X} \tag{1.1}$$

where V_s is the settling velocity of bioparticles, d_p is the diameter of the particle, SVI stands for sludge volume index, X is biomass concentration, α and β are two constant coefficients. It is obvious that the settling velocity of aerobic granules would be determined by the size of the granule, SVI, and biomass concentration of granules. A smaller SVI would lead to a higher settling velocity of bioparticles or an improved settleability.

Yang, Tay, and Liu (2005) found that cell surface hydrophobicity at steady state was positively correlated to the applied substrate N/COD ratio (figure 1.14). In fact,

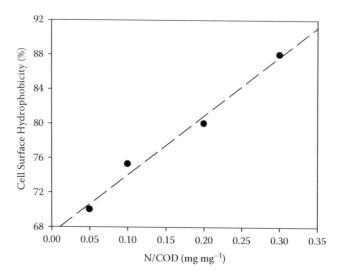

FIGURE 1.14 Cell surface hydrophobicity of aerobic granules versus the substrate N/COD ratio. (Data from Yang, S. F., Tay, J. H., and Liu, Y. 2005. *J Environ Eng* 131: 86–92.)

as revealed in figure 1.15, the activities of both ammonia oxidizer and nitrite oxidizer were proportionally related to the substrate N/COD ratio, that is, when the substrate N/COD ratio was increased, the nitrifying population would be enriched in the aerobic granules. Previous research reported that nitrifying bacteria would have a higher hydrophobic interaction than that of activated sludge (Sousa et al. 1997; Kim, Stabnikova, and Ivanov 2000). Thus, the enriched nitrifying population in aerobic granules cultivated at high substrate N/COD ratio would be responsible for the high cell surface hydrophobicity observed in figure 1.14.

Figure 1.16 exhibits the direct effect of substrate N/COD ratio on the ratio of sludge polysaccharides to sludge protein (PS/PN). It was found that the PS/PN ratio increased significantly along with the formation of aerobic granules. These results imply that extracellular polysaccharides could contribute, at least partially, to microbial aggregation. Meanwhile, increased substrate N/COD ratio would result in a low PS/PN ratio. In fact, other studies found a similar phenomenon, that a reduced substrate N/COD ratio could stimulate the production of extracellular polysaccharides, leading to an improved bacterial attachment to solid surfaces (Schmidt and Ahring 1996; Durmaz and Sanin 2001). There is evidence that cell carbohydrate content increased and protein content decreased in a very significant way as the substrate N/COD ratio decreased (Durmaz and Sanin 2001). It can be reasonably considered that nitrifying bacteria would produce much less extracellular polysaccharides than heterotrophic bacteria. For example, Tsuneda et al. (2001) used extracellular polysaccharides produced by heterotrophic bacteria to enhance the formation of nitrifying biofilm.

1.5.2 EFFECT OF N/COD RATIO ON POPULATION DISTRIBUTION

The activity of ammonium oxidizers and nitrite oxidizers as measured by specific ammonium oxygen utilization rate $(SOUR)_{NH4}$, specific nitrite oxygen utilization

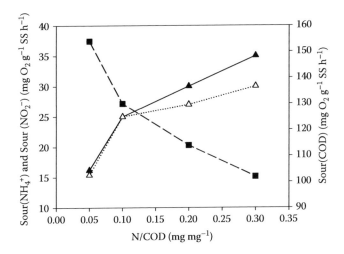

FIGURE 1.15 Activities of nitrifying population in terms of SOUR (NH_4^+) (▲) and SOUR (NO_2^-) (△) as well as the activity of heterotrophic population in terms of SOUR (COD) (■) of mature aerobic granules cultivated at different substrate N/COD ratios. (Data from Yang, S. F., Tay, J. H., and Liu, Y. 2005. *J Environ Eng* 131: 86–92.)

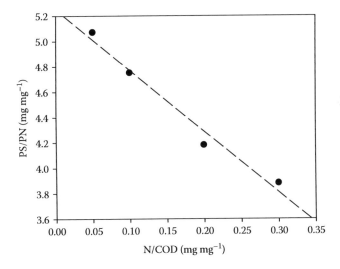

FIGURE 1.16 Effect of the substrate N/COD ratio on PS/PN ratio. (Data from Yang, S. F., Tay, J. H., and Liu, Y. 2005. *J Environ Eng* 131: 86–92.)

rate $(SOUR)_{NO2}$, and specific heterotrophic oxygen utilization rate $(SOUR)_H$ showed that higher $(SOUR)_{NH4}$ and $(SOUR)_{NO2}$ were obtained at the higher substrate N/COD ratios, while the $(SOUR)_H$ exhibited a decreasing trend with the increase in the substrate N/COD ratio (figure 1.16). It indicates that the nitrifying population in aerobic granules was greatly sustained at the high substrate N/COD ratios. In fact, a similar finding was also reported in biofilm reactors (Moreau et al. 1994; Ochoa et al. 2002). Evidence shows that bacteria can sense and move towards nutrients (Prescott,

Harley, and Klein 1999). As shown in figure 1.15, the activity of nitrifying population over heterotrophic population in the aerobic granules grown at the substrate N/COD ratio of 5/100 was very low as compared to the granules developed at high substrate N/COD ratios. At the high substrate N/COD ratio, competition between nitrifying and heterotrophic populations on nutrients would be significant.

It is clear that aerobic granules can be developed at a relatively wide range of substrate N/COD ratios. The substrate N/COD ratios had an insignificant effect on the formation of aerobic granules, but would determine microbial and physicochemical characteristics of aerobic granules.

1.6 CONCLUSIONS

Aerobic granulation is a gradual process from seed sludge to compact aggregates, further to granular sludge, and finally to mature granules. The formation of aerobic granules is insensitive to the substrate carbon source because granules can be formed with organic or inorganic carbon source, or toxic wastewater. However, substrate characteristics had a profound impact on the microbial structure and the diversity of aerobic granules. Aerobic granules can be formed in a relatively wide range of substrate concentrations and substrate N/COD ratios. This would make aerobic granular sludge technology applicable to low-strength wastewater, such as sewage, and to high-strength wastewater treatment, such as industrial waste.

REFERENCES

Adebowale, O. and Kiff, R. 1988. Operational trends in UASB reactor bed stability and during initiation of granulation. In *Proceedings of the Fifth International Symposium on Anaerobic Digestion*, 99–103. May 22–26, 1988. Bologna, Italy,

Arrojo, B., Mosquera-Corral, A., Garrido, J. M., and Mendez, R. 2004. Aerobic granulation with industrial wastewater in sequencing batch reactors. *Water Res* 38: 3389–3399.

Beeftink, H. H. 1987. Anaerobic bacterial aggregates. Ph.D. thesis, University of Amsterdam, Netherlands.

Beun, J.J., Hendriks, A., Van Loosdrecht, M.C.M., Morgenroth, E., Wilderer, P.A., and Heijnen, J. J. 1999. Aerobic granulation in a sequencing batch reactor. *Water Res* 33: 2283–2290.

Campos, C. M. M. and Anderson, G. K. 1992. The effect of the liquid upflow velocity and the substrate concentration on the start-up and steady-state periods of lab-scale UASB reactors. *Water Sci Technol* 25: 41–50.

Chen, J. and Lun, S. 1993. Study on mechanism of anaerobic sludge granulation in UASB reactors. *Water Sci Technol* 28: 171–178.

Chudoba, J. (1985) Control of activated sludge filamentous bulking. VI. Formulation of basic principles. *Water Res* 19: 1017–1022.

Crites, R. and Tchobanoglous, G. 1998. *Small and decentralized wastewater management systems*, 1st ed. Singapore, Indonesia: McGraw-Hill.

de Kreuk, M. K. and van Loosdrecht, M. C. M. 2006. Formation of aerobic granules with domestic sewage. *J Environ Eng* ASCE 132: 694–697.

Del Re, B., Sgorbati, B., Miglioli, M., and Palenzona, D. (2000) Adhesion, autoaggregation and hydrophobicity of 13 strains of Bifidobacterium longum. *Lett Appl Microbiol* 31: 438–442.

Durmaz, B. and Sanin, F. D. 2001. Effect of carbon to nitrogen ratio on the composition of microbial extracellular polymers in activated sludge. *Water Sci Technol* 44: 221–229.

Etterer, T. and Wilderer, P. A. 2001. Generation and properties of aerobic granular sludge. *Water Sci Technol* 43: 19–26.

Ghangrekar, M. M., Asolekar, S. R., Ranganathan, K. R., and Joshi, S. G. 1996. Experience with UASB reactor start-up under different operating conditions. *Water Sci Technol* 34: 421–428.

Grotenhuis, J. T. C., Kissel, J. C., Plugge, C. M., Stams, A. J. M., and Zehnder, A. J. B. 1991. Role of substrate concentration in particle-size distribution of methanogenic antigranulocytes sludge in UASB reactors. *Water Res* 25: 21–27.

Hulshoff Pol, L. W., Heijnekamp, K., and Lettinga, G. 1988. The selection pressure as a driving force behind the granulation of anaerobic sludge. In Lettinga, G., Zehnder, A. J. B., Grotenhuis, J. T. C., and Hulshoff Pol, L. W. (eds.), *Granular anaerobic sludge: Microbiology and technology*, 153–161. Wageningen, the Netherlands: Pudoc.

Hulshoff Pol, L. W., van de Worp, J. J. M., Lettinga, G., and Beverloo, W. A. 1986. Physical characteristics of anaerobic granular sludge. In EWPCA—Conference on Anaerobic Wastewater Treatment, 89–101. Amsterdam, the Netherlands, September 15–19, 1986.

Inizan, M., Freval, A., Cigana, J., and Meinhold, J. 2005. Aerobic granulation in a sequencing batch reactor (SBR) for industrial wastewater treatment. *Water Sci Technol* 52: 335–343.

Jiang, H. L., Tay, J. H., and Tay, S. T. L. 2002. Aggregation of immobilized activated sludge cells into aerobically grown microbial granules for the aerobic biodegradation of phenol. *Lett Appl Microbiol* 35: 439–445.

Jiang, H. L., Tay, J. H., Maszenan, A. I. M., and Tay, S. T. L. 2006. Enhanced phenol biodegradation and aerobic granulation by two coaggregating bacterial strains. *Environ Sci Technol* 40: 6137–6142.

Kim, I. S., Stabnikova, E. V., and Ivanov, V. N. 2000. Hydrophobic interactions within biofilms of nitrifying and denitrifying bacteria in biofilters. *Bioprocess Eng* 22: 285–290.

Kwok, W. K., Picioreanu, C., Ong, S. L., van Loosdrecht, M. C. M., Ng, W. J., and Heijnen, J. J. 1998. Influence of biomass production and detachment forces on biofilm structures in a biofilm airlift suspension reactor. *Biotechnol Bioeng* 58: 400–407.

Lan, H. X., Chen, Y. C., Chen, Z. H., and Chen, R. 2005. Cultivation and characters of aerobic granules for pentachlorophenol (PCP) degradation under microaerobic condition. *J Env Sci China* 17: 506–510.

Lettinga, G., Hulshoff Pol, L. W., Koster, I. W., Wiegant, W. M., de Zeeuw, W., Rinzema, A., Grin, P. C., Roersma, R. E., and Hobma, S. W. 1984. High-rate anaerobic wastewater treatment using the UASB reactor under a wide range of temperature conditions. *Biotechnol Genet Eng Rev* 2: 253–284.

Liu, Q. S. 2003. Aerobic granulation in sequencing batch reactor. Ph.D. thesis, Nanyang Technological University, Singapore.

Liu, Q. S., Tay, J. H., and Liu, Y. 2003. Substrate concentration-independent aerobic granulation in sequential aerobic sludge blanket reactor. *Environ Technol* 24: 1235–1242.

Liu, Y., Lin, Y. M., and Tay, J. H. 2005. The elemental compositions of P-accumulating microbial granules developed in sequencing batch reactors. *Process Biochem* 40: 3258–3262.

Liu, Y., Yang, S. F., Liu, Q. S., and Tay, J. H. 2003. The role of cell hydrophobicity in the formation of aerobic granules. *Curr Microbiol* 46: 270–274.

Liu, Y., Wang, Z. W., Liu, Y. Q., Qin, L., and Tay, J. H. 2005b. A generalized model for settling velocity of aerobic granular sludge. *Biotechnol Prog* 21: 621–626.

Moreau, M., Liu, Y., Capdeville, B., Audic, J. M., and Calvez, L. 1994. Kinetic behavior of heterotrophic and autotrophic biofilms in wastewater treatment processes. *Water Sci Technol* 29: 385–391.

Morgenroth, E., Sherden, T., Van Loosdrecht, M. C. M., Heijnen, J. J., and Wilderer, P. A. 1997. Aerobic granular sludge in a sequencing batch reactor. *Water Res* 31: 3191–3194.

Morvai, L., Mihaltz, P., and Czako, L. 1992. Kinetic basis of a new start-up method to ensure the rapid granulation of anaerobic sludge. *Water Sci Technol* 25: 113–122.

Morvai, L., Mihaltz, P., Czako, L., Peterfy, M., and Hollo, J. 1990. The influence of organic load on granular sludge development in an acetate-fed system. *Appl Microbiol Biotechnol* 33: 463–468.

Moy, B. Y. P., Tay, J. H., Toh, S. K., Liu, Y., and Tay, S. T. L. 2002. High organic loading influences the physical characteristics of aerobic sludge granules. *Lett Appl Microbiol* 34: 407–412.

Ochoa, J.C., Colprim, J., Palacios, B., Paul, E., and Chatellier, P. 2002. Active heterotrophic and autotrophic biomass distribution between fixed and suspended systems in a hybrid biological reactor. *Water Sci Technol* 46: 397–404.

Prescott, L., Harley, J., and Klein, D. 1999. *Microbiology*. Boston: McGraw-Hill.

Quarmby, J. and Forster, C. F. 1995. An examination of the structure of UASB granules. Water Res 29: 2449–2454.

Rosenberg, M., Gutnick, D., and Rosenberg, E. 1980. Adherence of bacteria to hydrocarbons: A simple method for measuring cell-surface hydrophobicity. *FEMS Microbiol Lett* 9: 29–33.

Schmidt, J. E. and Ahring, B. K. 1996. Granular sludge formation in upflow anaerobic sludge blanket (UASB) reactors. *Biotechnol Bioeng* 49: 229–246.

Schwarzenbeck, N., Erley, R., and Wilderer, P. A. 2004. Aerobic granular sludge in an SBR-system treating wastewater rich in particulate matter. *Water Sci Technol* 49: 41–46.

Schwarzenbeck, N., Borges, J. M., and Wilderer, P. A. 2005. Treatment of dairy effluents in an aerobic granular sludge sequencing batch reactor. *Appl Microbiol Biotechnol* 66: 711–718.

Sousa, M., Azeredo, J., Feijo, J., and Oliveira, R. 1997. Polymeric supports for the adhesion of a consortium of autotrophic nitrifying bacteria. *Biotechnol Tech* 11: 751–754.

Su, K. Z. and Yu, H. Q. 2005. Formation and characterization of aerobic granules in a sequencing batch reactor treating soybean-processing wastewater. *Environ Sci Technol* 39: 2818–2827.

Tay, J. H., Jiang, H. L., and Tay, S. T. L. 2004. High-rate biodegradation of phenol by aerobically grown microbial granules. *J Environ Eng* 130: 1415–1423.

Tay, J. H., Liu, Q. S., and Liu, Y. 2001. Microscopic observation of aerobic granulation in sequential aerobic sludge blanket reactor. *J Appl Microbiol* 91: 168–175.

Tay, J. H., Liu, Q. S., and Liu, Y. 2002. Characteristics of aerobic granules grown on glucose and acetate in sequential aerobic sludge blanket reactors. *Environ Technol* 23: 931936.

Tay, J. H., Xu, H. L., and Teo, K. C. 2000. Molecular mechanism of granulation. I. H+ translocation-dehydration theory. *J Environ Eng* 126: 403–410.

Tay, J. H., Yang, S. F., and Liu, Y. 2002. Hydraulic selection pressure-induced nitrifying granulation in sequencing batch reactors. *Appl Microbiol Biotechnol* 59: 332–337.

Tay, J. H., Pan, S., He, Y. X., and Tay, S. T. L. 2004. Effect of organic loading rate on aerobic granulation. I. Reactor performance. *J Environ Eng* 130: 1094–1101.

Tay, S. T. L., Zhuang, W. Q., and Tay, J. H. 2005. Start-up, microbial community analysis and formation of aerobic granules in a tert-butyl alcohol degrading sequencing batch reactor. *Environ Sci Technol* 39: 5774–5780.

Tijhuis, L., Hijman, B., Van Loosdrecht, M. C. M., and Heijnen, J. J. (1996) Influence of detachment, substrate loading and reactor scale on the formation of biofilms in airlift reactors. *Appl Microbiol Biotechnol* 45: 7–17.

Tsuneda, S., Park, S., Hayashi, H., Jung, J., and Hirata, A. 2001. Enhancement of nitrifying biofilm formation using selected EPS produced by heterotrophic bacteria. *Water Sci Technol* 43: 197–204.

Tsuneda, S., Nagano, T., Hoshino, T., Ejiri, Y., Noda, N., and Hirata, A. 2003. Characterization of nitrifying granules produced in an aerobic upflow fluidized bed reactor. *Water Res* 37: 4965–4973.

van Loosdrecht, M. C. M., Eikelboom, D., Gjaltema, A., Mulder, A., Tijhuis, L., and Heijnen, J. J. 1995. Biofilm structures. *Water Sci Technol* 32: 35–43.

Wang, S. G., Liu, X. W., Gong, W. X., Gao, B. Y., Zhang, D. H., and Yu, H. Q. 2007. Aerobic granulation with brewery wastewater in a sequencing batch reactor. *Bioresource Technol* 98: 2142–2147.

Wu, W. M. 1991. Technological and microbiological aspects of anaerobic granules. Ph.D. dissertation, East Lansing, MI: Michigan State University.

Yang, S. F. 2005. Effect of substrate N/COD ratio on the formation and characteristics of aerobic granules developed in sequencing batch reactors. Ph.D. thesis, Nanyang Technological University, Singapore.

Yang, S. F., Tay, J. H., and Liu, Y. 2005. Effect of substrate nitrogen/chemical oxygen demand ratio on the formation of aerobic granules. *J Environ Eng* 131: 86–92.

Yi, S., Zhuang, W. Q., Wu, B., Tay, S. T. L., and Tay, J. H. 2006. Biodegradation of p-nitrophenol by aerobic granules in a sequencing batch reactor. *Environ Sci Technol* 40: 2396–2401.

Zheng, Y. M., Yu, H. Q., Liu, S. H., and Liu, X. Z. 2006. Formation and instability of aerobic granules under high organic loading conditions. *Chemosphere* 63: 1791–1800.

Zhu, J. R. and Wilderer, P. A. 2003. Effect of extended idle conditions on structure and activity of granular activated sludge. *Water Res* 37: 2013–2018.

Zhuang, W. Q., Tay, J. H., Yi, S., and Tay, S. T. L. 2005. Microbial adaptation to biodegradation of tert-butyl alcohol in a sequencing batch reactor. *J Biotechnol* 118: 45–53.

2 Aerobic Granulation at Different Shear Forces

Qi-Shan Liu and Yu Liu

CONTENTS

2.1 INTRODUCTION

Shear force resulting from hydraulics and/or particle-particle collision has been considered as one of the most influencing factors in the formation, structure, and stability of biofilms (van Loosdrecht et al. 1995; Y. Liu and Tay 2001a, 2002). A higher shear force would result in a stronger and compact biofilm, whereas biofilm tends to become a heterogeneous, porous, and weaker structure when the shear force is weak (Chang et al. 1991; van Loosdrecht et al. 1995; Chen, Zhang, and Bott 1998). It has been shown that biofilm density increases with the increase of shear stress, while biofilm thickness exhibits a decreasing trend (Chang et al. 1991; Ohashi and Harada 1994; Kwok et al. 1998). Biofilm density correlates very closely with the self-immobilization strength of fixed bacteria, which is determined by the shear force imposed on the biofilms (Ohashi and Harada 1994; Chen, Zhang, and Bott 1998). It appears that a certain shear force in the biofilm system is necessary in order to produce a compact and stable biofilm structure, that is, higher shear force favors the formation of a smoother and denser biofilm.

In anaerobic granulation, it has been observed that granulation proceeded well at relatively high hydrodynamic shear condition in terms of high upflow liquid velocity, whereas anaerobic granulation was absent at a weak hydrodynamic shear force (Alphenaar, Visser, and Lettinga 1993; OFlaherty et al. 1997; Alves et al. 2000). These seem to indicate that shear force may also play an important role in the anaerobic granulation process. Thus, this chapter attempts to offer further insights into the role of shear force in aerobic granulation.

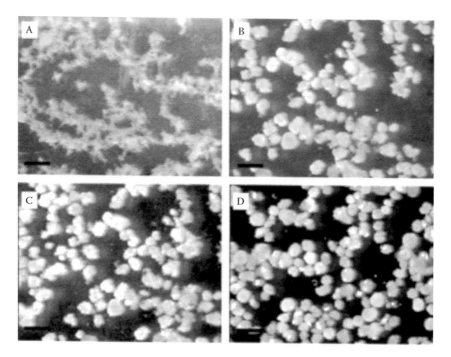

FIGURE 2.1 Sludge morphology in reactors with various superficial upflow air velocities at 2 weeks of operation. (A) 0.3 cm s^{-1}; (B) 1.2 cm s^{-1}; (C) 2.4 cm s^{-1}; (D) 3.6 cm s^{-1}. Bar: 1 mm. (From Liu, Q. S. 2003. Ph.D. thesis, Nanyang Technological University, Singapore. With permission.)

2.2 AEROBIC GRANULATION AT DIFFERENT SHEAR FORCES

In a column-type sequencing batch reactor (SBR) commonly employed for cultivation of aerobic granules, the superficial upflow air velocity (SUAV) has been known as a major cause of hydrodynamic turbulence and further hydraulic shear force (Chisti and Mooyoung 1989; Al-Masry 1999). Tay, Liu, and Liu (2001a) reported that shear force had a significant impact on the formation, structure, and metabolism of aerobic granules in the column SBR operated at different SUAV of 0.3 to 3.6 cm s^{-1}. It was shown that only typical bioflocs were observed in the reactor run at an SUAV of 0.3 cm s^{-1} during a time period of about 4 weeks, while aerobic granulation was observed in the reactors operated at SUAVs of 1.2, 2.6, and 3.6 cm s^{-1}, respectively. However, aerobic granules formed at the SUAV of 1.2 cm s^{-1} seemed unstable, and gradually disappeared 1 week after its formation. Figure 2.1 exhibits the morphology of biomass cultivated in each reactor operated at different SUAV after 2 weeks of operation. It can be seen that aerobic granules with a clear round outer shape and compact structure were developed in the SBRs operated at the SUAV higher than 1.2 m s^{-1} (figure 2.1C to D), whereas only loose and woolly structured bioflocs were observed in the reactor with SUAV of 0.3 m s^{-1} (figure 2.1A). In fact, another study by Tay, Liu, and Liu (2001b) also found that when the reactor was operated at a low SUAV of 0.8 cm s^{-1}, no granules were observed other than fluffy flocs (figure 2.2A). On the contrary, regular-shaped granules were successfully developed in the reactor operated at a high superficial air velocity of 2.5 cm s^{-1} (figure 2.2B).

FIGURE 2.2 Bioflocs cultivated at a superficial upflow air velocity of 0.008 m s⁻¹ (A); and granules formed at a superficial upflow air velocity of 0.025 m s⁻¹. (From Liu, Q. S. 2003. Ph.D. thesis, Nanyang Technological University, Singapore. With permission.)

Similarly, it has been reported that a low superficial air velocity did not lead to the formation of stable aerobic granules; however, at a relatively high superficial air velocity, granulation occurred and because of the high shear strength, smooth, dense, and stable aerobic granules formed (Beun et al. 1999; Wang et al. 2004). In addition, in the study by Shin, Lim, and Park (1992), conducted in an oxygen aerobic upflow sludge bed reactor, it was demonstrated that the granulation was governed by the physical stress exerted on the granular sludge. It is apparent that aerobic granulation would be a phenomenon associated very closely with the hydrodynamic conditions present in the SBR.

As shear force has an important role in aerobic granulation and granule stability, a minimum shear force seems necessary for aerobic granulation. It should be pointed out that high shear force in terms of upflow air velocity required for aerobic granulation will certainly increase the energy consumption for an aerobic granular sludge reactor. For example, if an upflow air velocity of 2.4 cm s⁻¹ is maintained in the system with a loading rate of 6.0 kg m⁻³.d, then about 400 m³ of air should be supplied per kilogram of COD removed, which is high as compared to air requirement of 20 to 50 m³ kg⁻¹ BOD for a conventional activated sludge process. This means that the operation cost for aeration in an aerobic granular sludge reactor would be several times higher than that of a conventional activated sludge process. In order to reduce

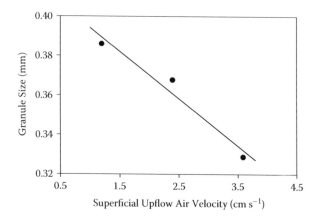

FIGURE 2.3 The effect of superficial upflow air velocity on granule size. (Data from Tay, J. H., Liu, Q. S., and Liu, Y. 2004. *Water Sci Technol* 49: 35–40.)

the operation cost for aeration in an aerobic granular sludge reactor, some counter-measures might have to be adopted, for example, optimizing air supply for minimum requirement of shear force, variable aeration, and so on.

2.3   EFFECT OF SHEAR FORCE ON GRANULE SIZE

The size of aerobic granules is strongly associated with the hydrodynamic shear force where smaller aerobic granules can be developed under higher shear conditions (Tay, Liu, and Liu 2001a, 2004). It was found that the mean size of aerobic granules tends to decrease with the increase of upflow air velocity (figure 2.3). It is evident that the size of aerobic granules is a net result of interaction between biomass growth and detachment, that is, the balance between growth and detachment would lead to a stable size. High hydrodynamic shear force would create more frequent collision and attrition among granules or particles, and subsequently high detachment (Gjaltema, van Loosdrecht, and Heijnen 1997). In fact, it has been observed that the thickness of biofilm is strongly associated with the hydrodynamic shear, for example, a thinner biofilm was developed under high shear conditions (Ohashi and Harada 1994; van Loosdrecht et al. 1995; Kwok et al. 1998; Wasche, Horn, and Hempel 2000; Y. Liu and Tay 2001a, 2001b). An example is given in figure 2.4 showing the effects of shear stress on biofilm thickness and density observed in a steady-state fluidized bed reactor. It can be seen that biofilm thickness decreased with the increase of shear stress.

2.4   EFFECT OF SHEAR FORCE ON GRANULE MORPHOLOGY

The morphology of aerobic granules can be described by aspect ratio or roughness. As shown in figure 2.5, both the aspect ratio and roundness of aerobic granules increased with the increase in the applied SUAV in the range of 1.2 to 3.6 cm s^{-1}. It is clear that aerobic granules became rounder and smoother at high applied shear force in terms of SUAV. As discussed earlier, rounder and regular aerobic granules obtained under higher shear conditions can be attributed to the more frequent

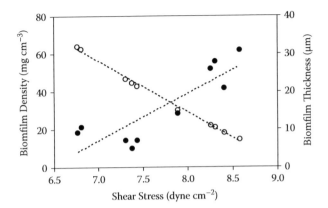

FIGURE 2.4 Effects of shear stress on biofilm thickness and density in a fluidized bed reactor: ○: thickness; ●: density. (Data from Chang, H. T. et al. 1991. *Biotechnol Bioeng* 38: 499–506.)

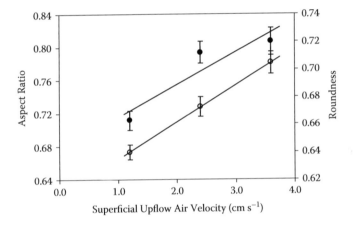

FIGURE 2.5 Effect of superficial upflow air velocity on granule morphology. ○: aspect ratio; ●: roundness. (Data from Liu, Q. S. 2003. Ph.D. thesis, Nanyang Technological University, Singapore.)

collision and attrition created by stronger upflow aeration. In fact, a heterogeneous, porous, and weaker biofilm was usually obtained when the shear force was weak, whereas smoother and denser biofilm can be obtained under high shear conditions (Chang et al. 1991; van Loosdrecht et al. 1995; Chen, Zhang, and Bott 1998; Kwok et al. 1998). These seem to indicate that a high shear would favor the formation of smoother and rounder aerobic granules or biofilm.

The growth of aerobic granules can be described by growth force and detachment force. In order to obtain a stable structure of aerobic granules, the growth force should be properly balanced with the detachment force. However, the effects of growth and detachment forces on aerobic granulation has often been studied independently, as discussed in chapter 7. A clear correlation of the interaction between growth and detachment forces to the metabolism and structure of aerobic granules

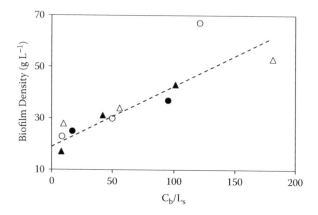

FIGURE 2.6   Effect of the ratio of basalt concentration (C_b) to organic loading rate (L_s) on biofilm density in a biofilm airlift suspension reactor. $L_s = 5$ (•); 10 ($\triangle$); 15 (o); 20 (▲) kg COD m^{-3} d^{-1}. (Data from Kwok, W. K et al. 1998. *Biotechnol Bioeng* 58: 400–407.)

is still lacking. As aerobic granules can be regarded as a special form of biofilm, the evidence coming from biofilm research may provide some in-depth insights into the above question. In this regard, the effect of the interaction between the growth and detachment forces on biofilm structure was discussed briefly in this section.

There is evidence that a dense biofilm is associated with a high detachment force (D_f), while at a low D_f or high growth force (G_f), a weak and porous biofilm structure is observed. These seem to indicate that the biofilm structure is the net result of the interaction between G_f and D_f, that is, if a stable biofilm is expected, G_f and D_f must be balanced. In addition, the growth and detachment forces cannot be considered independently in the biofilm process. It is a reasonable consideration that detachment force normalized to growth force, D_f/G_f ratio, can be used to describe the degree of balance of G_f and D_f (Y. Liu et al. 2003). This ratio indeed reflects the relative strength of detachment force acting on unit growth force. Y. Liu et al. (2003) thought that an equilibrium biofilm structure can be expected at a given D_f/G_f ratio. Figure 2.6 shows, using the D_f/G_f concept, the relationship between the ratio of carrier (basalt) concentration (C_b) to substrate loading rate (L_s) and biofilm density obtained at different carrier concentrations and organic loading rates in a biofilm airlift suspension reactor (Kwok et al. 1998). Obviously, in the biofilm airlift suspension reactor, detachment force is mainly due to particle-to-particle collision, which is proportional to the reactor's carrier concentration (C_b).

It appears that the biofilm density increased with the increase of the C_b/L_s ratio. This implies that a certain detachment force that is balanced with the growth force is necessary in order to produce and maintain a compact biofilm structure. In an open channel flow biofilm reactor, effective diffusivities increased with increasing glucose (substrate) concentration, but decreased with the increase in flow velocity that served as a major detachment force (Beyenal and Lewandowski 2000, 2002). High effective diffusivities at high substrate concentrations show lower biofilm densities, while reduced effective diffusivities at high flow velocities display higher biofilm densities (Tanyolac and Beyenal 1997). Beyenal and Lewandowksi (2002) hypothesized that biofilms, depending on the hydrodynamic shear force, could arrange their internal architecture to control the mechanical pliability needed to resist the shear stress exerted on them. It is obvious that structural arrangement of biofilms would be the result of changes in metabolic behaviors. In conclusion, it is the interaction between growth and detachment forces that governs the formation, structure, and metabolism of biofilms.

2.5 EFFECT OF SHEAR FORCE ON BIOMASS SETTLEABILITY

Figure 2.7 shows that the biomass settleability in terms of SVI can be improved markedly with increasing the SUAV. For example, an average biomass SVI value of 170 mL g^{-1} was obtained in the SBR with no successful granulation at SUAV of 0.3 cm s^{-1}, while the respective biomass SVI of 62, 55, and 46 mL g^{-1} were achieved in the SBRs operated at the SUAV of 1.2, 2.4, and 3.6 cm s^{-1}. The lowered SVI in turn implies that the physical structure of biomass becomes more compact and denser at higher applied shear force. Obviously, the shear force-associated aerobic granulation is mainly responsible for the observed improvement of sludge settleability.

The specific gravity of biomass represents the compactness of a microbial community. Figure 2.7 shows that biomass became denser and denser with the increase of the applied shear force, while the specific gravity of granular sludge was much higher than that of bioflocs. As presented in figure 2.4, biofilm density increased quasi-linearly with shear stress. Di Iaconic et al. (2005) also reported that the biomass density of aerobic granular sludge increased linearly with shear force in a sequencing batch biofilter reactor, and a very high biomass density of 70 to 110 g VSS L^{-1} biomass was obtained in the reactor (figure 2.8). Obviously, higher granule density can ensure a more efficient biosolid–liquid separation, which is essential for producing high-quality effluent.

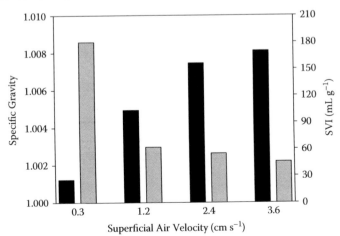

FIGURE 2.7 Sludge specific gravity (black) and SVI (gray) versus superficial upflow air velocity. (Data from Tay, J. H., Liu, Q.S., and Liu, Y. 2001a. *Appl Microbiol Biotechnol* 57: 227–233.)

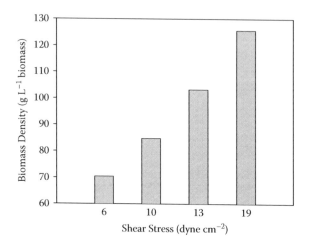

FIGURE 2.8 Effect of shear stress on biomass density of granular sludge in sequencing batch biofilter reactor. (Data from Di Iaconi, C. et al. 2005. *Environ Sci Technol* 39: 889–894.)

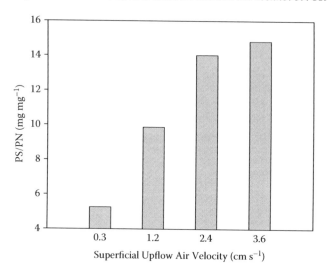

FIGURE 2.9 The effect of superficial upflow air velocity on the production of sludge polysaccharides (PS) normalized to sludge proteins (PN). (Data from Tay, J. H., Liu, Q.S., and Liu, Y. 2001a. *Appl Microbiol Biotechnol* 57: 227–233.)

2.6 EFFECT OF SHEAR FORCE ON THE PRODUCTION OF CELL POLYSACCHARIDES

Referring to chapter 10, extracellular polysaccharides can mediate both cohesion and adhesion of cells and play an essential role in maintaining the structural integrity of an immobilized community. It can be seen in figure 2.9 that the content of granule cellular polysaccharides normalized to the content of granule proteins tended to increase with the applied shear force up to a stable level. Higher shear force seems to enhance the production of cellular polysaccharides. This is confirmed

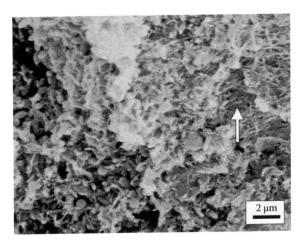

FIGURE 2.10 Extracellular polysaccharides surrounded the cells inside the granules observed by scanning electron microscope. (The arrow indicates the area of dense extracellular polysaccharides.) (From Liu, Q. S. 2003. Ph.D. thesis, Nanyang Technological University. With permission.)

by microscopic observation, as illustrated in figure 2.10, in which filaments of extracellular polysaccharides were visualized.

The shear force-stimulated production of extracellular polysaccharides has been widely reported in biofilm cultures (Trinet et al. 1991; Ohashi and Harada 1994; Chen, Zhang, and Bott 1998). It has been reported that the content of exopolysaccharides was fivefold greater for attached cells than for free-living cells (Vandevivere and Kirchman 1993), meanwhile colanic acid, an exopolysaccharide of *Escherichia coli* K-12, was found to be critical for the formation of the complex three-dimensional structure and depth of *E. coli* biofilms (Danese, Pratt, and Kolter 2000). These together imply that extracellular polysaccharides can make a great contribution to microbial self-immobilization. However, it should be pointed out that different views exist with regard to the relationship of extracellular polysaccharides to applied shear force. For example, Di Iaconic et al. (2005) found that both the content and composition of extracellular polymeric substances in aerobic granular sludge were not affected by hydrodynamic shear forces.

2.7   EFFECT OF SHEAR FORCE ON CELL HYDROPHOBICITY

Cell surface hydrophobicity can serve as an essential trigger of aerobic granulation (chapter 9). Figure 2.11 shows the effect of the applied shear force on cell hydrophobicity. The significant difference in cell hydrophobicity was observed before and after aerobic granulation. For example, in the SBR run at the highest SUAV of 3.6 cm s^{-1}, cell surface hydrophobicity increased from 54.3% in the period with no granulation to 81.2% after aerobic granulation. Similar trends were also observed in the other reactors with granulation, while it should be emphasized that there was no significant change in cell hydrophobicity in the SBR without granulation at the SUAV of 0.3 cm s^{-1}. The cell hydrophobicity of aerobic granules is nearly 50%

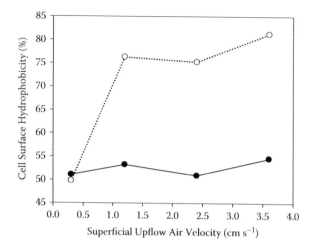

FIGURE 2.11   Comparison of cell surface hydrophobicity before (●) and after (○) granulation at different superficial upflow air velocities. (Data from Tay, J. H., Liu, Q. S., and Liu, Y. 2001a. *Appl Microbiol Biotechnol* 57: 227–233.)

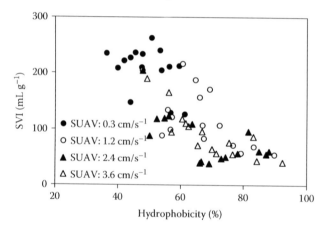

FIGURE 2.12   The relationship between sludge volume index and cell surface hydrophobicity. (From Tay, J. H., Liu, Q. S., and Liu, Y. 2001a. *Appl Microbiol Biotechnol* 57: 227–233.)

higher than that of seed sludge. These provide experimental evidence showing that aerobic granulation seems to be closely associated with an increase in cell hydrophobicity. It can be seen in figure 2.12 that the sludge SVI decreased almost linearly with the increase of cell surface hydrophobicity, that is, high cell hydrophobicity results in a more strengthened cell-to-cell interaction and, further, a compact and dense structure.

2.8   CONCLUSIONS

Hydrodynamic conditions caused by upflow aeration served as the main shear force in the column-type reactor commonly employed for the cultivation of aerobic granules.

Shear force in terms of superficial upflow air velocity was important in the aerobic granulation process. The microbial structure and metabolism of microorganisms are also influenced by shear force. Higher shear force led to more compact, denser, rounder and smaller granules. Shear force has a positive effect on the production of cell polysaccharides, while shear force associated cell polysaccharides production could affect the formation and stability of aerobic granules. It is reasonable to consider that hydrophobicity might act as an inducing force for the cell immobilization and further strengthen cell-cell interaction, while shear-stimulated extracellular polysaccharide production may play an important role in building up and maintaining the architecture of granular sludge.

REFERENCES

Al-Masry, W. A. 1999. Effect of scale-up on average shear rates for aerated non-Newtonian liquids in external loop airlift reactors. *Biotechnol Bioeng* 62: 494–498.

Alphenaar, P., Visser, A., and Lettinga, G. 1993. The effect of liquid upward velocity and hydraulic retention time on granulation in UASB reactors treating wastewater with a high sulphate content. *Bioresource Technol* 43: 249–258.

Alves, M., Cavaleiro, A. J., Ferreira, E. C., Amaral, A. L., Mota, M., da Motta, M., Vivier, H., and Pons, M.N. 2000. Characterisation by image analysis of anaerobic sludge under shock conditions. *Water Sci Technol* 41: 207–214.

Beun, J.J., Hendriks, A., Van Loosdrecht, M. C. M., Morgenroth, E., Wilderer, P. A., and Heijnen, J. J. 1999. Aerobic granulation in a sequencing batch reactor. *Water Res* 33: 2283–2290.

Beyenal, H. and Lewandowski, Z. (2000) Combined effect of substrate concentration and flow velocity on effective diffusivity in biofilms. *Water Res* 34: 528–538.

Beyenal, H. and Lewandowski, Z. 2002. Internal and external mass transfer in biofilms grown at various flow velocities. *Biotechnol Prog* 18: 55–61.

Chang, H. T., Rittmann, B. E., Amar, D., Heim, R., Ehlinger, O., and Lesty, Y. 1991. Biofilm detachment mechanisms in a liquid-fluidized bed. *Biotechnol Bioeng* 38: 499–506.

Chen, M. J., Zhang, Z., and Bott, T. R. 1998. Direct measurement of the adhesive strength of biofilms in pipes by micromanipulation. *Biotechnol Tech* 12: 875–880.

Chisti, Y. and Mooyoung, M. 1989. On the calculation of shear rate and apparent viscosity in airlift and bubble column bioreactors. *Biotechnol Bioeng* 34: 1391–1392.

Danese, P. N., Pratt, L. A., and Kolter, R. 2000. Exopolysaccharide production is required for development of *Escherichia coli* K-12 biofilm architecture. *J Bacteriol* 182: 3593–3596.

Di Iaconi, C., Ramadori, R., Lopez, A., and Passino, R. 2005. Hydraulic shear stress calculation in a sequencing batch biofilm reactor with granular biomass. *Environ Sci Technol* 39: 889–894.

Gjaltema, A., van Loosdrecht, M. C. M., and Heijnen, J. J. 1997. Abrasion of suspended biofilm pellets in airlift reactors: Effect of particle size. *Biotechnol Bioeng* 55: 206–215.

Kwok, W. K., Picioreanu, C., Ong, S. L., van Loosdrecht, M. C. M., Ng, W. J., and Heijnen, J. J. 1998. Influence of biomass production and detachment forces on biofilm structures in a biofilm airlift suspension reactor. *Biotechnol Bioeng* 58: 400–407.

Liu, Q. S. 2003. Aerobic granulation in sequencing batch reactor. Ph.D. Thesis, Nanyang Technology University, Singapore.

Liu, Y. and Tay, J. H. 2001a. Detachment forces and their influence on the structure and metabolic behaviour of biofilms. *World J Microbiol Biotechnol* 17: 111–117.

Liu, Y. and Tay, J. H. 2001b. Metabolic response of biofilm to shear stress in fixed-film culture. *J Appl Microbiol* 90: 337–342.

Liu, Y. and Tay, J. H. 2002. The essential role of hydrodynamic shear force in the formation of biofilm and granular sludge. *Water Res* 36: 1653–1665.

Liu, Y., Lin, Y. M., Yang, S. F., and Tay, J. H. 2003. A balanced model for biofilms developed at different growth and detachment forces. *Process Biochem* 38: 1761–1765.

OFlaherty, V., Lens, P. N. L., deBeer, D., and Colleran, E. 1997. Effect of feed composition and upflow velocity on aggregate characteristics in anaerobic upflow reactors. *Appl Microbiol Biotechnol* 47: 102–107.

Ohashi, A. and Harada, H. 1994. Adhesion strength of biofilm developed in an attached-growth reactor. *Water Sci Technol* 29: 281–288.

Shin, H. S., Lim, K. H., and Park, H. S. 1992. Effect of shear stress on granulation in oxygen aerobic upflow sludge bed reactors. *Water Sci Technol* 26: 601–605.

Tanyolac, A. and Beyenal, H. 1997. Prediction of average biofilm density and performance of a spherical bioparticle under substrate inhibition. *Biotechnol Bioeng* 56: 319–329.

Tay, J. H., Liu, Q. S., and Liu, Y. 2001a. The effects of shear force on the formation, structure and metabolism of aerobic granules. *Appl Microbiol Biotechnol* 57: 227–233.

Tay, J. H., Liu, Q. S., and Liu, Y. 2001b. Microscopic observation of aerobic granulation in sequential aerobic sludge blanket reactor. *J Appl Microbiol* 91: 168–175.

Tay, J. H., Liu, Q. S., and Liu, Y. 2004. The effect of upflow air velocity on the structure of aerobic granules cultivated in a sequencing batch reactor. *Water Sci Technol* 49: 35–40.

Trinet, F., Heim, R., Amar, D., Chang, H. T., and Rittmann, B. E. 1991. Study of biofilm and fluidization of bioparticles in a three-phase liquid-fluidized-bed reactor. *Water Sci Technol* 23: 1347–1354.

Vandevivere, P. and Kirchman, D. 1993. Attachment stimulates exopolysaccharide synthesis by a bacterium. *Appl Environ Microbiol* 59: 3280–3286.

van Loosdrecht, M. C. M., Eikelboom, D., Gjaltema, A., Mulder, A., Tijhuis, L., and Heijnen, J. J. 1995. Biofilm structures. *Water Sci Technol* 32: 35–43.

Wang, F., Liu, Y. H., Yang, F. L., Zhang, X. W., and Zhang, H. M. 2004. Study on the stability of aerobic granules in SBAR: Effect of superficial upflow air velocity and carbon source. Paper presented at IWA Workshop on Aerobic Granular Sludge, September 27–28, Munich, Germany.

Wasche, S., Horn, H., and Hempel, D. C. 2000. Mass transfer phenomena in biofilm systems. *Water Sci Technol* 41: 357–360.

3 Aerobic Granulation at Different SBR Cycle Times

Zhi-Wu Wang and Yu Liu

CONTENTS

3.1 INTRODUCTION

It appears from the preceding chapters that a number of operating parameters of a sequencing batch reactor (SBR) can influence aerobic granulation. This chapter looks into another SBR operating parameter, cycle time and its effect on aerobic granulation, as well as on the characteristics of aerobic granules. Cycle time is associated with the washout frequency of SBR, which can be regarded as a kind of hydraulic selection pressure. In fact, hydraulic selection pressure has been shown to be important for the formation of anaerobic granules in an anaerobic SBR (Hulshoff Pol et al. 1982; Shizas and Bagley 2002). Definitely, a sound understanding of the role of SBR cycle time in aerobic granulation would be helpful for the optimization and design of large-scale aerobic granular sludge SBR.

3.2 EFFECT OF CYCLE TIME ON AEROBIC GRANULATION

Tay, Yang, and Liu (2002) investigated the formation of nitrifying granules at different cycles times of 3 to 24 hours in SBRs. A complete washout of the sludge occurred and led to a failure of nitrifying granulation in the SBR run at the shortest cycle time of 3 hours, while only typical bioflocs were cultivated in SBR operated at the longest cycle time of 24 hours (figure 3.1a). After 2 weeks of operation, tiny nitrifying granules with mean diameters of 0.22 and 0.24 mm appeared in SBRs run at the cycle times of 12 and 6 hours, respectively. In comparison with the seed sludge that had a very loose and irregular structure, nitrifying granules developed showed a compact structure with a clear outer shape; moreover, the nitrifying granules formed at the cycle time of 6 hours were found to be smoother and denser than those developed at a cycle time of 12 hours (figures 3.1b and c).

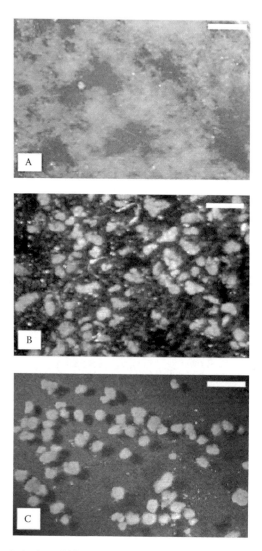

FIGURE 3.1 Morphologies of bioparticles cultivated in SBRs run at the cycle times of 24 (a), 12 (b), and 6 (c) hours; scale bar: 1 mm. (From Tay, J. H., Yang, S. F., and Liu, Y. 2002. *Appl Microbiol Biotechnol* 59: 332–337. With permission.)

One outstanding characteristic of aerobic granules compared to bioflocs is their relatively large particle size. Figure 3.2 shows that a short cycle time of SBR favors the development of large nitrifying granules. A similar phenomenon was also observed in an upflow anaerobic sludge blanket (UASB) reactor, that is, when the Hydraulic Retention Time (HRT) was decreased from 10 days to 1.5 days, the diameter of the UASB granules increased from 0.56 mm to 0.89 mm (Lin, Chang, and Chang 2001).

The basis of aerobic and anaerobic granulation is the continuous selection of sludge particles, that is, light and dispersed sludge is washed out, while heavier components

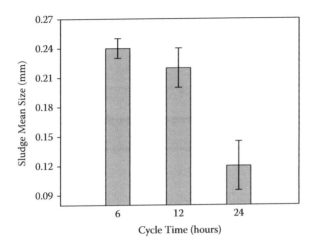

FIGURE 3.2   Mean size of bioparticles cultivated in SBRs run at different cycle times. (Data from Tay, J. H., Yang, S. F., and Liu, Y. 2002. *Appl Microbiol Biotechnol* 59: 332–337.)

are retained in the system. In an SBR, hydraulic selection pressure may result from the cycle time (Shizas and Bagley 2002). This is due to the fact that the SBR cycle time represents the frequency of solid discharge through effluent withdrawal, and it is related to the HRT. This implies that the longest cycle time would result in the lowest selection pressure. As a result, no nitrifying granulation was observed in the SBR run at the cycle time of 24 hours. In fact, the absence of anaerobic granulation was observed when the hydraulic selection pressure was very weak (Alphenaar, Visser, and Lettinga 1993; OFlaherty et al. 1997). Nitrifying granules developed well in SBRs with cycle times of 12 and 6 hours. It is evident that a relatively short cycle time should suppress the growth of suspended sludge because of frequent washout of the poorly settleable sludge. However, if the SBR is run at an extremely short cycle time, for example, 3 hours, the sludge loss due to hydraulic washout from the system cannot be compensated for by the growth of nitrifying bacteria. In this case, biomass cannot be retained in the system, and a complete washout of sludge blanket occurs and eventually leads to a failure of nitrifying granulation. A similar phenomenon was also reported in a UASB reactor (Alphenaar, Visser and Lettinga 1993).

The effect of the cycle time of SBR on the formation of heterotrophic aerobic granules was also reported in the literature. Wang et al. (2005) used sucrose as the sole carbon source to cultivate aerobic granules at the cycle times of 3 and 12 hours. Round aerobic granules first appeared in the SBR run at the cycle time of 3 hours after 30 cycles of operation, while irregular small granules were observed in the SBR operated at the cycle time of 12 hours after 120 cycles (figure 3.3). So far, all the evidence points to the fact that in order to achieve a rapid aerobic granulation in SBR, SBR cycle time needs to be controlled at a relatively low level.

Pan et al. (2004) initiated five column SBRs using precultivated glucose-fed aerobic granules with a mean size of 0.88 mm as seed, and operated them at the cycle times of 1, 2, 6, 12, and 24 hours, respectively. It was found that biomass in the SBR run at a cycle time of 1 hour was entirely washed out soon after the reactor start-up, and this in

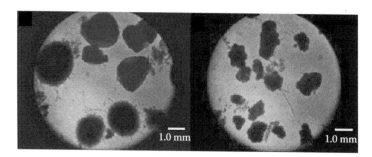

FIGURE 3.3 Morphologies of sucrose-fed aerobic granules cultivated in SBRs run at cycle times of 3 hours (left) and 12 hours (right), respectively. (From Wang, F. et al. 2005. World *J Microbiol Biotechnol* 21: 1379–1384. With permission.)

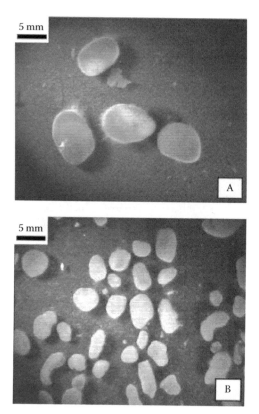

FIGURE 3.4 Morphologies of sludge cultivated in SBRs run at cycle times of 2 (a), 6 (b), 12 (c), and 24 (d) hours. (From Pan, S. et al. 2004. *Lett Appl Microbiol* 38: 158–163. With permission.)

turn resulted in reactor failure. In the SBR with cycle times of 2 to 24 hours, the seed granules were finally stabilized after 10 to 20 days of operation (figure 3.4). Figure 3.4a to c shows that at the short cycle time, aerobic granules tended to grow into large granules with fewer bioflocs in the bulk solution, while in the SBR run at the longest cycle time of 24 hours, a mixture of irregular granules and abundant suspended bioflocs

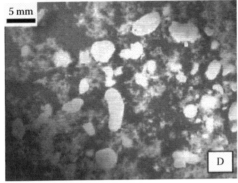

FIGURE 3.4 (continued)

was cultivated (figure 3.4d). It can be further seen in figure 3.5 that the size of stable granules turned out to be inversely related to the applied cycle time.

3.3   EFFECT OF CYCLE TIME ON PROPERTIES OF AEROBIC GRANULES

Pan et al. (2004) reported that a long cycle time does not favor improvement in sludge settleability. For instance, very poor settleability sludge with a high sludge volume index (SVI) of 110 mL g^{-1} was cultivated in the SBR operated at the cycle time of 24 hours. In contrast, excellent sludge with the lower SVI of 50 mL g^{-1} were harvested in the SBRs operated at the cycle times of 2, 6, and 12 hours (figure 3.6). It seems certain that a short cycle time can selectively retain bioparticles with good settleability. By virtue of those excellent settleability sludge retained in SBR, high volatile suspended solids concentrations (VSS), up to 13 g VSS L^{-1}, were achieved in SBRs run at short cycle times (figure 3.7). However, only 4 g VSS L^{-1} was finally achieved in the SBR run at the cycle time of 24 hours. In addition, the quality of effluent from SBR run at short cycle times was found to be much better than that from those operated at long cycle times (figure 3.8).

Compared to bioflocs with loose structure, aerobic granules always have a relatively high biomass density. The specific gravity of sludge reflects the compactness of the sludge structure. Figure 3.9 shows the specific gravity of nitrifying sludge

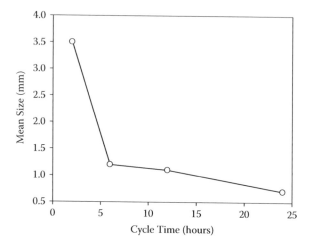

FIGURE 3.5   Sludge mean size in SBRs run at different cycle times. (Data from Pan, S. et al. 2004. *Lett Appl Microbiol* 38: 158–163.)

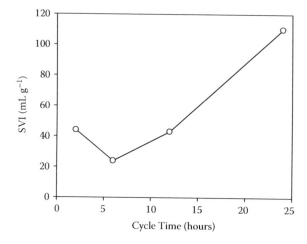

FIGURE 3.6   Effect of cycle time on sludge settleability. (Data from Pan, S. et al. 2004. *Lett Appl Microbiol* 38: 158–163.)

cultivated in SBRs run at different cycle times. As pointed out earlier, poorly settling sludge would be washed out, and only bioparticles with good settleability would be retained in SBRs run at short cycle times. According to the well-known Stokes law, a more compact particle would have a better settleability. This may explain the phenomenon shown in figure 3.9, that is, the specific gravity of nitrifying granules cultivated at a short cycle time is indeed much higher than those cultivated at long cycle times. Pan et al. (2004) also reported results similar to figure 3.9, that is, a short cycle time favors the development of aerobic granules with high specific gravity (figure 3.10). In addition, a high sludge integrity coefficient was obtained at short cycle times, indicating that aerobic granules with high mechanical strength can be cultivated at short cycle times (figure 3.10 and figure 3.11).

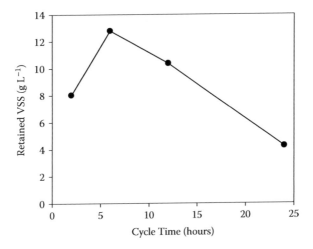

FIGURE 3.7 Effect of cycle time on suspended solids concentration retained in SBRs. (Data from Pan, S. et al. 2004. *Lett Appl Microbiol* 38: 158–163.)

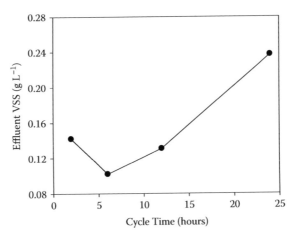

FIGURE 3.8 Effect of cycle time on effluent suspended solids concentrations from SBRs. (Data from Pan, S. et al. 2004. *Lett Appl Microbiol* 38: 158–163.)

As discussed in chapter 9, cell surface hydrophobicity plays an important role in microbial aggregation. The effect of cycle time on cell surface hydrophobicity is shown in figure 3.12. As can be seen, the cell surface hydrophobicity was improved as the cycle time was shortened. Higher cell surface hydrophobicity was developed in SBRs run at cycle times of 6 and 12 hours as compared to that found in the SBR run at the cycle time of 24 hours. Figure 3.1 and figure 3.12 together seem to suggest that the formation of nitrifying granules is associated with the cell surface hydrophobicity induced by the short cycle time of SBR. In fact, the importance of cell surface hydrophobicity in biofilm and biogranulation processes is demonstrated in chapter 9.

Similar results to figure 3.12 were also reported by Pan et al. (2004). It was observed that aerobic granules with high cell surface hydrophobicity were obtained at

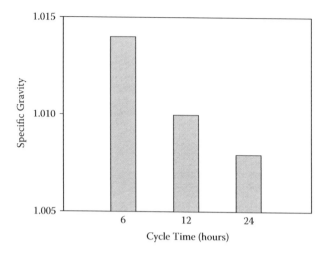

FIGURE 3.9 Specific gravity of sludge developed at different cycle times. (Data from Tay, J. H., Yang, S. F., and Liu, Y. 2002. *Appl Microbiol Biotechnol* 59: 332–337.)

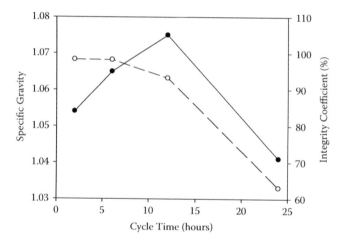

FIGURE 3.10 Specific gravity (●) and integrity coefficient (○) of sludge cultivated in SBRs run at different cycle times. (Data from Pan, S. et al. 2004. *Lett Appl Microbiol* 38: 158–163.)

short cycle times, for example, the cell surface hydrophobicity at a cycle time of 2 hours was nearly two times higher than that at the cycle time of 24 hours (figure 3.13).

It seems certain that a short cycle time would enhance cell surface hydrophobicity. Wilschut and Hoekstra (1984) proposed that the strong repulsive hydration interaction was the main force keeping the cells apart, and when bacterial surfaces were strongly hydrophobic, irreversible adhesion would occur. According to the thermodynamics theory, an increase in cell surface hydrophobicity would cause a corresponding decrease in the excess Gibbs energy of the cell surface, which promotes cell-to-cell interaction and further serves as a driving force for bacteria to

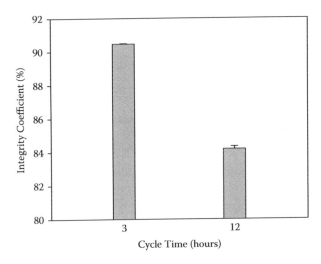

FIGURE 3.11   Integrity coefficient of sludge cultivated in SBRs run at different cycle times. (Data from Wang, F. et al. 2005. *World J Microbiol Biotechnol* 21: 1379–1384.)

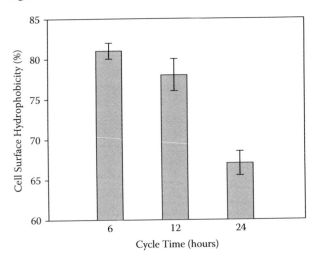

FIGURE 3.12   Effect of cycle time of SBR on cell surface hydrophobicity of nitrifying bacteria. (Data from Tay, J. H., Yang, S. F., and Liu, Y. 2002. *Appl Microbiol Biotechnol* 59: 332–337.)

self-aggregate out of liquid phase, that is, a high cell surface hydrophobicity would result in a strengthened cell-to-cell interaction, leading to the formation of dense and stable aerobic granule structures (see chapter 9).

Extracellular polysaccharides can mediate both cohesion and adhesion of cells and play a crucial role in building and further maintaining structure integrity in a community of immobilized cells (see chapter 10). Figure 3.14 displays the effect of cycle time on the ratio of cell polysaccharides (PS) to proteins (PN) of nitrifying granules. It is apparent that a short cycle time would stimulate the production

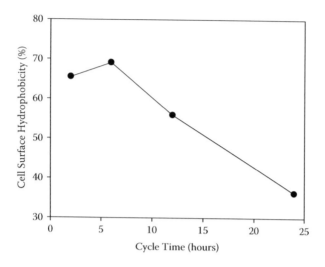

FIGURE 3.13 Effect of cycle time of SBR on cell surface hydrophobicity of glucose-fed aerobic granules. (Data from Pan, S. et al. 2004. *Lett Appl Microbiol* 38: 158–163.)

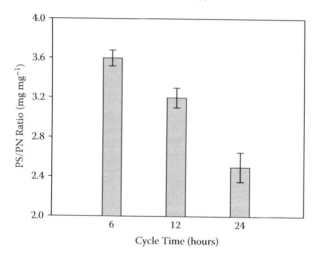

FIGURE 3.14 Effect of cycle time on ratio of extracellular polysaccharide (PS) to protein (PN) of the sludge fed by ammonia. (Data from Tay, J. H., Yang, S. F., and Liu, Y. 2002. *Appl Microbiol Biotechnol* 59: 332–337.)

of cell polysaccharides over proteins in nitrifying granules. In fact, heterotrophic bacteria were also found to overproduce PS at short cycle times (figure 3.15). In addition, the PS/PN ratio in heterotrophic granules cultivated at the same cycle time was nearly two times higher than that produced by nitrifying granules (figure 3.14). This is mainly due to the fact that nitrifying bacteria cannot utilize organic carbon for their growth, and only 11% to 27% of the energy generated goes to biosynthesis (Laudelout, Simonart, and van Droogenbroeck 1968), while the heterotrophic bacteria is able to convert up to 70% of the substrate energy into biosynthesis as well

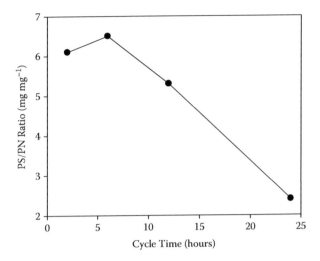

FIGURE 3.15   Effect of cycle time on the ratio of extracellular polysaccharide (PS) to protein (PN) of the sludge fed by glucose. (Data from Pan, S. et al. 2004. *Lett Appl Microbiol* 38: 158–163.)

as PS production (Rittmann and McCarty 2001). Besides, it has been also reported that an increased C/N ratio favors the production of cellular polysaccharides, leading to improved bacterial attachment to solid surfaces (Schmidt and Ahring 1996). In fact, extracellular polysaccharides are the key bounding material that interconnects individual cells into attached growth and are further responsible for the structural integrity of the granular matrix (see chapter 10).

The microbial activity of nitrifying bacteria can be quantified by the specific nitrification oxygen utilization rate (SNOUR) (Tay, Yang, and Liu 2002). The relationship between SNOUR and cycle time is presented in figure 3.16. It was found that the SNOUR was inversely related to the hydraulic selection pressure in terms of the SBR cycle time, that is, a shortened cycle time could stimulate the respiration activity of nitrifying bacteria.

The metabolic network of cells includes interrelated catabolic and anabolic reactions. The catabolic activity of microorganisms is directly correlated with the electron transport system activity, which can be described by the specific oxygen utilization rate. As can be seen in figure 3.16, the SNOUR was proportionally related to the cycle time of the SBR. A higher selection pressure or a short cycle time results in an increased SNOUR. This may imply that the nitrifying community can respond metabolically to changes in the hydraulic selection pressure. Figure 3.17 further shows that the PS/PN ratio increases with the increase in SNOUR. The SNOUR is in fact correlated with the electron transport system activity that determines catabolic activity of microorganisms. Therefore, it appears that when the hydraulic selection pressure imposed on the microbial community is increased, much of the energy generated by the catabolism is used for the production of cell polysaccharides rather than for growth.

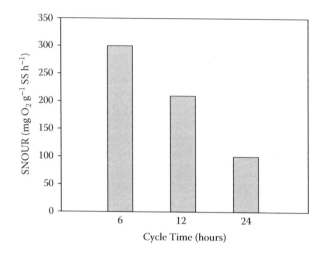

FIGURE 3.16 Specific nitrification oxygen utilization rate (SNOUR) of sludge cultivated in different cycle SBRs. (Data from Tay, J. H., Yang, S. F., and Liu, Y. 2002. *Appl Microbiol Biotechnol* 59: 332–337.)

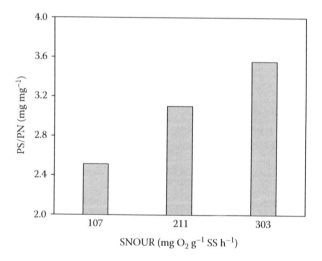

FIGURE 3.17 The relationship of SNOUR to PS/PN ratio. (Data from Tay, J. H., Yang, S. F., and Liu, Y. 2002. *Appl Microbiol Biotechnol* 59: 332–337.)

It is very likely that at a high hydraulic selection pressure, in order to avoid being washed out from the system, the nitrifying community has to regulate its metabolic pathway so as to maintain a balance with the external pressure via consuming non-growth-associated energy for polysaccharide production. In fact, a decreased biomass yield at shortened cycle time has been reported by Pan et al. (2004). Figure 3.18 shows that the biomass yield obtained at a cycle time of 2 hours was only half of that observed at the cycle time of 24 hours. Robinson, Trulear, and Characklis (1984) thought that most of the nongrowth-associated energy was attributable to polysaccharide

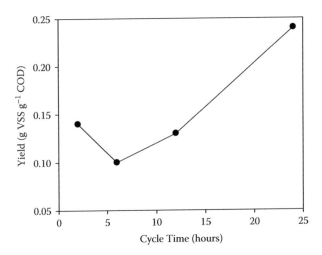

FIGURE 3.18 Effect of cycle time on biomass yield of aerobic granules. (Data from Pan, S. et al. 2004. *Lett Appl Microbiol* 38: 158–163.)

production, while extracellular redox activity was found to be associated with the production of extracellular polymeric substances (Wuertz et al. 1998). In fact, it is well known that the inhibition of the energy-generating function may prevent the development of competence for cell aggregation (Calleja 1984).

3.4 CONCLUSIONS

The nature of SBR is its cycle operation, thus cycle time of SBR can serve as a hydraulic selection pressure imposed on the microbial community. Complete wash-outs of biomass occur at short cycle times, and lead to failure of aerobic granulation in SBRs. On the other hand, aerobic granulation could not be developed at long cycle times because of microbial decay and weak hydraulic selection pressure. Evidence shows that short cycle time favors the cultivation of strong aerobic granules with excellent settleability, high specific gravity, and strengthened structure. The enhanced cell surface hydrophobicity and production of extracellular polysaccharides at short cycle time also contribute to rapid and successful aerobic granulation.

REFERENCES

Alphenaar, P., Visser, A., and Lettinga, G. 1993. The effect of liquid upward velocity and hydraulic retention time on granulation in UASB reactors treating wastewater with a high sulphate content. *Bioresource Technol* 43: 249–258.
Calleja, G. B. 1984. *Microbial aggregation*. Boca Raton, FL: CRC Press.
Hulshoff Pol, L. W., de Zeeuw, W. J., Velzeboer, C. T. M., and Lettinga, G. 1982. Granulation in UASB-reactors. *Water Sci Technol* 15: 291–304.
Laudelout, H., Simonart, P. C., and van Droogenbroeck, R. 1968. Calorimetric measurement of free energy utilization by Nitrosomonas and Nitrobacter. *Arch Mikrobiol* 63: 256–277.

Lin, C. Y., Chang, F. Y., and Chang, C. H. 2001. Treatment of septage using an upflow anaerobic sludge blanket reactor. *Water Environ Res* 73: 404408.

OFlaherty, V., Lens, P. N. L., deBeer, D., and Colleran, E. 1997. Effect of feed composition and upflow velocity on aggregate characteristics in anaerobic upflow reactors. *Appl Microbiol Biotechnol* 47: 102–107.

Pan, S., Tay, J. H., He, Y. X., and Tay, S. T. L. 2004. The effect of hydraulic retention time on the stability of aerobically grown microbial granules. *Lett Appl Microbiol* 38: 158–163.

Rittmann, B. E. and McCarty, P. L. 2001. *Environmental biotechnology: Principles and applications.* Boston: McGraw-Hill.

Robinson, J. A., Trulear, M. G., and Characklis, W. G. 1984. Cellular reproduction and extracellular polymer formation by Pseudomonas aeruginosa in continuous culture. *Biotechnol Bioeng* 26: 1409–1417.

Schmidt, J. E. and Ahring, B. K. 1996. Granular sludge formation in upflow anaerobic sludge blanket (UASB) reactors. *Biotechnol Bioeng* 49: 229–246.

Shizas, L. and Bagley, D. M. 2002. Improving anaerobic sequencing batch reactor performance by modifying operational parameters. *Water Res* 36: 363–367.

Tay, J. H., Yang, S. F., and Liu, Y. 2002. Hydraulic selection pressure-induced nitrifying granulation in sequencing batch reactors. *Appl Microbiol Biotechnol* 59: 332–337.

Wang, F., Yang, F. L., Zhang, X. W., Liu, Y. H., Zhang, H. M., and Zhou, J. 2005. Effects of cycle time on properties of aerobic granules in sequencing batch airlift reactors. *World J Microbiol Biotechnol* 21: 1379–1384.

Wilschut, J. and Hoekstra, D. 1984. Membrane fusion: From liposome to biological membrane. *Trends Biochem Sci* 9: 479–483.

Wuertz, S., Pfleiderer, P., Kriebitzsch, K., Spath, R., Griebe, T., Coello-Oviedo, D., Wilderer, P. A., and Flemming, H. C. 1998. Extracellular redox activity in activated sludge. *Water Sci Technol* 37: 379–384.

4 Aerobic Granulation at Different Settling Times

Lei Qin and Yu Liu

CONTENTS

4.1 INTRODUCTION

The selection pressure in terms of upflow liquid velocity has been demonstrated to be a driving force of anaerobic granulation in upflow anaerobic sludge blanket (UASB) reactors (Hulshoff Pol, Heijnekamp, and Lettinga 1988; Alphenaar, Visser, and Lettinga 1993). Although aerobic granulation is now studied extensively in SBRs, it is not yet clear how the aerobically grown granules are formed in SBR. The main feature of a column SBR is its successive cycle operation, and each cycle consists of filling, aeration, settling, and discharging. At the end of each cycle, settling of biomass takes place before effluent is withdrawn and sludge that cannot settle down within a given settling time is washed out of the reactor together with effluent through a fixed discharge port. As aerobic granules are much denser than suspended flocs, they require less time to settle than flocs do.

It appears that in SBR the settling time is likely to exert a selection pressure on the sludge particles, that is, only particles that can settle down below the discharge point within the given settling time are retained in the reactor; otherwise, they are discharged. This chapter aims to offer in-depth insights into the role of settling time in aerobic granulation in SBR. Such information would be useful for further setting up a practical guideline for successful aerobic granulation in SBR.

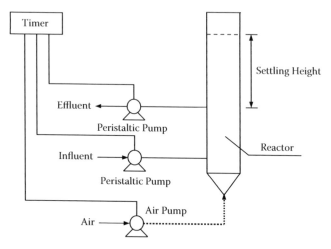

FIGURE 4.1   Schematic diagram of experimental system. (From Qin, L. 2006. Ph.D. thesis, Nanyang Technological University, Singapore. With permission.)

TABLE 4.1

Operation Strategies of R1 to R4

Operation Time (min)	R1	R2	R3	R4
Feed	5	5	5	5
Aeration	210	215	220	230
Settling	20	15	10	5
	5	2	1	—
Discharge	5	5	5	5
Total cycle time	240	240	240	240
Minimal settling velocity (m h^{-1})	1.89	2.52	3.78	7.56

Source: Qin, L. (2006) Ph.D. thesis, Nanyang Technological University, Singapore. With permission.

4.2   EFFECT OF SETTLING TIME ON THE FORMATION OF AEROBIC GRANULES

Qin, Liu, and Tay (2004a) investigated the effect of settling time on aerobic granulation in four column reactors, namely R1, R2, R3, and R4, each with a working volume of 2.5 liters, which were operated in sequencing batch mode (figure 4.1). R1 to R4 were run at settling times of 20, 15, 10, and 5 minutes, respectively, while the other operation parameters were kept the same. The duration of different operation stages and operation conditions applied for different reactors are shown in table 4.1. Effluent was discharged at the middle point of each SBR, which gives a volume exchange ratio of 50%. The sequential operation of the reactors was automatically controlled by timers, while two peristaltic pumps were employed for influent feeding and effluent withdrawal. In order to look into the effect of settling time on aerobic

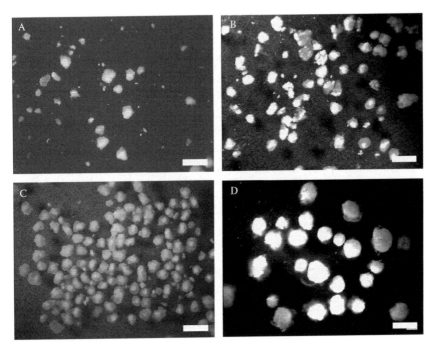

FIGURE 4.2 Morphology of aerobic granules developed in R1 (a), R2 (b), R3 (c), and R4 (d). Bar: 2 mm. (From Qin, L., Liu, Y., and Tay, J. H. 2004a. *Biochem Eng J* 21: 47–52. With permission.)

granulation, in the first phase of the study, R1 to R4 were run at respective settling times of 20, 15, 10, and 5 minutes.

The seed sludge had a mean floc size of 0.11 mm, and a sludge volume index (SVI) value of 230 mL g^{-1}. After 7 days of operation, aerobic granules were first observed in R4 operated at the settling time of 5 minutes. On day 10, tiny aggregates appeared in R1 to R3 run at respective settling times of 20, 15, and 10 minutes. After 3 weeks of operation, the four reactors reached steady state. The respective biomass concentrations in R1 to R4 at steady state were 5.3, 4.9, 5.5, and 5.4 g L^{-1}. Figure 4.2 shows that aerobic granules had a very regular and spherical outer shape, and the size of mature aerobic granules seems to increase gradually with the decrease of the settling time. Kim et al. (2004) also reported that granules cultivated with a minimum settling velocity of 0.7 m h^{-1} had a mean size of 1 to 1.35 mm, whereas granule size varied from 0.1 to 0.5 mm and rarely exceeding 1 mm when cultivated with a lower minimum settling velocity of 0.6 m h^{-1}. Other studies also showed that settling time employed would have an impact on the formation, size, and structure of aerobic granules at steady state (Beun, van Loosdrecht, and Heijnen 2002; McSwain, Irvine, and Wilderer 2004).

One of the prominent differences between aerobic granules and suspended flocs is the magnitude of the micropellets. It is observed that in R1, R2, and R3 aerobic granules coexisted with suspended flocs, whereas in R4 large aerobic granules became dominant over suspended flocs. The fractions of aerobic granules in steady-state R1 to R4 are shown in figure 4.3. It is obvious that only in R4 run at the shortest

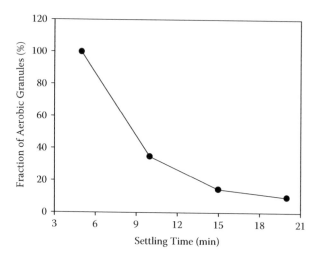

FIGURE 4.3 Fraction of aerobic granules developed at different settling times. (Data from Qin, L., Liu, Y., and Tay, J. H. 2004a. *Biochem Eng J* 21: 47–52.)

settling time of 5 minutes were aerobic granules the dominant form of growth; whereas the fraction of aerobic granules was only about 10% in R1, 15% in R2, and 35% in R3. These results clearly indicate that a mixture of aerobic granules and suspended sludge developed in R1 to R3 instead of a pure aerobic granular sludge blanket as observed in R4. The fractions of aerobic granules in the reactors seem to be related to the settling times. McSwain, Irvine, and Wilderer (2004) also observed a similar phenomenon in two SBRs operated at different settling times of 2 and 10 minutes, respectively. At a longer settling time, poorly settling flocs cannot be effectively withdrawn, and they may outcompete granule-forming bioparticles. As a result, the longer settling time would lead to failure of aerobic granulation due to the absence of strong selection pressure.

4.3   EFFECT OF SETTLING TIME ON THE SETTLEABILITY OF SLUDGE

SVI has been commonly used to describe the settleability and compactness of activated sludge in the field of environmental engineering. Figure 4.4 shows the relationship between the settling time and SVI observed in steady-state R1 to R4. It was found that the SVI was closely related to the settling time, that is, a more compact microbial structure of the aerobic granules could be expected at a shorter settling time. The SVI decreased from 230 mL g^{-1} in seed sludge to 49 mL g^{-1} in R4 after the formation of aerobic granules. However, in SBRs with partial aerobic granulation (R1 to R3), the SVI was much higher than that in R4. In consideration of the fraction of aerobic granules in each reactor (figure 4.3), it is reasonable to consider that the SVI is determined by the degree of aerobic granulation as well as the size and density of aerobic granules. McSwain, Irvine, and Wilderer (2004) reported that aerobic granules developed in the SBR operated at a settling time of 2 minutes had an SVI of 47 mL g^{-1}, while an SVI of 115 mL g^{-1} was found for the flocculent SBR

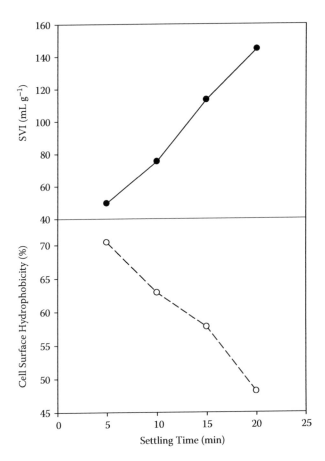

FIGURE 4.4   Effect of settling time on SVI (●) and cell surface hydrophobicity (○). (Data from Qin, L., Liu, Y., and Tay, J. H. 2004a. *Biochem Eng J* 21: 47–52.)

operated at a settling time of 10 minutes. The improvement of settling ability with decrease of settling time can be attributed to the increase of size and number or so-called fraction of aerobic granules in the reactors as flocs are effectively washed out at short settling time.

4.4   EFFECT OF SETTLING TIME ON CELL SURFACE HYDROPHOBICITY

Figure 4.4 shows the effect of settling time on cell surface hydrophobicity. A low cell surface hydrophobicity was found to be associated with a long settling time. The cell surface hydrophobicity tended to increase from 20% for the seed sludge to a stable value of 48% in R1, 58% in R2, 63% in R3, and 72% in R4. Likely, the cell surface hydrophobicity is inversely related to the settling time, that is, the microbial community developed at short settling time exhibits a high cell surface hydrophobicity. As shown in figure 4.3, the partial aerobic granulation was observed in R1, R2, and

R3, whereas aerobic granules were dominant in R4. It appears that the selection pressure-induced change in cell surface hydrophobicity contributes to cell-to-cell aggregation. In fact, it has been well known that cell surface hydrophobicity highly contributes to the formation of biofilm and anaerobic granules (see chapter 9).

Evidence shows that bacteria can change their surface hydrophobicity under some stressful conditions (see chapter 9). The cell surface hydrophobicity of the seed sludge was about 20%; however, after the appearance of aerobic granules in R1 to R4, the cell surface hydrophobicity was greatly improved (figure 4.4). In R4 dominated by aerobic granules, the cell surface hydrophobicity was much higher than those in R1 to R3. The settling time seems to induce changes in cell surface hydrophobicity, and a shorter settling time or a stronger hydraulic selection pressure results in a more hydrophobic cell surface. Research on anaerobic granulation also showed that anaerobic granular sludge in UASB reactors was more hydrophobic than the nongranular sludge washed out (Mahoney et al. 1987). It seems that microbial association has to adapt its surface properties to resist being washed out from the reactors through microbial self-aggregation at short settling time.

4.5   EFFECT OF SETTLING TIME ON PRODUCTION OF EXTRACELLULAR POLYSACCHARIDES

Extracellular polysaccharides (PS) are produced by most bacteria out of cell wall with the purpose of providing cells with the ability to compete in a variety of environments, providing a mode for adhesion to surface or self-immobilization (see chapter 10). Figure 4.5 shows that a shortened settling time would stimulate the production of PS, for example, an increase from 60.0 to 166.2 mg g^{-1} volatile solids (VS) was observed in the mature granules with the decrease of settling time in R1 to R4, whereas the production of extracellular proteins (PN) was not significantly influenced by the settling time, ranging from 16.5 to 25.0 mg g^{-1} VS. It appears

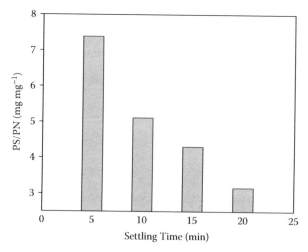

FIGURE 4.5   Effect of settling time on PS/PN ratio. (Data from Qin, L., Liu, Y., and Tay, J. H. 2004b. *Process Biochem* 39: 579–584.)

from figure 4.5 that the PS/PN ratio was inversely correlated to the settling time, that is, a shorter settling time would stimulate cells to produce more polysaccharide. Together with figure 4.3, these seem to suggest that extracellular polysaccharides play an essential role in the formation and further maintaining the structure and stability of aerobic granules.

The PS/PN ratios in the aerobic granules cultivated in R2 to R4 are much higher than that in the seed sludge (about 0.5 mg mg^{-1}). This is consistent with the earlier finding by Vandevivere and Kirchman (1993) that the content of extracellular polysaccharides for attached cells was five times higher than for free-living cells. The failure of aerobic granulation in SBR was also observed due to the inhibition of the production of extracellular polysaccharides (Yang, Tay, and Liu 2004), while the disappearance of aerobic granules in SBR was found to be tightly coupled to a drop of extracellular polysaccharides (Tay, Liu, and Liu 2001). It has been reported that high shear force can induce both aerobic biofilms and granules to secrete more extracellular polysaccharides, leading to a balanced structure of biofilm or granules under given hydrodynamic conditions (Ohashi and Harada 1994; Tay, Liu, and Liu 2001; Liu and Tay 2002). In fact, there is controversial report with regard to the essential role of extracellular polysaccharides in aerobic granulation (chapter 10).

4.6 EFFECT OF SETTLING TIME ON MICROBIAL ACTIVITY OF AEROBIC GRANULES

Microbial activity can be quantified by the specific oxygen utilization rate (SOUR) in terms of milligrams of oxygen consumed per milligram of volatile biomass per hour. To reflect the microbial activity of aerobic granules, aerobic granules were sampled just during the half hour of reaction period, and SOUR was measured immediately after sampling (Qin, Liu, and Tay 2004a). The correlation between the SOUR and settling time is presented in figure 4.6. The SOUR was found to be inversely related

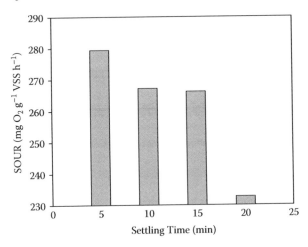

FIGURE 4.6 Effect of settling time on microbial activity in terms of SOUR. (Data from Qin, L., Liu, Y., and Tay, J. H. 2004b. *Process Biochem* 39: 579–584.)

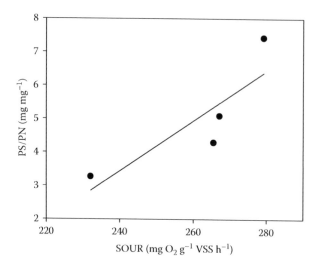

FIGURE 4.7 Relationships between PS/PN and SOUR. (Data from Qin, L., Liu, Y., and Tay, J. H. 2004b. *Process Biochem* 39: 579–584.)

to the settling time, that is, a shorter settling time would significantly stimulate the respirometric activity of microorganisms. These results may imply that bacteria may regulate their energy metabolism in response to the changes in hydraulic selection pressure exerted on them.

The catabolic activity of microorganisms is directly correlated to the electron transport system activity, which can be described by SOUR. As shown in figure 4.6, the SOUR was closely related with the hydraulic selection pressure in terms of settling time, for example a shorter settling time results in a remarkable increase of SOUR. This may indicate that the microbial community responds metabolically to changes in hydraulic selection pressure. As pointed out earlier, shorter settling time may trigger the production of extracellular polysaccharides. The correlation between the PS/PN ratio and SOUR is further shown in figure 4.7. More extracellular polysaccharides were secreted at higher SOUR. It is most likely that when the microbial community is exposed to an increased hydraulic selection pressure, much energy produced through the catabolism would go for the synthesis of extracellular polysaccharides rather than for growth, that is, under a high selection pressure, the microbial community would have to regulate its metabolic pathway in order to maintain a balance with the external forces through consuming nongrowth-associated energy for the production of polysaccharides and the improvement of cell surface hydrophobicity.

4.7 ACCUMULATION OF POLYVALENT CATIONS IN AEROBIC GRANULES

The contents of polyvalent cations (Ca, Mg, Fe, and Al) in aerobic granules cultivated in R1 to R4 are shown in table 4.2. The calcium content increased significantly at the shorter settling times, while the total content of Mg, Fe, and Al in aerobic granules did not show much difference at various settling times (figure 4.8). The increased

TABLE 4.2
Metal Content in Aerobic Granules in Percent by Dry Weight

	R1	R2	R3	R4	AS[a]	AG[b]
Ca	2.039	2.130	7.287	18.757	0.975	14.556
Mg	0.256	0.264	0.238	0.262	0.187	1.904
Fe	0.3813	0.3261	0.3533	0.4826	1.4420	0.087
Al	0.0041	0.0053	0.0049	NA[c]	0.4440	NA[c]
Microelement[d]	0.0446	0.0481	0.0408	0.0439	0.9038	NA[c]

[a] Activated sludge.
[b] Anaerobic granules (data from Fukuzak et al. 1991).
[c] Not available.
[d] Microelements including Co, Cu, Mn, Ni, and Zn.
Source: Data from Qin, L. (2006) Ph.D. thesis, Nanyang Technological University, Singapore.

calcium content of aerobic granules would result in a decrease of the ratio of volatile solids (VS) to total solids (TS) from 0.88 to 0.53. It appears that aerobic granules tend to selectively accumulate calcium that could play a part in the initiation and development of aerobic granules. In fact, it has been generally believed that multivalent positive ions, especially calcium, can favor both anaerobic and aerobic granulation (Schmidt and Ahring 1996; Teo, Xu, and Tay 2000; Yu, Tay, and Fang 2001; Jiang et al. 2003). Accumulation of calcium content in aerobic granules has been observed in aerobic granules cultivated under short settling times of 1 to 3 minutes and an organic loading rate of 4.8 kg chemical oxygen demand (COD) m^{-3} day^{-1} (Wang, Du, and Chen 2004).

Figure 4.8 clearly shows that the calcium content of aerobic granules in R4 operated at the shortest settling time of 5 minutes is about 18% of dry weight, which is much higher than those in the granule-suspended sludge mixtures cultivated in R1 to R3. However, the total contents of iron, magnesium, and aluminum in aerobic granules are minor and independent of the selection pressure as compared to the calcium, that is, the microbial community prefers to accumulate calcium instead of iron, magnesium, and aluminum. In fact, it was observed that the accumulation of calcium was accompanied by a rapid increase in granule size, while a nucleus was observed in the aerobic granule with high calcium content.

The selective accumulation of calcium would be a defensive strategy of the microbial community to selection pressure to increase its settleability to resist washout from the reactor. According to the proton translocation-dehydration theory developed for anaerobic granulation, Teo, Xu, and Tay (2000) proposed a biological explanation for the selective calcium accumulation in anaerobic granulation, and they considered that the positive effect of calcium on anaerobic granulation was probably due to the calcium-induced dehydration of bacterial cell surfaces, which was observed by Xu, Jiao, and Liu (1993), that is, the calcium-induced cell fusion might initiate the formation of a cell cluster, which acts as a microbial nucleus for further granulation.

It has been reported that the calcium content in anaerobic granules was about 14.6% by dry weight (Fukuzak et al. 1991). In fact, calcium is a constituent of

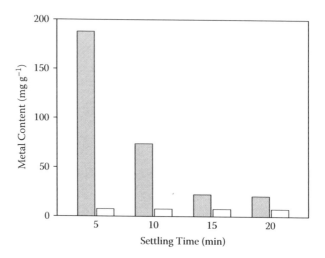

FIGURE 4.8 The accumulation of polyvalent cations in aerobic granules developed at various settling times, Ca (gray) and total Mg, Fe, and Al (white). (Data from Qin, L., Liu, Y., and Tay, J. H. 2004a. *Biochem Eng J* 21: 47–52.)

extracellular polysaccharides and/or proteins, which are used as adsorbing and linking materials in the anaerobic granulation process (Morgan, Evison, and Forster 1991). However, different views exist regarding the role of calcium in biogranulation, for example calcium has been thought not to induce granulation, and the contribution of calcium to anaerobic granulation was overestimated (Guiot et al. 1988; Thiele et al. 1990). As presented in chapter 13, the accumulation of calcium in aerobic granules may not be a prerequisite of microbial granulation. In R1 to R4, the VS/TS ratio of aerobic granules declined from 88% to 53% when the calcium content in aerobic granules increased from 20.4 to 187.6 mg g^{-1} TS. It is obvious that calcium and calcium-related compounds would be mainly responsible for the reduced VS content in aerobic granules. As a result, aerobic granules are substantially mineralized at high calcium contents.

4.8 EFFECT OF SHIFT OF SETTLING TIME ON AEROBIC GRANULATION

After the stabilization of the four reactors, the settling times in R1 to R3 were further shortened from 20 to 5, 15 to 2, and 10 to 1 minutes, respectively, without changing the other operation parameters. As shown in figure 4.3, the fraction of aerobic granules is in the range of 10% to 35% in R1 to R3 operated at respective settling times of 20, 15, and 10 minutes. In order to confirm the effect of settling time or hydraulic selection pressure on aerobic granulation, the settling times in steady-state R1, R2, and R3 were shifted from 20 to 5, 15 to 2, and 10 to 1 minutes on day 60 accordingly. This led to immediate washout of the light and dispersed sludge from the reactors, while only heavier granules remained. Two weeks after the shift of settling time, R1 to R3 gradually restabilized, and aerobic granules completely replaced suspended sludge and became dominant in R1 to R3. Figure 4.9 shows a comparison

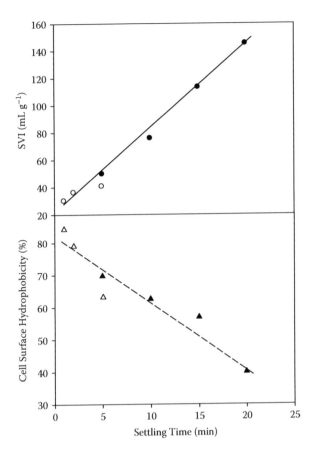

FIGURE 4.9　SVI before (●) and after (○) shift of settling time; cell surface hydrophobicity before (▲) and after (△) shift of settling time. (Data from Qin, L., Liu, Y., and Tay, J. H. 2004a. *Biochem Eng J* 21: 47–52.)

of cell surface hydrophobicity and microbial settleability before and after the shift of settling time in R1 to R3. It can be seen that both cell surface hydrophobicity and microbial settleability were improved significantly after the settling time was shortened. The production of extracellular polysaccharides was increased after the settling time was shortened. Moreover, the bed volume of aerobic granules in R1 to R3 reached nearly 100% because the sludge bed had been converted from partial to complete aerobic granulation after the settling time was shortened.

As can be seen in figure 4.9, cell surface hydrophobicity was increased significantly after the settling times were shortened in the reactors, while microbial settleability in terms of SVI was highly improved due to the formation of aerobic granules. It should be noted that microbial responses to the settling time in terms of cell hydrophobicity and settleability are subject to the parallel patterns no matter how the settling time was manipulated. This might mean that the state of the microbial community in a reactor is determined mainly by the external selection pressure exerted on it. Figure 4.9 also implies that the initial state of the seed sludge has less

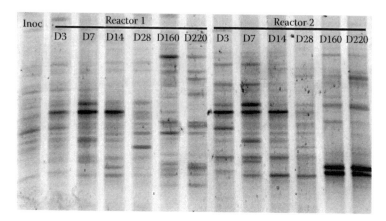

FIGURE 4.10 DGGE gel showing samples from R1 (10 minutes of settling time) and R2 (2 minutes of settling time) taken over time (for example, D3 means on day 3). Arrows indicate differences in dominated bands during start-up; boxes indicate dominant bands at steady state. (From McSwain, B. S., Irvine, R. L., and Wilderer, P. A. 2004. *Water Sci Technol* 50: 195–202. With permission.)

influence on the final granulation process as well as the physical characteristics of the mature granules; in other words, the characteristics of the mature granules are highly dependent on the selection pressure applied. It is reasonably considered that self-aggregation of bacteria would be a natural response of a microbial community to strong selection pressure in order to resist being washed out from the system. These results point to the fact that aerobic granules would be driven by hydraulic selection pressure in terms of settling time.

4.9 EFFECT OF SETTLING TIME ON MICROBIAL POPULATION

The effect of settling time on species selection has also been reported. By using polymerase chain reaction-denaturing gradient gel electrophoresis (PCR-DGGE) techniques, McSwain, Irvine, and Wilderer (2004) found that a more diverse microbial community was identified in the steady-state sludge harvested from longer-settling-time SBRs than sludge from shorter-settling-time reactors (figure 4.10). This observation might suggest settling time could cause species selection during the cultivation of aerobic granules. It seems that further investigation into the species selection with the effect of settling time is still needed in order to obtain more detailed information.

4.10 RATIONALE BEHIND SETTLING TIME-INITIATED AEROBIC GRANULATION

Figure 4.3 shows that aerobic granules are dominant only when the settling time is as short as 5 minutes, and a mixture of aerobic granules and suspended sludge is developed at longer settling times. In SBR, a short settling time preferentially selects for the growth of good settling bacteria, and the sludge with a poor settleability is washed out. Obviously, the selection of good settling sludge is crucial for aerobic granulation. In fact, absence of anaerobic granulation in the UASB reactors was observed at very weak hydraulic selection pressure in terms of liquid upflow velocity (OFlaherty et al. 1997). As compared to the operation of aerobic SBR, it is extremely difficult to quantitatively describe the hydraulic selection pressure in the UASB reactor due to the eventual production of biogas.

According to the well-known Stokes formula, the settling velocity of a particle can be estimated as:

$$V_s = \frac{g(\rho_p - \rho)d_p^{\,2}}{18\mu} \tag{4.1}$$

where V_s is the settling velocity of a particle, d_p is the diameter of the particle, ρ_p and ρ are the respective densities of particle and solution, and μ is the viscosity of the solution. The equation shows that the settling velocity of a particle is the function of the density and diameter of the aggregates. If the distance for mixed liquor to travel to the discharge port is L, the corresponding traveling time of bioparticles can be calculated as:

$$\text{Traveling time to the discharge port} = \frac{L}{V_s} \tag{4.2}$$

Equation 4.2 shows that a higher settling velocity (V_s) results in a shorter traveling time for particles to the discharge port. Thus, the bioparticles with a low settling velocity and hence a traveling time longer than the designed settling time are discharged from the reactor, that is, a minimum settling velocity, $(V_s)_{min}$, for bioparticles to be retained in the reactor, is only defined with a given L and settling time:

$$(V_s)_{min} = \frac{L}{\text{settling time}} \tag{4.3}$$

Only those bioparticles with a settling velocity greater than $(V_s)_{min}$ can be retained in the system. Equation 4.3 indeed shows that $(V_s)_{min}$ is determined by the settling time as well as traveling distance. The traveling distance (L) for a given reactor with a certain height is affected by the height of the discharging port, or more specifically, the volume exchange ratio (see chapter 5).

McSwain, Irvine, and Wilderer (2004) and McSwain et al. (2005) also investigated the influence of settling time on the formation of aerobic granules. Flocculent and granular sludge were formed at different settling times of 10 and 2 minutes, respectively. The sludges cultivated in these two SBRs had completely different morphologies (figure 4.11), while it was found that the properties of the sludge, Mixed Liquor Suspended Solids (MLSS) content, and settling characteristics were all different in two SBRs operated at the respective settling times of 10 and 2 minutes. According to McSwain, Irvine, and Wilderer (2004), at the shorter settling time of 2 minutes, granular sludge was dominant, with an average biomass concentration of 8.8 g L^{-1} and an SVI of 50 mL g^{-1}, compared to flocculent sludge with a biomass concentration of 3.0 g L^{-1} and an SVI of 120 mL g^{-1} at the longer settling time of

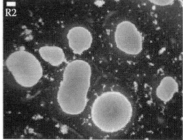

FIGURE 4.11 Steady-state sludge (day 200) from reactors 1 and 2 (scale = 1 mm) with different settling times of 10 and 2 minutes, respectively. (From McSwain, B. S. et al. 2005. *Appl Environ Microbiol* 71: 1051–1057. With permission.)

TABLE 4.3
Comparison of Sludge Characteristics with the Settling Time

Settling time (min)	$(V_s)_{min}$ (m h⁻¹)	Sludge size (mm)	SVI (ml g⁻¹)	References
2	12[a]	NA[b]	47	McSwain et al. 2004
4–5	10	2.5	NA[b]	Beun et al. 2002
5	7.56	2.15	49	Qin 2006
10	2.4[a]	NA[b]	115	McSwain et al. 2004
20	1.89	0.4	144	Qin 2006
75	0.7	1–1.3	50	Kim et al. 2004
75	0.6	0.1–0.5	85	Kim et al. 2004

[a] The minimum settling velocities were obtained from McSwain et al. (2005).
[b] Data of sludge size are not available.

10 minutes. As shown in equation 4.3, the minimum settling velocity, $(V_s)_{min}$ can be estimated as 2.4 m h⁻¹ for settling time of 10 minutes and 12 m h⁻¹ for 2 minutes (McSwain et al. 2005).

Kim et al. (2004) reported that granules cultivated with a minimum settling velocity of 0.7 m h⁻¹ have a mean size of 1 to 1.35 mm, whereas granule sizes varied from 0.1 to 0.5 mm and rarely exceeding 1 mm when cultivated with a lower minimum settling velocity of 0.6 m h⁻¹. The characteristics of sludge associated with settling time and minimum settling velocity are summarized in table 4.3. In most cases the reactor was operated mainly as a COD removal reactor to study the effect of settling time on aerobic granulation (Beun, van Loosdrecht, and Heijnen 2002; McSwain, Irvine, and Wilderer 2004), while the study for COD and nitrogen removal by selection of minimum settling velocity of granules was also carried out (Kim et al. 2004). It is apparent that the settling time has a significant impact on the sludge size and settleability during the cultivation of aerobic granules from activated sludge. Both granule size and settleability were increased by applying a shorter settling time in the SBR. Most significantly, when the settling time is shortened, the

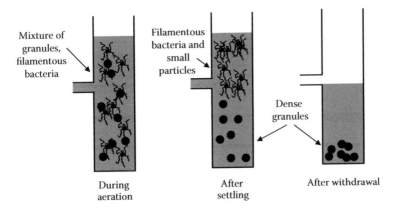

FIGURE 4.12 Selection of well-settling, dense granules in SBAR. (Adapted from Beun, J. J., van Loosdrecht, M. C. M., and Heijnen, J. J. 2002. *Water Res* 36: 702–712.)

time required for granulation and the time required for reaching steady state are all reduced. This implies that an appropriate selection of applied settling time is an effective strategy towards rapid and stable aerobic granulation. Most applied settling times reported thus far fall into a range of 2 to 20 minutes, with the exception of the study of Kim et al. (2004), in which settling time was kept as long as 75 minutes, leading to a small settling velocity of 0.6 to 0.7 m h^{-1}.

It is thought that short settling time selects for fast-settling flocs and granules, so as to enhance granule formation. Based on the difference in settling velocity between granules and the flocs, Beun, van Loosdrecht, and Heijnen (2002) used a short settling time of 4 to 5 minutes to effectively select for particles with a settling velocity larger than 10 m h^{-1} in a sequencing batch airlift reactor (SBAR). During the formation period of aerobic granules, a mixture of granules, filaments, and flocs presented in the reactor, while selection of the biomass granules from the mixture occurred in the period of settling, as illustrated in figure 4.12.

4.11 CONCLUSIONS

It appears from this chapter that settling time is a decisive parameter for the formation of aerobic granules in SBR. Aerobic granules can be successfully cultivated and become dominant only in the SBR operated at a short settling time, while a mixture of aerobic granules and suspended sludge is developed at a relatively long settling time. It seems certain that weak selection pressure does not favor aerobic granulation, and relatively strong selection pressure is essential for rapid and stable aerobic granulation in SBR. It is most likely that aerobic granulation would be an effective defensive or protective strategy of a microbial community against external selection pressure.

REFERENCES

Alphenaar, P., Visser, A., and Lettinga, G. 1993. The effect of liquid upward velocity and hydraulic retention time on granulation in UASB reactors treating wastewater with a high sulphate content. *Bioresource Technol* 43: 249–258.

Beun, J. J., van Loosdrecht, M. C. M., and Heijnen, J. J. 2002. Aerobic granulation in a sequencing batch airlift reactor. *Water Res* 36: 702–712.

Fukuzak, S., Chang, Y.-J., Nishio, N., and Nagai, S. 1991. Characteristics of granular methanogenic sludge grown on lactate in a UASB reactor. *J Ferment Bioeng* 72: 465–472.

Guiot, S. R., Gorur, S. S., Bourque, D., and Samson, R. 1988. Metal effect on microbial aggregation during upflow anaerobic sludge bed-filter (UBF) reactor start-up. *In Granular anaerobic sludge: Microbiology and technology*, eds. Lettinga, G., Zehnder, A. J. B., Grotenhuis, J. T. C., and Hulshoff Pol, L. W Wageningen, 187–194. The Netherlands: Purdoc.

Hulshoff Pol, L. W., Heijnekamp, K., and Lettinga, G. 1988. The selection pressure as a driving force behind the granulation of anaerobic sludge. In Lettinga, G., Zehnder, A. J. B., Grotenhuis, J. T. C., and Hulshoff Pol, L. W. (eds.), *Granular anaerobic sludge: Microbiology and technology*, 153–161. Wageningen, the Netherlands: Purdoc.

Jiang, H. L., Tay, J. H., Liu, Y., and Tay, S. T. L. 2003. Ca2+ augmentation for enhancement of aerobically grown microbial granules in sludge blanket reactors. *Biotechnol Lett* 25: 95–99.

Kim, S. M., Kim, S. H., Choi, H. C., and Kim, I. S. 2004. Enhanced aerobic floc-like granulation and nitrogen removal in a sequencing batch reactor by selection of settling velocity. *Water Sci Technol* 50: 157–162.

Liu, Y. and Tay, J. H. 2002. The essential role of hydrodynamic shear force in the formation of biofilm and granular sludge. *Water Res* 36: 1653–1665.

Mahoney, E. M., Varangu, L. K., Cairns, W. L., Kosaric, N., and Murray, R. G. E. 1987. The effect of calcium on microbial aggregation during UASB reactor start-up. *Water Sci Technol* 19: 249–260.

McSwain, B. S., Irvine, R. L., and Wilderer, P. A. 2004. The influence of settling time on the formation of aerobic granules. *Water Sci Technol* 50: 195–202.

McSwain, B. S., Irvine, R. L., Hausner, M., and Wilderer, P. A. 2005. Composition and distribution of extracellular polymeric substances in aerobic flocs and granular sludge. *Appl Environ Microbiol* 71: 1051–1057.

Morgan, J. W., Evison, L. M., and Forster, C. F. 1991. Examination into the composition of extracellular polymers extracted from anaerobic sludges. *Process Safety and Environmental Protection: Transactions of the Institution of Chemical Engineers, Part B* 69: 231–236.

OFlaherty, V., Lens, P. N. L., deBeer, D., and Colleran, E. 1997. Effect of feed composition and upflow velocity on aggregate characteristics in anaerobic upflow reactors. *Appl Microbiol Biotechnol* 47: 102–107.

Ohashi, A. and Harada, H. 1994. Adhesion strength of biofilm developed in an attached-growth reactor. *Water Sci Technol* 29: 281–288.

Qin, L. 2006. Development of microbial granules under alternating aerobic-anaerobic conditions for carbon and nitrogen removal. Ph.D. thesis, Nanyang Technological University, Singapore.

Qin, L., Liu, Y., and Tay, J. H. 2004a. Effect of settling time on aerobic granulation in sequencing batch reactor. *Biochem Eng J* 21: 47–52.

Qin, L., Liu, Y., and Tay, J. H. 2004b. Selection pressure is a driving force of aerobic granulation in sequencing batch reactors. *Process Biochem* 39: 579–584.

Schmidt, J. E. and Ahring, B. K. 1996. Granular sludge formation in upflow anaerobic sludge blanket (UASB) reactors. *Biotechnol Bioeng* 49: 229–246.

Tay, J. H., Liu, Q. S., and Liu, Y. 2001. The effects of shear force on the formation, structure and metabolism of aerobic granules. *Appl Microbiol Biotechnol* 57: 227–233.

Teo, K. C., Xu, H. L., and Tay, J. H. 2000. Molecular mechanism of granulation. II. Proton translocating activity. *J Environ Eng* 126: 411–418.

Thiele, J. H., Wu, W. M., Jain, M. K., and Zeikus, J. G. 1990. Ecoengineering high rate anaerobic digestion systems. Analysis of improved syntrophic biomethanation catalysts. *Biotechnol Bioeng* 35: 990–999.

Vandevivere, P. and Kirchman, D. 1993. Attachment stimulates exopolysaccharide synthesis by a bacterium. *Appl Environ Microbiol* 59: 3280–3286.

Wang, Q., Du, G. C., and Chen, J. 2004. Aerobic granular sludge cultivated under the selective pressure as a driving force. *Process Biochem* 39: 557–563.

Xu, H. L., Jiao, X. M., and Liu, S. S. 1993. Fluorescence measurement of surface dielectric constant of cll membrane. *Acta Biophys Sin* 9.

Yang, S. F., Tay, J. H., and Liu, Y. 2004. Inhibition of free ammonia to the formation of aerobic granules. *Biochem Eng J* 17: 41–48.

Yu, H. Q., Tay, J. H., and Fang, H. H. P. 2001. The roles of calcium in sludge granulation during UASB reactor start-up. *Water Res* 35: 1052–1060.

5 Roles of SBR Volume Exchange Ratio and Discharge Time in Aerobic Granulation

Zhi-Wu Wang and Yu Liu

CONTENTS

5.1 INTRODUCTION

It appears from the preceding chapters, among all the operation parameters that have been discussed so far, only settling time can serve as an effective selection pressure for aerobic granulation. However, a basic question to be addressed is if there are still other parameters that can also play the roles of selection pressure in aerobic granulation other than the identified settling time. The answer to such a question is essential for developing the design and operation strategy for rapid and stable aerobic granulation in both small- and large-scale sequencing batch reactors (SBRs). This chapter looks into two other potential candidate parameters that may act as selection pressures in aerobic granulation in SBR, namely SBR volume exchange ratio and discharge time.

5.2 THE ROLE OF SBR VOLUME EXCHANGE RATIO IN AEROBIC GRANULATION

According to figure 5.1, the mixed liquor volume exchange ratio, or volume exchange ratio for short, is the volume of effluent that is withdrawn after a preset settling time divided by the total working volume of a column SBR:

$$\text{Volume exchange ratio} = \frac{\pi r^2 L}{\pi r^2 H} = \frac{L}{H} \tag{5.1}$$

in which r is the radius of a column SBR, and H is the working height of the column SBR. This equation clearly shows that the volume exchange ratio is proportionally related to L. To look into the potential role of volume exchange ratio in aerobic granulation, Wang, Liu, and Tay (2006) designed and ran four identical column SBRs at different volume exchange ratios of 20% to 80% (figure 5.1), while the other operating conditions were all maintained at the same levels.

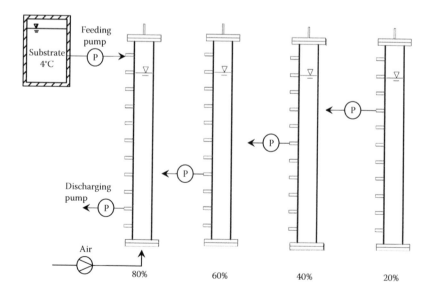

FIGURE 5.1 Schematics of four SBRs operated at the respective volume exchange ratios of 80%, 60%, 40%, and 20%. (From Wang, Z.-W. 2007. Ph.D. thesis, Nanyang Technological University, Singapore. With permission.)

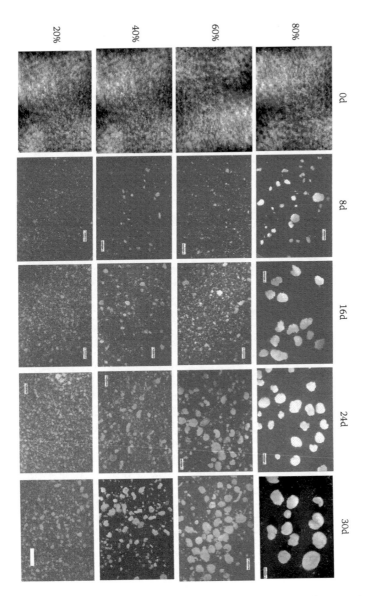

FIGURE 5.2 Morphologies of sludge cultivated at different volume exchange ratios in the course of aerobic granulation in SBRs; scale bar: 6 mm.

5.3 EFFECT OF VOLUME EXCHANGE RATIO ON AEROBIC GRANULATION

Wang, Liu, and Tay (2006) investigated the effect of volume exchange ratio on the formation of acetate-fed aerobic granules. The evolution of sludge morphology in the course of SBR operation at different volume exchange ratios is shown in figure 5.2. Morphologies of aerobic granules formed in four reactors appeared to be closely correlated with the applied volume exchange ratio; that is, only 8 days after reactor startup,

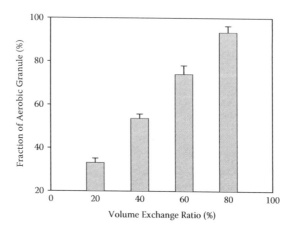

FIGURE 5.3 Fraction of aerobic granules in four SBRs run at volume exchange ratios of 20% to 80%. (Data from Wang, Z.-W., Liu, Y., and Tay, J.-H. 2006. *Chemosphere* 62: 767–771.)

aerobic granules first appeared in the SBR operated at the highest volume exchange ratio of 80%, while aerobic granules were subsequently observed at the volume exchange ratios of 60%, 40%, and 20%, respectively, 6, 12, and 20 days later. It can be seen in figure 5.2 that the larger and more spherical aerobic granules were formed at the higher volume exchange ratio of 80%, whereas bioflocs were cultivated and became predominant in the SBR operated at the lower volume exchange ratio of 20%.

It is apparent from figure 5.2 that only a mixture of bioflocs and aerobic granules was cultivated at small volume exchange ratios. Analysis of the fraction of aerobic granules formed in each SBR reveals that nearly a pure aerobic granular sludge blanket was indeed developed at the volume exchange ratio of 80% (figure 5.3). In contrast, almost no aerobic granulation was found at the volume exchange of 20%, indicating a failed granulation (figure 5.3). It is thus reasonable to consider that the SBR volume exchange ratio can play an essential role in aerobic granulation, and a high SBR volume exchange ratio facilitates rapid and successful aerobic granulation in SBR.

As presented in the preceding chapters, aerobic granules can be simply distinguished from bioflocs by their large particle size. The mean size of aerobic granules cultivated at different volume exchange ratios are presented in figure 5.4. The size of the aerobic granules tended to increase with the increase in the SBR volume exchange ratio, for example, the size of aerobic granules developed at the volume exchange ratio of 20% was smaller than 1 mm, whereas aerobic granules as large as about 3.8 mm were obtained at the volume exchange ratio of 80%.

In the operation of nitrogen-removal SBRs, Kim et al. (2004) also manipulated the SBR discharge height so as to impose on microorganisms two slightly different selection pressures in terms of minimum settling velocity of 0.6 and 0.7 m h[-1]. Even such a marginal difference in the minimum settling velocity could also result in distinct morphologies of cultivated sludge. For example, large bioparticles of 1.0 to 2.0 mm were harvested at the $(V_s)_{min}$ of 0.7 m h[-1], while only small bioparticles of 0.1 to 0.5 mm were cultivated at the $(V_s)_{min}$ of 0.6 m h[-1] (Kim et al. 2004). Microscopic observation further revealed that the high volume exchange ratio SBR favored the

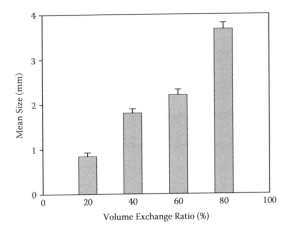

FIGURE 5.4 Comparison of mean size of aerobic granules developed at volume exchange ratios of 20% to 80%. (Data from Wang, Z.-W., Liu, Y., and Tay, J.-H. 2006. *Chemosphere* 62: 767–771.)

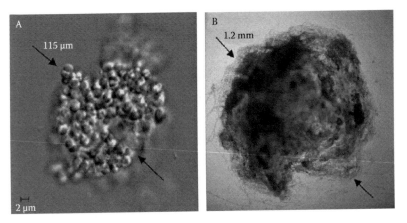

FIGURE 5.5 Morphology of steady-state granules obtained at different minimum settling velocities, (a): 0.6 m h^{-1} and (b): 0.7 m h^{-1}, respectively. (From Kim, S. M. et al. 2004. *Water Sci Technol* 50: 157–162. With permission.)

cultivation of spherical granular sludge (figure 5.5b), while only bioflocs instead of granular sludge were developed in the SBR run at the low volume exchange (figure 5.5a). Similar to the findings by Wang, Liu, and Tay (2006), the time for formation of aerobic granules and to reach steady state was significantly shortened at a high volume exchange ratio (Kim et al. 2004).

5.4 EFFECT OF VOLUME EXCHANGE RATIO ON SLUDGE SETTLEABILITY

In the field of biological wastewater treatment, sludge volume index (SVI) has been used commonly as a good indicator of microbial sludge settleability. Figure 5.6

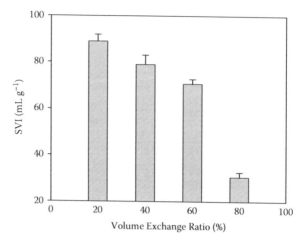

FIGURE 5.6 Sludge volume index (SVI) versus volume exchange ratios in SBRs. (Data from Wang, Z.-W., Liu, Y., and Tay, J.-H. 2006. *Chemosphere* 62: 767–771.)

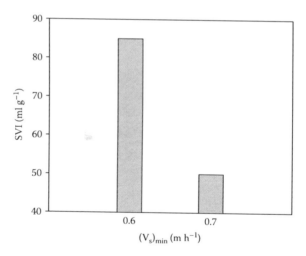

FIGURE 5.7 Sludge volume index (SVI) versus minimum settling velocities $(V_s)_{min}$ determined from the volume exchange ratios. (Data from Kim, S. M. et al. 2004. *Water Sci Technol* 50:157–162.)

shows comparison of the settleability of sludge cultivated at different volume exchange ratios in SBRs. It can be seen that the sludge SVI was inversely correlated to the volume exchange ratio, that is, a sludge with excellent settleability would be developed at a high volume exchange ratio. For example, the settleability of sludge cultivated at the volume exchange ratio of 80% is almost three times superior to that harvested at the volume exchange ratio of 20%. Kim et al. (2004) also reported similar results showing that high volume exchange ratio SBR corresponding to a high $(V_s)_{min}$ could promote the development of sludge with excellent settleability, indicated by a low SVI of 50 mL g^{-1} (figure 5.7).

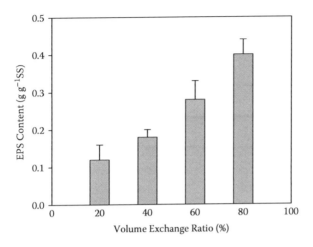

FIGURE 5.8 Extracellular polysaccharide production at different volume exchange ratios. (Data from Wang, Z.-W., Liu, Y., and Tay, J.-H. 2006. *Chemosphere* 62: 767–771.)

5.5 EFFECT OF VOLUME EXCHANGE RATIO ON PRODUCTION OF EXTRACELLULAR POLYSACCHARIDES

Extracellular polysaccharides (PS) are a kind of bioglue that interconnects individual cells into the three-dimensional structure of attached-growth microorganisms (see chapter 10). A high applied volume exchange ratio was found to stimulate cells to produce more PS (figure 5.8). As discussed in chapter 10, PS indeed is not an essential cell component under normal living conditions, and its production is only necessary when microbial cells are subjected to stressful conditions. Figure 5.8 seems to indicate that the high SBR volume exchange ratio can impose a pressure on microbial sludge, leading to an enhanced production of PS.

5.6 EFFECT OF VOLUME EXCHANGE RATIO ON CALCIUM ACCUMULATION IN AEROBIC GRANULES

Calcium ion was accumulated significantly in aerobic granules developed at high volume exchange ratio, for example, the calcium content in granules cultivated at the volume exchange ratio of 80% was almost three times higher than that obtained at the volume exchange ratio of 20% (figure 5.9). Figure 5.10 shows further that the mean size of the aerobic granules tended to increase with the calcium content, while an inverse trend was found for SVI. According to Stokes law, the increase in particle size will improve the settling ability of particles, and this in turn results in a lowered SVI (figure 5.10). The improved settleability of bioparticles can effectively prevent them from being washed out of the SBR at a high volume exchange ratio (figure 5.9). Thus, it is most likely that the selective accumulation of calcium would be a defensive strategy of microbial aggregates to resist the hydraulic discharge from the reactor through the calcium-promoted increases in their size and settleability in terms of SVI (figure 5.10). In fact, it is generally believed that calcium may facilitate

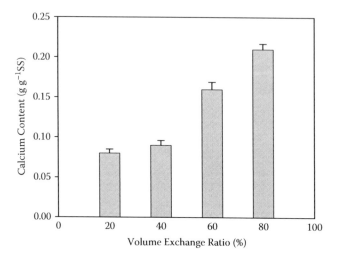

FIGURE 5.9 Calcium content of sludge cultivated at different volume exchange ratios in SBRs. (Data from Wang, Z.-W., Liu, Y., and Tay, J.-H. 2006. *Chemosphere* 62: 767–771.)

anaerobic granulation (Schmidt and Ahring 1996; Yu, Tay, and Fang 2001), while evidence also shows that the removal of calcium from the anaerobic granule matrix results in lowered strength of upflow anaerobic sludge blanket (UASB) granules (Pereboom 1997). Consequently, a certain amount of calcium in biogranules would improve their long-term stability.

5.7 VOLUME EXCHANGE RATIO IS A SELECTION PRESSURE FOR AEROBIC GRANULATION

It appears from chapter 4 that the settling time of SBR can serve as a selection pressure for aerobic granulation, for example, at a short settling time, bioparticles with poor settleability would be washed out according to the minimum settling velocity:

$$(V_s)_{min} = \frac{L}{t_s} \qquad (5.2)$$

in which t_s is settling time and L is the traveling distance of the bioparticles above the discharge port, which is proportionally correlated to the volume exchange ratio of SBR (figure 5.11).

At a designed settling time and discharge height, bioparticles with a settling velocity less than $(V_s)_{min}$ are washed out of the reactor, while those with a settling velocity greater than $(V_s)_{min}$ are retained (figure 5.11). It is obvious that the selection pressure in terms of minimum settling velocity $(V_s)_{min}$ is not only a function of settling time (t), but also depends on the discharge height (L), which can be translated to the volume exchange ratio as given in equation 5.2. This means that the volume exchange ratio can be another essential selection pressure for successful aerobic granulation.

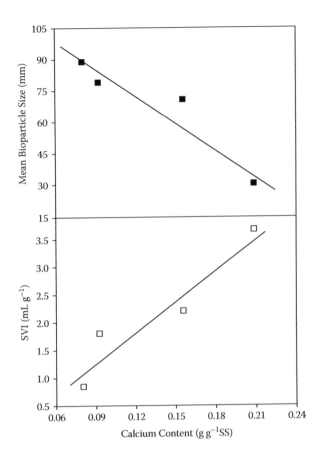

FIGURE 5.10 Correlations among size of bioparticles (■), SVI (□), and calcium content. (Data from Wang, Z.-W., Liu, Y., and Tay, J.-H. 2006. *Chemosphere* 62: 767–771.)

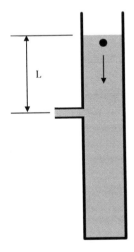

FIGURE 5.11 Schematic diagram of a column SBR.

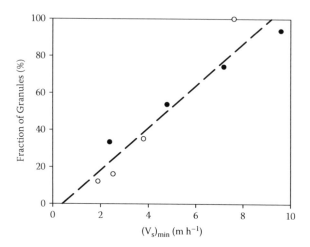

FIGURE 5.12 Fraction of aerobic granules versus $(V_s)_{min}$, obtained from studies of volume exchange ratio (●) and settling time (○). (Data on volume exchange ratio from Wang, Z.-W., Liu, Y., and Tay, J.-H. 2006. *Chemosphere* 62: 767–771; data on settling time from Qin, L., Liu, Y., and Tay, J. H. 2004. *Biochem. Eng. J.* 21: 47–52.)

According to equation 5.2, the minimum settling velocity is the function of settling time and discharge height or volume exchange ratio for an SBR with a given diameter. By controlling $(V_s)_{min}$, bioparticles can be effectively selected according to their respective settleability. This means that selection of bioparticles indeed can be realized by manipulating settling time and volume exchange ratio. To examine the collective effects of SBR volume exchange ratio and settling velocity on aerobic granulation, figure 5.12 shows the correlation of the fractions of aerobic granules in SBRs to $(V_s)_{min}$ calculated from various settling times (see chapter 6) and volume exchange ratios. As expected, the degree of aerobic granulation in an SBR is determined by $(V_s)_{min}$.

It appears from figure 5.12 that at a $(V_s)_{min}$ less than 4 m h^{-1}, only a partial aerobic granulation can be achieved in SBR, and the growth of suspended sludge seems to be promoted in this case. The typical settling velocity of conventional activated sludge is generally less than 5 m h^{-1} (Giokas et al. 2003). This implies that for an SBR operated at a $(V_s)_{min}$ below the settling velocity of conventional activated sludge, suspended sludge cannot be effectively withdrawn. In this case, suspended sludge will easily out compete aerobic granules, which will lead to the instability and even failure of aerobic granular sludge SBRs. Now it is clear that suspended sludge will take over the entire reactor at low $(V_s)_{min}$, as shown in figure 5.12. To achieve rapid and enhanced aerobic granulation in SBRs, the minimum settling velocity $(V_s)_{min}$ must be controlled at a level higher than the settling velocity of suspended sludge (see chapter 6).

5.8 EFFECT OF DISCHARGE TIME ON FORMATION OF AEROBIC GRANULES

As illustrated in figure 5.13, discharge time of SBR (t_d) is defined as the time preset to withdraw the volume of the mixed liquor above the discharge port of the SBR, and

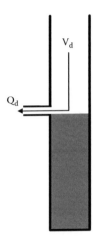

FIGURE 5.13 Illustration of the discharge time.

it can be expressed as the ratio of the discharge volume of SBR (V_e) to the discharge flow rate of the SBR (Q_e):

$$t_d = \frac{V_d}{Q_d} \tag{5.3}$$

Wang (2007) studied the potential effect of discharge time on aerobic granulation in SBRs. For this purpose, four identical SBRs were operated at different discharge times of 5 to 20 minutes, while all other operating conditions were kept at the same levels. Figure 5.14 shows the morphologies of bioparticles developed at the various discharge times. It can be seen that smooth, round aerobic granules were successfully cultivated at a short discharge time of 5 minutes, and only floc-like bioparticles were observed in the SBR operated at the longest discharge time of 20 minutes. Moreover, figure 5.15 shows that the mean size of the bioparticles was inversely related to the applied discharge time. This seems to indicate that a prolonged discharge time would delay or prevent the formation of aerobic granules in SBR even though both settling time and volume exchange ratio are properly controlled.

As discussed in chapter 4, the fraction of aerobic granules over the whole sludge blanket in an SBR represents the degree of aerobic granulation that can be achieved under given operating conditions. Figure 5.16 shows that the fraction of aerobic granules decreased as the applied discharge time was prolonged from 5 to 20 minutes, for example, in the SBR run at the discharge time of 20 minutes, almost no aerobic granules were formed. Similar to settling time and volume exchange ratio, the observed failure of aerobic granulation at the long discharge time may imply that this parameter could also serve as a kind of selection pressure for aerobic granulation.

5.9   EFFECT OF DISCHARGE TIME ON SETTLEABILITY OF BIOPARTICLES

As presented in the preceding chapters, settleability of bioparticles can be evaluated by a simple parameter, namely the SVI. Figure 5.17 shows a comparison of the

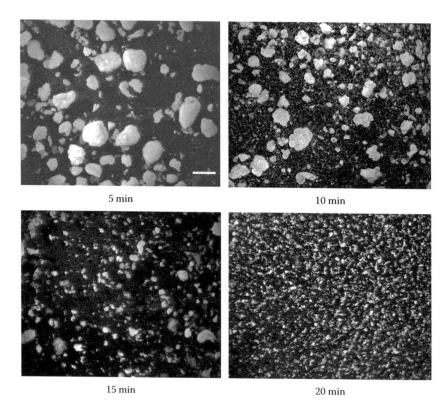

FIGURE 5.14 Morphology of the steady-state sludge cultivated at different discharge times; scale bar: 3 mm. (From Wang, Z.-W. 2007. Ph.D. thesis, Nanyang Technological University, Singapore. With permission.)

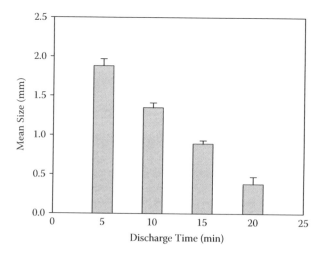

FIGURE 5.15 Mean size of bioparticles versus different discharge times observed in SBRs. (Data from Wang, Z.-W. 2007. Ph.D. thesis, Nanyang Technological University, Singapore.)

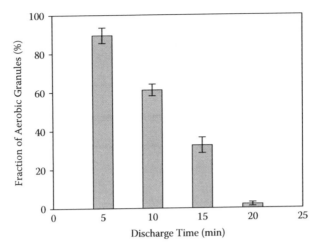

FIGURE 5.16 Fraction of aerobic granules cultivated at different discharge times. (Data from Wang, Z.-W. 2007. Ph.D. thesis, Nanyang Technological University, Singapore.)

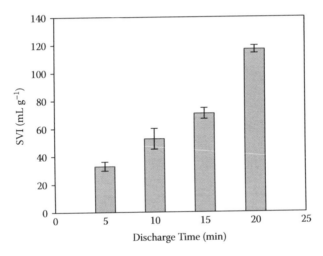

FIGURE 5.17 Sludge volume index (SVI) of sludge cultivated at different discharge times. (Data from Wang, Z.-W. 2007. Ph.D. thesis, Nanyang Technological University, Singapore.)

SVI of bioparticles harvested in SBRs run at different discharge times. High SVI values were obtained at long discharge times. This implies that bioparticles with poorer settleability would be cultivated at the longer discharge time. It should be realized that the SVI of bioparticles obtained from an SBR with a 20-minute discharge time was similar to that of typical activated sludge flocs. This observation is in line with the results presented in figure 5.15. Arrojo et al. (2004) also reported that a long discharge time would lead to a high concentration of total suspended solids in effluent, for example the concentration of total suspended solids in the effluent from an SBR operated at a discharge time of 3 minutes was four times higher than that at 0.5 minutes of discharge time, indicating that the sludge with poor settleability

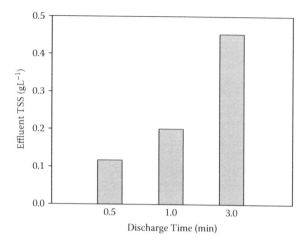

FIGURE 5.18 Total suspended solids (TSS) in the effluents from SBRs operated at different discharge times. (Data from Arrojo, B. et al. 2004. *Water Res* 38: 3389–3399.)

was cultivated at long discharge time (figure 5.17). The poor settleability sludge is actually a practical indicator of unsuccessful aerobic granulation in the SBR. It seems certain that a prolonged SBR discharge time would prevent aerobic granulation in the SBR.

5.10 EFFECT OF DISCHARGE TIME ON CELL SURFACE HYDROPHOBICITY

The effect of discharge time on cell surface hydrophobicity is shown in figure 5.19. A low hydrophobicity was observed at a long discharge time, for example, the cell surface hydrophobicity was increased from 26% for the seed sludge to a stable value of 71%, 65%, 52%, and 39% in SBRs run at discharge times of 5 to 20 minutes. This implies that a microbial community developed at short discharge time would exhibit a high cell surface hydrophobicity. As shown in figure 5.16, incomplete aerobic granulation was observed in SBRs run at the discharge times of 10, 15, and 20 minutes, while successful aerobic granulation was only achieved at the discharge time of 5 minutes. This may be partially attributed to the difference in cell surface hydrophobicity, as detailed in chapter 9.

5.11 EFFECT OF DISCHARGE TIME ON PRODUCTION OF EXTRACELLULAR POLYSACCHARIDES

As shown in the preceding chapters, the production of extracellular polysaccharides (PS) is closely associated with the operation conditions in the SBR. Figure 5.20 shows the effect of discharge time on the sludge PS content. An inverse correlation of the PS production to the applied discharge time was observed, that is, a shorter discharge time would stimulate cells to produce more PS. Microscopic observation

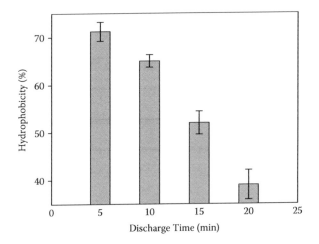

FIGURE 5.19 Cell surface hydrophobicity at different discharge times. (Data from Wang, Z.-W. 2007. Ph.D. thesis, Nanyang Technological University, Singapore.)

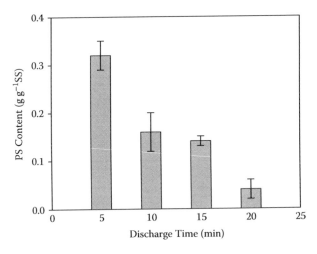

FIGURE 5.20 Polysaccharide content of sludge cultivated at different discharge times. (Data from Wang, Z.-W. 2007. Ph.D. thesis, Nanyang Technological University, Singapore.)

further reveals that bacteria were tightly connected and embedded in the PS matrix of aerobic granules cultivated at the shorter discharge time of 5 minutes (figure 5.21).

5.12 CONCLUSIONS

This chapter provides experimental evidence showing that the volume exchange ratio and discharge time of the SBR are two decisive parameters that highly influence aerobic granulation, and can serve as effective selection pressures for aerobic granulation in an SBR. A high volume exchange ratio favors aerobic granulation, and a short discharge time has the same function.

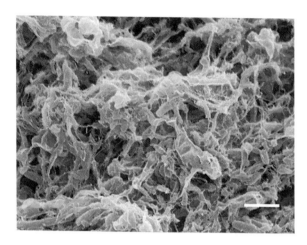

FIGURE 5.21   Filamentous PS observed in aerobic granules, scale bar = 3 μm.

The essential role of volume exchange ratio in aerobic granulation in SBRs can be reasonably interpreted by the concept of minimum settling velocity, while the mechanism by which discharge time can influence aerobic granulation will be further discussed in chapter 6.

REFERENCES

Arrojo, B., Mosquera-Corral, A., Garrido, J. M., and Mendez, R. 2004. Aerobic granulation with industrial wastewater in sequencing batch reactors. *Water Res* 38: 3389–3399.
Giokas, D. L., Daigger, G. T., von Sperling, M., Kim, Y., and Paraskevas, P. A. 2003. Comparison and evaluation of empirical zone settling velocity parameters based on sludge volume index using a unified settling characteristics database. *Water Res* 37: 3821–3836.
Kim, S. M., Kim, S. H., Choi, H. C., and Kim, I. S. 2004. Enhanced aerobic floc-like granulation and nitrogen removal in a sequencing batch reactor by selection of settling velocity. *Water Sci Technol* 50: 157162.
Pereboom, J. H. F. 1997. Strength characterisation of microbial granules. *Water Sci Technol* 36: 141–148.
Qin, L., Liu, Y., and Tay, J. H. 2004. Effect of settling time on aerobic granulation in sequencing batch reactor. *Biochem Eng J* 21: 47–52.
Schmidt, J. E. and Ahring, B. K. 1996. Granular sludge formation in upflow anaerobic sludge blanket (UASB) reactors. *Biotechnol Bioeng* 49: 229–246.
Wang, Z.-W. 2007. Insights into the mechanism of aerobic granulation in sequencing batch reactor. Ph.D. thesis. Nanyang Technological University, Singapore.
Wang, Z.-W., Liu, Y., and Tay, J.-H. 2006. The role of SBR mixed liquor volume exchange ratio in aerobic granulation. *Chemosphere* 62: 767–771.
Yu, H. Q., Tay, J. H., and Fang, H. H. P. 2001. The roles of calcium in sludge granulation during UASB reactor start-up. *Water Res* 35: 1052–1060.

6 Selection Pressure Theory for Aerobic Granulation in Sequencing Batch Reactors

Yu Liu and Zhi-Wu Wang

CONTENTS

6.1 INTRODUCTION

It appears from the preceding chapters that almost all research on aerobic granulation has been conducted in sequencing batch reactors (SBRs), while no successful aerobic granulation has been observed in continuous microbial culture. As shown in chapter 5, it is believed that aerobic granules form through self-immobilization of bacteria when suitable selection pressure is provided in the SBR. Compared to continuous microbial culture, the unique feature of an SBR is its ability to be operated in a cyclic mode, and a cycle of SBR for aerobic granulation may comprise filling, aeration, settling, and sludge discharge.

A number of parameters have been known to influence the properties of aerobic granules formed in SBRs. Basically, contributing parameters include substrate composition, organic loading, hydrodynamic shear force, feast-famine regimen, feeding strategy, dissolved oxygen, reactor configuration, solids retention time, cycle time, settling time, and volume exchange ratio, while only the parameters associated with selection pressure on sludge particles can contribute to the formation of aerobic granules, as shown in chapters 4 and 5. In an SBR, three major selection pressures have been identified, and they are settling time, volume exchange ratio, and discharge time (see chapters 4 and 5). In fact, selection pressure in terms of upflow velocity has been shown to serve as an effective driving force towards successful anaerobic granulation in upflow anaerobic sludge blanket (UASB) reactors (Hulshoff Pol, Heijnekamp, and Lettinga 1988; Alphenaar, Visser, and Lettinga 1993). The key to successful aerobic granulation is to identify and model selection pressures; thus, this chapter attempts to offer an overview of the major selection pressures for aerobic granulation in SBRs, and subsequently a unified selection pressure theory is described as well.

6.2   IS AEROBIC GRANULATION INDUCIBLE?

In view of modern molecular biology, the information on aerobic granulation may reside in the genetic makeup of the microbial species involved. According to Calleja (1984), "the deposition of structural and regulatory genes may determine whether the aggregation function of cells is constitutive or inducible." If the capability of microorganisms for aerobic granulation is constitutive, that is, whatever stage the cell is in with regard to its cell cycle or its life cycle, aerobic granulation will be present, provided the environmental conditions allow it to occur. On the contrary, if it is inducible, it will be present only when the cells are physiologically competent under given conditions.

Aerobic granulation has been observed in the cultures of different microbial species. Such experimental evidence indeed supports the view that aerobic granulation is a microbial self-aggregation induced by environmental conditions through changing microbial surface properties and metabolic behaviors, as shown in the preceding chapters. Therefore, it is reasonable to consider that aerobic granulation would be species-independent, and represents an inducible rather than constitutive microbiological phenomenon (Y. Liu et al. 2005b).

6.3   EARLIER UNDERSTANDING OF AEROBIC GRANULATION

Based on microscopic observations, Beun et al. (1999) proposed a schematic mechanism of aerobic granulation in SBR (figure 6.1). This mechanism shows that the growth of filamentous fungi is a prerequisite of aerobic granulation. After reactor seeding, these fungi can easily form pellets with a compact structure under hydrodynamic shear conditions, and then the fungal pellet can settle quickly, while other bacteria without this property are washed out of the SBR. Obviously, these pellets may provide a protective matrix in which bacteria can further grow into colonies up to a diameter of 5 to 6 mm. Subsequently, the large pellets break up due to microbial lysis in their inner parts, probably caused by oxygen and nutrient limitations. Because of their good settleability, bacterial colonies produced from the breakup of the fungal pellets are easily retained in the SBR, and further grow to aerobic granules.

Development of aerobic granules requires aggregation of microorganisms. For bacteria in a culture to aggregate, a number of conditions have to be met. So far, it has been believed that intercellular communication and multicellular coordination are crucial for bacteria to achieve an organized spatial structure. According to research on cell-to-cell communication in biofilms (Davies et al. 1998; Pratt and Kolter 1998), it is a reasonable consideration that a cell-to-cell signaling mechanism would also be involved in the formation of aerobic granules, as well as in the organization of the spatial structure of granule-associated bacteria. In the study of aerobic granulation by two coaggregating bacterial strains, it was found that the coaggregating bacterial strains could produce autoinducer-like signals during aerobic granulation (Jiang et al. 2006). The benefits of an organized microbial structure include more efficient proliferation, access to resources and niches that cannot be utilized by isolated cells, collective defense against antagonists that eliminate isolated cells, and optimization of population survival by differentiation into distinct cell types (Shapiro 1998). Obviously, a sound understanding of the cell-to-cell communication in aerobic granulation is essential.

Y. Liu and Tay (2002) proposed a generic four-step model for aerobic granulation.

Step 1: Physical movement to initiate bacterium-to-bacterium contact or bacterial attachment onto nuclei

Step 2: Initial attractive forces to keep stable multicellular contacts

Step 3: Microbial forces to make cell aggregation mature

Step 4: Steady-state three-dimensional structure of microbial aggregate shaped by hydrodynamic shear forces

The microbial aggregates would be finally shaped by hydrodynamic shear force to form a certain structured community. The outer shape and size of microbial aggregates are determined by the interactive strength/pattern between aggregates and of hydrodynamic shear force, microbial species, and substrate loading rate. This four-step model for aerobic granulation, as well as that shown in Figure 10.1, still cannot explain what is the key driving force of aerobic granulation. In this regard, a more profound understanding of the mechanisms responsible for aerobic granulation is needed.

6.4   BRIEF REVIEW OF PARAMETERS CONTRIBUTING TO AEROBIC GRANULATION

Aerobic granulation is the gathering together of cells through cell-to-cell immobilization to form a fairly stable and multicellular association. Evidence shows that aerobic granulation is a gradual process from seed sludge to compact aggregates, further to granular sludge and finally to mature granules (see chapter 1). Obviously, for cells in

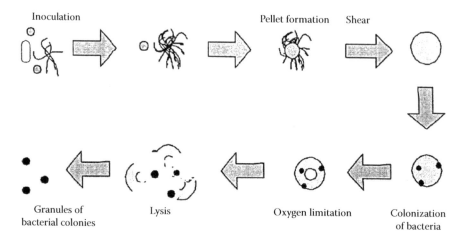

FIGURE 6.1 Illustration of aerobic granulation proposed by Beun et al., (1999).

a culture to aggregate, a number of conditions have to be fulfilled. The focus of this section is thus to identify the main driving forces of aerobic granulation in SBR.

6.4.1 SUBSTRATE COMPOSITION AND LOADING

As shown in chapter 1, aerobic granules have been cultivated successfully for treating a wide variety of wastewaters. It is evident that the formation of aerobic granules is independent of or insensitive to the characteristics of feed wastewater, while the microbial structure and diversity of mature aerobic granules are closely related to the type of wastewater (chapter 1).

The essential role of organic loading in the formation of aerobic granules was discussed in chapter 1. It has been found that relatively high organic loading facilitates the formation of anaerobic granules in upflow anaerobic sludge blanket (UASB) reactors (Hulshoff Pol, Heijnekamp, and Lettinga 1988; Kosaric et al. 1990). This is due mainly to the fact that the high organic loading-enhanced biogas production results in an increased upflow liquid velocity known as the major selection pressure for anaerobic granulation in the UASB reactor (Hulshoff Pol, Heijnekamp, and Lettinga 1988). In contrast to anaerobic granulation, it appears from chapter 1 that aerobic granules can form across a very wide range of organic loading rates from 2.5 to 15 kg COD m^{-3} day^{-1}, while nitrifying and P-accumulating granules can also be developed at a very wide range of ammonia-nitrogen and phosphate loadings. These indicate that the substrate loading in the range studied so far is not a determinant of aerobic granulation in SBRs. As concluded in chapter 1, aerobic granulation in SBRs would be substrate concentration-independent, but the kinetic behavior of aerobic granules is related to the applied substrate loading (see chapter 7).

6.4.2 HYDRODYNAMIC SHEAR FORCE

In a bubble column or airlift SBR, hydrodynamic shear force is created mainly by aeration that can be described roughly by the upflow air velocity (see chapter 2).

A higher shear force favors the formation of more compact and denser aerobic granules (chapter 2). Similar to the formation of biofilms, aerobic granules can form at different levels of hydrodynamic shear forces. It is believed that the structure of mature aerobic granules is determined by hydrodynamic shear force, but there is no concrete evidence to show that shear force is a primary inducer of aerobic granulation in SBRs.

6.4.3 FEAST-FAMINE REGIMEN

An SBR is operated in a sequencing cycle of feeding, aeration, settling, and discharge of supernatant. In SBR, the aeration period consists of two phases: a degradation phase in which the substrate is depleted to a minimum, followed by an aerobic starvation phase in which the external substrate is no longer available. It is likely that microorganisms in the SBR are subject to a periodic feast and famine regimen, called periodic starvation (Tay, Liu, and Liu 2001). There is evidence showing that bacteria become more hydrophobic under the periodic feast-famine conditions, and high cell hydrophobicity in turn facilitates microbial aggregation (Bossier and Verstraete 1996). In fact, the periodic feast-famine regimen in SBRs can be regarded as a kind of microbial selection pressure that may alter the surface properties of the cell. However, it has been revealed in the preceding chapters that aerobic granules cannot successfully be developed if the settling time in the SBR is not properly controlled even though the periodic feast-famine regimen was present. As shown in chapter 14, short-term C, N, P, and K starvations reduce granule extracellular polysaccharide content, inhibit microbial activity, weaken structural integrity, and subsequently worsen settleability of aerobic granules. So far, no solid experimental evidence shows that starvation can act as a trigger of aerobic granulation in SBRs.

6.4.4 FEEDING STRATEGY

McSwain, Irvine, and Wilderer (2004) reported that intermittent feeding is an effective operating strategy for enhancing aerobic granulation in SBRs. For this purpose, different filling times were applied to SBRs, resulting in different degrees of feast-famine to microorganisms. A high feast-famine ratio or pulse feeding to the SBR was found to be favorable for the formation of compact and dense aerobic granules. This seems to indicate that the feeding strategy may influence the characteristics of aerobic granules formed in an SBR, but it is unlikely to play the role of a trigger of aerobic granulation.

6.4.5 DISSOLVED OXYGEN

Dissolved oxygen (DO) concentration is an important parameter in the operation of aerobic wastewater treatment processes. Evidence shows that aerobic granules can form at DO concentrations as low as 0.7 to 1.0 mg L^{-1} in an SBR (Peng et al. 1999; Tokutomi 2004), while they can also be successfully developed at relatively high DO concentrations of 2 to 6 mg L^{-1} (Tsuneda et al. 2003; Yang, Tay, and Liu 2003; Qin, Liu, and Tay 2004a). Obviously, if an aerobic condition is maintained by sufficient aeration, the DO concentration would not be a decisive parameter of aerobic granulation.

6.4.6 Reactor Configuration

In a column SBR a higher ratio of reactor height (H) to diameter (D) can ensure a circular flow trajectory, which in turn creates a more effective hydraulic attrition to microbial aggregates. On the other hand, a high H/D ratio also improves oxygen transfer. Q. S. Liu (2003) looked into aerobic granulation in a column-type continuous activated sludge reactor, and found that aerobic granulation failed, while Pan (2003) showed that aerobic granules could be developed in SBRs with various H/D ratios. These studies indicate that aerobic granulation may not be associated with the H/D ratio.

6.4.7 Solids Retention Time

Y. Li (2007) systematically investigated the role of solids retention time (SRT) in aerobic granulation in SBR, and found that SRT up to 40 days had no significant influence on aerobic granulation (figure 6.2). It is apparent that a complete aerobic granular sludge blanket was not developed over the SRT range of 3 to 40 days if selection pressures were too weak in the SBR (Y. Li 2007). In fact, in the past 100 years of research and application history of the conventional activated sludge process, aerobic granulation has never been reported in the processes operated in an extremely wide range of SRT. Thus, there is no reason to believe that SRT would be an inducer of aerobic granulation in SBR.

6.4.8 Cycle Time

If an SBR is run at an extremely short cycle time, microbial growth should be suppressed by insufficient reaction time for bacteria to break down substrates. As a result, the sludge loss due to hydraulic washout cannot be compensated for by

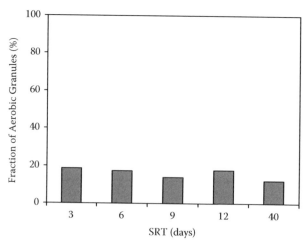

FIGURE 6.2 Fraction of aerobic granules versus solids retention time (SRT) in SBRs operated at extremely low selection pressures. (Data from Li, Y. 2007. Ph.D. thesis, Nanyang Technological University, Singapore.)

bacterial growth. For example, a complete washout of the sludge blanket and subsequent failure of nitrifying granulation was observed in an SBR run at a very short cycle time (see chapter 3). On the contrary, if the cycle time is kept much longer than that required for complete degradation of substrates, hydrolysis or decay of biomass occurs and eventually causes a negative effect on microbial aggregation (chapter 3).

Pan et al. (2004) reported that at the shortest HRT of 1 hour, the strong hydraulic pressure triggered biomass washout and led to reactor failure, while at the longest HRT of 24 hours, aerobic granules were gradually substituted by bioflocs. Therefore, it seems reasonable to consider that the cycle time of SBRs should be short to suppress biomass hydrolysis, but long enough for biomass growth and accumulation in the system. However, even for SBRs operated at the optimum cycle time, aerobic granulation still failed if the settling time was kept longer than 15 minutes (see chapter 3). Consequently, cycle time is not a decisive factor in aerobic granulation in SBR.

6.4.9 Settling Time

In a column SBR, wastewater is treated in successive cycles of a few hours. At the end of a cycle, settling of the biomass takes place before the effluent is withdrawn. Sludge that cannot settle down within the preset settling time is washed out of the reactor through a fixed discharge port, as illustrated in figure 6.3. Basically, a short settling time preferentially selects for the growth of fast-settling bioparticles. Thus, the settling time acts as a major hydraulic selection pressure exerted on the microbial community. As discussed in chapter 4, aerobic granules were successfully cultivated and became dominant only in the SBRs operate at a settling time of less than 5 minutes, while a mixture of aerobic granules and suspended sludge was developed in the SBRs run at the longer settling times. So far, a short settling time has been commonly practiced as an effective means of control to enhance aerobic

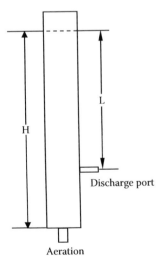

FIGURE 6.3 Schematic of a column-type SBR for aerobic granulation.

granulation in SBRs (Jiang, Tay, and Tay2002; Lin, Liu, and Tay 2003; Q. S. Liu, Tay and Liu 2003; Y. Liu, Yang, and Tay 2003; Yang, Tay. and Liu 2003; Wang, Du, and Chen2004; Hu et al. 2005). At a long settling time, poorly settling bioflocs cannot be withdrawn effectively, and they may in turn out compete granule-forming bioparticles (see chapter 7). This points to the fact that settling time can be regarded as a decisive factor in aerobic granulation in SBRs.

6.4.10 Exchange Ratio

The exchange ratio in an SBR is defined as the liquid volume withdrawn at the end of the given settling time over the total reactor working volume (see chapter 5). For column SBRs with the same diameter, the exchange ratio is proportionally related to the height (L) of the discharge port from the water surface (figure 6.3). A larger exchange ratio is associated with a higher L. The fraction of aerobic granules in the total biomass was found to be proportionally related to the exchange ratio, for example, only in the SBRs run at the higher exchange ratios of 60% and 80% were aerobic granules dominant, and a mixture of aerobic granules and bioflocs instead of pure aerobic granules developed at smaller exchange ratios of 40% and 20% (see chapter 5). It appears that aerobic granulation is highly dependent on the exchange ratio of the SBR.

6.4.11 Discharge Time

The essential role of discharge time in aerobic granulation in SBRs has been demonstrated in chapter 5. A prolonged discharge time results in a failure of aerobic granulation even though both settling time and volume exchange ratio were properly controlled, that is, the discharge time of effluent from the SBR is one of the key parameters that determine aerobic granulation in an SBR. To develop a unified theory for aerobic granulation in SBRs, the role of discharge time in aerobic granulation should be taken into account seriously.

6.5 MAIN SELECTION PRESSURES OF AEROBIC GRANULATION

Aerobic granulation is a microbial phenomenon that is induced by selection pressure through changing microbial surface properties and metabolic behavior, as documented in the preceding chapters. Compared to continuous microbial culture, SBR is a fill-and-draw process that is fully mixed during the batch reaction step. The sequential steps of aeration and clarification in an SBR occur in the same tank (Metcalf and Eddy 2003). The operation of nearly all SBRs employed for aerobic granulation comprises four steps: feeding, aeration, settling, and discharge (figure 6.4).

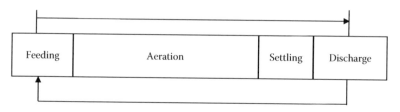

FIGURE 6.4   Cycle operation of an SBR for aerobic granulation.

It appears from the discussion in section 6.4, that settling time, volume exchange ratio, and discharge constitute the main selection pressures on aerobic granulation in SBR, that is, no matter how other variables are manipulated, aerobic granulation would not be successful without proper control of these three main selection pressures in the SBR. This means that optimization and scale up of an aerobic granular sludge SBR must obviously take account of these selection pressures.

6.6 A SELECTION PRESSURE THEORY FOR AEROBIC GRANULATION IN SBRS

It is now clear that the settling time, exchange ratio, and discharge time in SBRs are the most effective selection pressures for aerobic granulation. Successful and stable aerobic granulation in SBRs closely depends on those applied selection pressures. Y. Liu, Wang, and Tay (2005) proposed a selection pressure theory by which the three identified key parameters can be unified into an easy concept of minimum settling velocity of bioparticles. Following is a discussion of this approach.

In present operation of a column SBR for aerobic granulation, the effluent is discharged at a discharge outlet (figure 6.3), that is, the volume of mixed liquor above the discharge port is withdrawn immediately at the end of the preset settling time. As shown in figure 6.3, for an SBR with a given diameter, the volume exchange ratio translates to the suspension discharge depth. According to the well-known Stokes formula, the settling velocity of a particle can be calculated as follows:

$$V_s = \frac{g(\rho_p - \rho)d_p^2}{18\mu} \tag{6.1}$$

in which V_s is the settling velocity of the particle, d_p is the diameter of the particle, ρ_p is the density of the particle, ρ is the density of the solution, and μ is the viscosity of the solution. Equation 6.1 shows that the settling velocity of the particle is determined mainly by the density and diameter of aggregates in an SBR.

For a column SBR (figure 6.5a) with the effluent discharged at an outlet located at depth L, that is, at the end of the designed settling time (t_s), the volume of suspension above the discharge port will be withdrawn during the preset discharge time (t_d). If the distance for bioparticles to travel to the discharge port is L, the corresponding travel time of the bioparticles is given by:

$$\text{Traveling time to the discharge port} = \frac{L}{V_s} \tag{6.2}$$

in which V_s is the settling velocity of the bioparticles. As can be seen in figure 6.5a, L is proportionally related to the volume exchange ratio.

Equation 6.2 shows that a high V_s results in short travel time of bioparticles to the discharge port. This implies that bioparticles with a travel time that is longer than the designed settling time will be discharged out of the SBR. Thus, a minimum settling

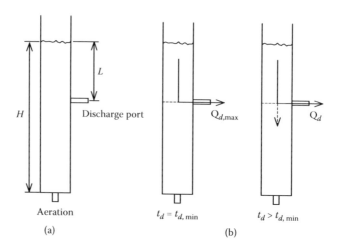

FIGURE 6.5   (a) Schematic of a column SBR; (b) hypothetical flows during discharge.

velocity, $(V_s)_{min}$ exists for the bioparticles to be retained in the reactor. According to Y. Liu, Wang, and Tay(2005), $(V_s)_{min}$ can be defined as:

$$(V_s)_{min} = \frac{L}{\text{effective settling time}} \tag{6.3}$$

As discussed in chapter 9, there appears to exist a minimum discharge time $t_{d,min}$ at which the fraction of aerobic granules in the SBR is close to 100%, that is, a full granular sludge blanket can be developed at $t_{d,min}$. If the discharge time (t_d) is set to be longer than $t_{d,min}$, a portion of the liquor above the discharge port will continue to settle during discharge time t_d, and this will eventually lower the effective selection pressure on microorganisms (figure 6.5b). Therefore, for $t_d > t_{d,min}$, the settling time should be calibrated in order to account for the effect of the longer discharge time. According to Y. Liu, Wang, and Tay (2005), the effective settling time involved in equation 6.3 can be expressed as follows:

$$\text{Effective settling time} = \text{settling time preset } (t_s)$$
$$+\text{relaxation of settling time due to } t_d \tag{6.4}$$

Y. Liu, Wang, and Tay (2005) further thought that if the discharge flow rates at t_d and $t_{d,min}$ are Q_d and $Q_{d,max}$, respectively, they can be calculated in a way such that:

$$Q_{d,max} = \frac{V_e}{t_{d,min}} \tag{6.5}$$

$$Q_d = \frac{V_e}{t_d} \tag{6.6}$$

V_e in equation 6.5 is the exchange volume above the discharge port, as shown in figure 6.5b, and the hypothetical flow that can settle is $Q_{d,max} - Q_d$ (figure 6.5b). Thus, the relaxation of settling time due to t_d can be given as:

$$\text{Relaxation of settling time due to } t_d = \frac{Q_{d,max} - Q_d}{Q_{d,max}} \times \left(t_d - t_{d,min}\right) \qquad (6.7)$$

Substitution of equations 6.5 and 6.6 in equation 6.7 yields the following equation:

$$\text{Relaxation of setting time due to } t_d = \left(1 - \frac{t_{d,min}}{t_d}\right)\left(t_d - t_{d,min}\right) = \frac{\left(t_d - t_{d,min}\right)^2}{t_d} \qquad (6.8)$$

In this case, equation 6.4 becomes:

$$\text{Effective settling time} = t_s + \frac{\left(t_d - t_{d,min}\right)^2}{t_d} \qquad (6.9)$$

Combining equation 6.9 with equation 6.3 leads to:

$$\left(V_s\right)_{min} = \frac{L}{t_s + \dfrac{\left(t_d - t_{d,min}\right)^2}{t_d}} \qquad (6.10)$$

Equation 6.10 integrates the three major selection pressures (i.e., t_s, t_d, and L) in SBRs into an easy concept of the minimum settling velocity required for successful aerobic granulation. Basically, fast-settling bioparticles are heavy spherical aggregates, whereas the slow-settling particles are small, light, and have irregular shapes. Clearly, bioparticles can be selected according to their settling velocity, and this has been confirmed in laboratory-scale aerobic granular sludge SBRs (Y. Liu et al. 2005a). Equation 6.10 reasonably explains why t_s, L, and t_d in SBRs can serve as the effective selection pressures and the way that they determine aerobic granulation.

The decisive effect of $(V_s)_{min}$ on aerobic granulation in stable SBRs operated at different selection pressures is shown in figure 6.6. It can be seen that the fraction of aerobic granules expressed as the ratio of biomass of aerobic granules to the total biomass tends to increase with the increase of $(V_s)_{min}$. At $(V_s)_{min}$ values smaller than 1.0 m·h^{-1}, only suspended bioflocs are cultivated and no aerobic granules are developed. As the $(V_s)_{min}$ increases above 1.0 m·h^{-1}, an aerobic granular sludge blanket starts to appear. At the $(V_s)_{min}$ value of 4.0 m·h^{-1}, aerobic granules prevail over suspended flocs (figure 6.6). As the typical settling velocity of suspended activated sludge is generally less than 3 to 5 m·h^{-1} (Giokas et al. 2003), figure 6.6 seems to indicate that if the SBR is operated at a $(V_s)_{min}$ value below that of suspended flocs, suspended sludge will not be effectively washed out of the reactor.

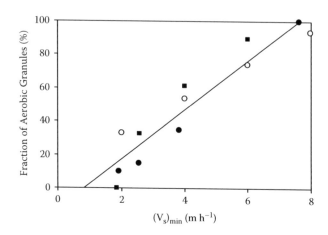

FIGURE 6.6 Relationship between the mass fraction of aerobic granules and $(V_s)_{min}$. (●) at different settling times (Qin, Liu, and Tay 2004a); (○) at various volume exchange ratios (Z.-W. Wang, Liu, and Tay 2006); (■) at different discharge times (Z.-W. Wang 2007).

The specific growth rates and growth yield of aerobic granules are lower than that of suspended activated sludge (see chapter 7), that is, suspended sludge can easily out compete aerobic granules. This in turn represses the formation and growth of aerobic granules and eventually leads to disappearance of the aerobic granular sludge blanket in the SBR if suspended sludge is not effectively withdrawn. It is the main reason $(V_s)_{min}$ must be controlled at a level higher than the settling velocity of suspended sludge, otherwise successful aerobic granulation will not be achieved and maintained stably. Equation 6.10 indicates that enhanced selection of bioparticles for rapid aerobic granulation can be realized through properly controlling and adjusting settling time, discharge time, and the volume exchange ratio (or depth of discharge port) in SBRs (Y. Liu, Wang, and Tay 2005). However, compared to the exchange ratio and discharge time, control of the settling time is more flexible in terms of manipulation in a full-scale SBR operation.

It also appears from Equation 6.10 that the H/D ratio of an SBR may not serve as a selection pressure for aerobic granulation; nevertheless, a larger H/D ratio may be desirable in the design of a full-scale SBR because it may allow more space for operation engineers to manipulate L and subsequent $(V_s)_{min}$ according to needs. In addition, as shown in the preceding chapters, selection pressures have a profound effect on the surface properties of aerobic granules in terms of cell surface hydrophobicity and extracellular polysaccharides (EPS), which in turn favor the formation of aerobic granules in an SBR. Similarly, the selection pressure-associated cell surface hydrophobicity and EPS production was also observed in anaerobic granulation in an upflow anaerobic sludge blanket (UASB) reactor (Mahoney et al. 1987; Schmidt and Ahring 1996). As equation 6.1 shows, the settling velocity of particles is closely related to the diameter of aggregates. It is likely that microbial granulation induced by selection pressures is an effective microbial survival strategy that enables the bacteria to aggregate into big granules and consequently to avoid being discharged out of the reactor.

FIGURE 6.7 Morphology of seed sludge. (From Liu, Q. S. 2003. Ph.D. thesis, Nanyang Technological University, Singapore. With permission.)

6.7 FAILURE OF AEROBIC GRANULATION IN CONTINUOUS MICROBIAL CULTURE

The selection pressure theory for aerobic granulation is further supported by experimental observations in a continuous activated sludge reactor in which selection pressure in terms of the minimum settling velocity (equation 6.10) is absent or extremely weak. Q. S. Liu (2003) reported the failure of aerobic granulation in a column-type continuous activated sludge reactor in which hydraulic selection pressure was almost negligible. The seed sludge had an average size of 0.07 mm, and exhibited a typical morphology of conventional activated sludge flocs (figure 6.7). After 3 weeks of operation, aerobic granules appeared in the SBR. However, only bioflocs with a size of 0.1 mm prevailed in the continuous activated sludge reactor over the whole experimental period. The morphology of bioparticles present in both the reactors on day 23 is shown in figure 6.8 and figure 6.9. It can be seen that aerobic granules with a clear, round-shaped structure were successfully cultivated in the SBR (figure 6.8), but only a fluffy, irregular, loose-structured sludge was developed in the continuous activated sludge reactor (figure 6.9).

Figure 6.10 shows further changes in sludge size as a function of operation time in both the continuous activated sludge reactor and the SBR. It was found that the sludge size in the SBR gradually increased up to a relatively stable value of 0.50 mm after 120-cycle operation; however, the sludge size fluctuated at the level of around 0.1 mm and no granulation was observed in a continuous activated sludge reactor. The sludge settling property in terms of sludge volume index (SVI) improved along with aerobic granulation in the SBR (figure 6.11), for example, the SVI value dropped from 190 mL g^{-1} at the beginning to an average value of 56 mL g^{-1} after aerobic granulation. In contrast, the SVI value in the continuous activated sludge reactor fluctuated at 200 mL g^{-1}, which was similar to that of seed sludge.

The continuous activated sludge process has been used to treat an extremely wide variety of wastewaters for the removal of organics, nitrogen, and phosphorus

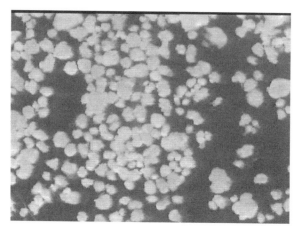

FIGURE 6.8   Morphology of aerobic granule observed on day 23 in an SBR. (From Liu, Q. S. 2003. Ph.D. thesis, Nanyang Technological University, Singapore. With permission.)

FIGURE 6.9   Morphology of bioflocs observed on day 23 in a continuous activated sludge reactor. (From Liu, Q. S. 2003. Ph.D. thesis, Nanyang Technological University, Singapore. With permission.)

for more than a century; however, up to now, no successful aerobic granulation has been reported in this process. So far, nearly 100% of aerobic granules are produced in sequencing batch reactors only. As shown in figures 6.10 and 6.11, the feasibility and efficiency of a continuous activated sludge reactor in the development of aerobic granular sludge was not sufficiently demonstrated. In the sense of reaction operation, the SBR and complete-mix continuous activated sludge reactor (figure 6.12) have very different behaviors in terms of selection pressure. The selection pressure in an SBR can be created and further manipulated as elaborated by the concept of the minimum settling velocity (equation 6.10), while in a complete-mix continuous activated sludge reactor, it is almost impossible to generate such a minimum settling velocity by which bioparticles would be selected according to their settleability. This is due mainly to the fact that in the complete-mix continuous activated sludge

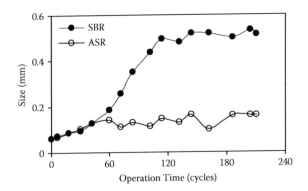

FIGURE 6.10 Change in sludge size versus operation time in an SBR and continuous activated sludge reactor (ASR). (From Liu, Q. S. 2003. Ph.D. thesis, Nanyang Technological University, Singapore. With permission.)

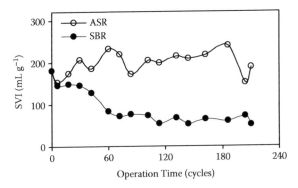

FIGURE 6.11 Change in sludge volume index (SVI) observed in an SBR and continuous activated sludge reactor (ASR). (From Liu, Q. S. 2003. Ph.D. thesis, Nanyang Technological University, Singapore. With permission.)

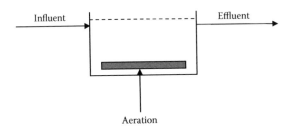

FIGURE 6.12 Schematic illustration of a complete-mix continuous activated sludge reactor.

reactor, sludge is discharged out of the reactor in a complete-mix state, that is, quick- and slow-settling bioparticles would be equally thrown away without any selection.

The selection pressure theory for aerobic granulation clearly shows that selective washout of suspended flocs according to their settling velocities is essential for rapid and successful aerobic granulation in an SBR. Through regular removal of completely mixed bioflocs from the aerobic granular sludge SBR, stable aerobic granules with good settleability can be produced and maintained (Z. H. Li, Kuba, and Kusuda 2006).

The upflow anaerobic sludge blanket (UASB) reactor is operated in a continuous mode, and it was thought that anaerobic granulation in the UASB reactor is driven by hydraulic selection pressure in terms of the liquid upflow velocity, V_{up} (Hulshoff Pol, Heijnekamp, and Lettinga 1988; Alphenaar, Visser, and Lettinga 1993). The liquid upflow velocity has a dual effect on anaerobic granulation, that is, low V_{up} does not encourage or enhance anaerobic granulation, whereas if V_{up} is too high, sludge washout from the UASB reactor can occur, and eventually lead to failure of anaerobic granulation. Theoretically, the critical liquid upflow velocity that can be applied to a UASB reactor should be equal to V_s. In practice, in order to maintain the long-term stability of the UASB reactor and to avoid eventual washout of the anaerobic granular sludge blanket, the liquid upflow velocity applied to the UASB reactor must be much lower than V_s. Thus, information on the settling velocity is critical in the determination of the allowable upflow velocity of the anaerobic granular sludge in UASB reactors. This means that the proposed selection pressure theory may also be applicable for anaerobic granulation.

6.8   UPSCALING AEROBIC GRANULAR SLUDGE SBRS

Aerobic granules have been developed successfully in laboratory-scale SBRs with aspect ratios of 1.9 to 20 (Morgenroth et al. 1997; Beun et al. 1999; Tay, Liu, and Liu 2001; Yang, Tay, and Liu 2003; McSwain, Irvine, and Wilderer 2004; Qin, Liu, and Tay 2004a; Schwarzenbeck, Erley, and Wilderer 2004; Zheng, Yu, and Sheng 2005). It has been proposed that SBRs should have a high aspect ratio to improve selection of granules that settle well (Beun, van Loosdrecht, and Heijnen 2002). However, from equation 6.10, aspect ratio does not appear to influence selection pressure in an SBR. Nevertheless, a large aspect ratio is desirable because it offers more flexibility for operators to manipulate L and, therefore, $(V_s)_{min}$.

In upscaling an aerobic granular sludge SBR, settling time, discharge time, and volume exchange ratio (or depth of the outlet port) must be collectively controlled according to equation 6.10. Compared to the volume exchange ratio and discharge time, control of the settling time is more flexible in manipulating the operation of a full-scale SBR. To avoid initial washout of biomass, settling time should be gradually shortened, for example, from 20 minutes to 2 minutes (Lin, Liu, and Tay 2003; Qin, Liu, and Tay 2004a, 2004b; Tay et al. 2004; Hu et al. 2005). According to figure 6.4, $(V_s)_{min}$ for enhanced aerobic granulation should not be less than 8 m·h^{-1}. Successful aerobic granulation has been obtained at settling velocities of 10 and 16.2 m·h^{-1} (Beun, van Loosdrecht, and Heijnen 2000, 2002).

An essential aspect of the design of an aerobic granular sludge SBR is the estimation of the discharge time. Discharge time greatly influences the formation of aerobic granules (equation 6.10) and determines the discharge pumping rate (equation 6.5) that relates to energy consumption. In practice, engineers and operators have limited space for manipulating the volume exchange ratio or depth of the outlet port of the reactor. Indeed, most laboratory-scale aerobic granular sludge SBRs are actually operated at a fixed volume exchange ratio (Morgenroth et al. 1997; Tay, Liu, and

Liu 2001; Lin, Liu, and Tay 2003; McSwain, Irvine, and Wilderer 2004; Q. Wang, Du, and Chen 2004). In practice it may be preferable to control the settling time and discharge time in order to achieve the minimum settling velocity required for aerobic granulation; hence, equation 6.10 can be rewritten as follows:

$$t_s = \frac{L}{\left(V_s\right)_{min}} - \frac{\left(t_d - t_{d,min}\right)^2}{t_d} \tag{6.11}$$

The following example considers a full-scale column-type SBR with a diameter of 4 m and a working height of 8 m, and a volume exchange ratio of 50% is assumed. This corresponds to a discharge depth (L) of 4 m and a minimum discharge time of 5 minutes. The latter value is based on laboratory studies as presented in chapter 5. If the minimum settling velocity is increased in steps of 2 m·h^{-1} from an initial value of 8 m·h^{-1}, the corresponding relationship of settling time to discharge time can be determined by equation 6.11; that is,

At $(V_s)_{min} = 8$ m·h^{-1}, $t_s = 30 - \dfrac{\left(t_d - 5\right)^2}{t_d}$

At $(V_s)_{min} = 10$ m·h^{-1}, $t_s = 24 - \dfrac{\left(t_d - 5\right)^2}{t_d}$

At $(V_s)_{min} = 12$ m·h^{-1}, $t_s = 20 - \dfrac{\left(t_d - 5\right)^2}{t_d}$

At $(V_s)_{min} = 16$ m·h^{-1}, $t_s = 15 - \dfrac{\left(t_d - 5\right)^2}{t_d}$

At $(V_s)_{min} = 20$ m·h^{-1}, $t_s = 12 - \dfrac{\left(t_d - 5\right)^2}{t_d}$

The above equations show relationships of the settling time to the discharge time at different desired minimum settling velocities (figure 6.13). The salient points from figure 6.13 are as follows:

1. Any pair of t_s and t_d that satisfies the above t_s - t_d relationship would result in the same desirble $(V_s)_{min}$ for aerobic granulation, that is, a longer t_d would be compensated for by a shorter t_s and vice versa.
2. For any given settling time, the discharge time required can be computed.
3. The longest settling time for the preset $(V_s)_{min}$ can be estimated.
4. The longest discharge time allowed for achieving preset $(V_s)_{min}$ can be obtained.

These pieces of information are essential for rational scale-up of an aerobic granular sludge SBR.

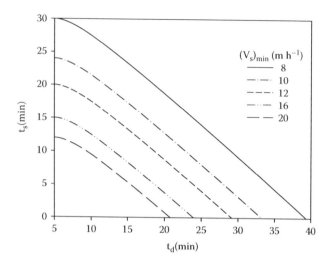

FIGURE 6.13 Relationship between settling time and discharge time for a desired $(V_s)_{min}$. (From Liu, Y., Wang, Z.-W., and Tay, J. H. 2005. *Biotechnol Adv* 23: 335–344. With permission.)

It should be realized that the discharge time determines the effluent pumping rate that relates to the energy cost, that is, a short discharge time results in a higher effluent pumping rate. From the point of view of the SBR operation, a long discharge time is preferable for reducing the pumping power. However, a long discharge time can be obtained only by shortening the settling time (figure 6.13). As demonstrated in chapter 4, optimal settling time for successful aerobic granulation should be less than 5 minutes. If a similar settling time applies to the above full-scale SBR, the discharge time needed for achieving a desired $(V_s)_{min}$ for aerobic granulation can be easily estimated, for example, 34.4 min for a $(V_s)_{min}$ of 8 m·h^{-1}. Consequently, equation 6.10 offers a guide for design and operation engineers to manipulate the aerobic granulation process through collective adjustments of the settling time, discharge time, and the volume exchange ratio; however, compared to the exchange ratio and discharge time, control of the settling time seems more flexible in manipulating the full-scale SBR.

6.9 PREDICTION OF SETTLING VELOCITY OF BIOPARTICLES

As discussed in section 6.4, to enhance aerobic granulation in SBR, bioparticles with settling velocity smaller than the minimum are effectively washed out. This implies that estimating the settling velocity of the bioparticles is essential for the optimal control and operation of an aerobic granular sludge SBR for wastewater treatment. In this regard, Y. Liu et al. (2005a) developed a generalized equation for the settling velocity of aerobic granules, which is also applicable for anaerobic granules.

In study of the settling velocity of activated sludge, the role of particle size is often ignored and is not quantitatively reflected in the original Vesilind equation and the modified Vesilind equations (Hartel and Pöpel 1992; Akca, Kinaci, and Karpuzcu 1993; Wahlberg and Keinath 1995; Renko 1998; Giokas et al. 2003). This is because the size of activated sludge flocs is very small, normally less than 100 μm, that is, activated sludge is relatively homogeneous. However, in an aerobic granular sludge SBR, the size of the aerobic granules may vary in a very wide range of 0.25 to 9 mm. In this case, the effect of the granule size on V_s must be taken into account. The well-known Stokes formula (equation 6.1) shows that the settling velocity of a particle is determined mainly by the density and diameter of aggregates. In the study of aerobic granulation, SVI has been commonly measured to reflect the compactness of microbial association. Y. Liu et al. (2005a) assumed that the term $(\rho_p - \rho)$ in equation 6.1 that represents the wet density of aerobic granules can be roughly estimated as:

$$\rho_p - \rho = \frac{1}{SVI} \tag{6.12}$$

Such an assumption indeed is reasonable, as shown below.

According to the SVI definition (APHA 1998), its inverse can be expressed as:

$$\frac{1}{SVI} = \frac{M_{ps}}{V_{ps} + V_{pw} + V_v} \tag{6.13}$$

in which M_{ps} is the dry mass of the bioparticles, V_{ps} is the dry volume of the bioparticles, V_{pw} is the volume of water retained in the bioparticles, and V_v is the void space between settled bioparticles. In the SVI test of aerobic granules, V_v is often very small or negligible as compared to the whole volume of settled biomass, thus equation 6.13 is reduced to

$$\frac{1}{SVI} = \frac{M_{ps}}{V_{ps} + V_{pw}} \tag{6.14}$$

On the other hand, the water density can be expressed as

$$\rho_w = \frac{M_{pw}}{V_{pw}} = 1 \tag{6.15}$$

Likewise, the density of the bioparticles can be analytically defined in a way such that:

$$\rho_p = \frac{M_{ps} + M_{pw}}{V_{ps} + V_{pw}} \tag{6.16}$$

in which M_{pw} is the mass of water contained in the bioparticles. Combining equations 6.15 and 6.16 yields the following relationship:

$$\rho_p - \rho_w = \frac{M_{ps}}{V_{ps} + V_{pw}} + \frac{V_{pw}}{V_{ps} + V_{pw}} - 1 \tag{6.17}$$

It appears from equations 6.15 and 6.16 that the second term on the right-hand side of equation 6.17 can be rewritten as:

$$\frac{V_{pw}}{V_{ps} + V_{pw}} = \rho_p \frac{M_{pw}}{M_{pw} + M_{ps}} \tag{6.18}$$

In fact, the term $\rho_p \dfrac{M_{pw}}{M_{pw} + M_{ps}}$ in equation 6.18 is the product of bioparticle density and water content of the bioparticles.

It is well known that the density of aerobic granules generally falls into a range of 1.01 to 1.06 kg L^{-1}, and its water content is as high as 95% to 98%. As a consequence, the value of the term $\rho_p \dfrac{M_{pw}}{M_{pw} + M_{ps}}$ in Equation 6.18 should be very close to 1. In this case, equation 6.17 becomes:

$$\rho_p - \rho_w \approx \frac{M_{ps}}{V_{ps} + V_{pw}} \tag{6.19}$$

Substitution of equation 6.19 into equation 6.14 gives:

$$\frac{1}{SVI} \approx \rho_p - \rho_w \tag{6.20}$$

Substitution of equation 6.12 into equation 6.1 gives:

$$V_s = \frac{g}{18\mu} \frac{d_p^2}{SVI} \tag{6.21}$$

Existing evidence shows that the viscosity of mixed liquor in a biological reactor is an exponential function of the concentration of biomass (Manem and Sanderson 1996; Hasar et al. 2004), that is,

$$\mu = ae^{\beta X} \tag{6.22}$$

in which X is the biomass concentration, and a and β are constant coefficients. Combining equations 6.21 and 6.22 yields:

$$V_s = \alpha \frac{d_p^2}{SVI} e^{-\beta X} \tag{6.23}$$

n which α is a constant coefficient and equals g/18a. Equation 6.23 shows that the settling velocity of aerobic granular sludge is determined by the size of the granules, SVI, and biomass concentration of the granules.

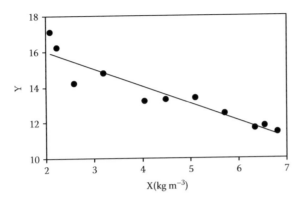

FIGURE 6.14 Settling velocity of microbial aggregates as a function of sludge volume index, mean size, and biomass concentration observed in the course of aerobic granulation in SBR; equation 6.23 prediction with respective α and β values of 5.94×10^7 m^2 kg^{-1} h^{-1} and 0.98 m^3 kg^{-1}, and correlation coefficient = 0.94. (From Liu, Y. et al. 2005a. *Biotechnol Prog* 21: 621–626. With permission.)

In order to determine the constant involved, equation 6.23 is linearized as follows:

$$Y = \ln \alpha - \beta X \tag{6.24}$$

in which:

$$Y = \ln\left(\frac{V_s \times SVI}{d_p^2}\right) \tag{6.25}$$

Thus, plotting Y versus X gives a straight line with a slope of $-\beta$ and an intercept of lnα. Figure 6.14 and figure 6.15 show changes in size, biomass concentration, SVI, and corresponding settling velocity of microbial aggregates in the course of aerobic granulation in SBRs. It can be seen that equation 6.23 can provide a satisfactory description to the experimental data obtained, indicated by a respective correlation coefficient of 0.94 and 0.89. In fact, Moon et al. (2003) also reported that the zone settling velocity of activated sludge was related to its size.

In order to further confirm the proposed equation (equation 6.23), additional experiments were conducted with aerobic granules having defined mean size in the range of 0.23 to 2.4 mm, while the biomass concentration was kept constant at 1.2 kg m^{-3}. In this case, equation 6.23 reduces to:

$$V_s = \gamma \frac{d_p^2}{SVI} \tag{6.26}$$

in which:

$$\gamma = \alpha e^{-\beta X} \tag{6.27}$$

Figure 6.16 shows a plot of d_p^2/SVI against V_s. A good agreement between the equation 6.26 prediction and the experimental data is obtained, indicated by a

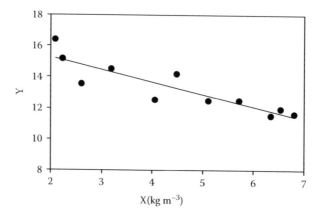

FIGURE 6.15 Settling velocity of microbial aggregates as a function of sludge volume index, mean size, and biomass concentration observed in the course of aerobic granulation in an SBR; equation 6.23 prediction is shown by a solid curve with respective α and β values of 2.19×10^7 m^2 kg^{-1} h^{-1} and 0.81 m^3 kg^{-1}, and correlation coefficient = 0.89. (From Liu, Y. et al. 2005a. *Biotechnol Prog* 21: 621–626. With permission.)

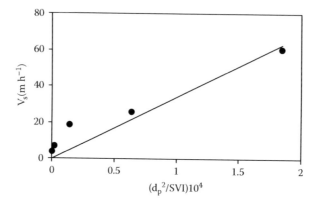

FIGURE 6.16 Settling velocity of aerobic granules as a function of sludge volume index, mean size of aerobic granules at a constant biomass concentration observed in the course of aerobic granulation in an SBR; equation 6.23 prediction is shown by a solid curve with γ value of 3.4×10^5 m^2 kg^{-1} h^{-1}. (From Liu, Y. et al. 2005a. *Biotechnol Prog* 21: 621–626. With permission.)

correlation coefficient of 0.94. It is apparent that equation 6.23 can offer a satisfactory prediction to the settling velocity of bioparticles.

As discussed earlier, among all three key hydraulic selection pressures identified, settling time is the strongest one and it offers high flexibility in control and operation of an aerobic granular sludge SBR. There is solid evidence that settling time must be set as short as possible in order to enhance and accelerate the aerobic granulation process, that is, a short settling time preferentially selects for the growth of good settling bacteria, and the sludge with a poor settleability is washed out. However, a basic question remaining unanswered so far is how to determine the critical settling time for a rapid aerobic granulation. Equation 6.2 shows that the higher V_s results in a shorter traveling time of particles to the discharge port, that is, the bioparticles with a traveling time longer than the settling time chosen for the SBR system will be washed out of the reactor. Thus, the critical settling time for aerobic granulation in an SBR would be defined as follows:

$$\text{Critical settling time} = \frac{L}{V_s} \quad (6.28)$$

Substitution of equation 6.23 into equation 6.28 gives:

$$\text{Critical settling time} = \frac{L}{\alpha} \frac{SVI \times e^{\beta X}}{d_p^2} \quad (6.29)$$

It can be seen that the critical settling time for rapid aerobic granulation is a function of the SVI, mean size of the bioparticles, and biomass concentration. Equation 6.29 provides theoretical guidelines for selecting and adjusting the settling time according to the parameters describing settleability in terms of particle size and SVI. In fact, this strategy has been applied successfully to accelerate aerobic granulation in SBRs by adjusting the settling time according to changes in the settleability of bioparticles (Qin, Liu, and Tay 2004b, 2004a).

6.10 CONCLUSION

This chapter shows that aerobic granulation in SBRs is driven by selection pressures exerted on microorganisms. The major selection pressures identified include settling time, volume exchange ratio, and discharge time, and they can all be unified to a single and easy concept of the minimal settling velocity of bioparticles that ultimately determines aerobic granulation in SBRs. One may expect to manipulate the formation and characteristics of aerobic granules by properly controlling the minimum settling velocity, which is a function of the settling time, exchange ratio, and discharge time in SBRs. It was demonstrated that the proposed selection pressure theory by Y. Liu, Wang, and Tay (2005) can offers a useful guide for manipulating and optimizing the formation and characteristics of aerobic granules in SBRs.

REFERENCES

Akca, L., Kinaci, C., and Karpuzcu, M. 1993. A model for optimum design of activated sludge plants. *Water Res* 27: 1461–1468.

Alphenaar, P., Visser, A., and Lettinga, G. 1993. The effect of liquid upward velocity and hydraulic retention time on granulation in UASB reactors treating wastewater with a high sulphate content. *Bioresource Technol* 43: 249–258.

APHA. 1998. Standard methods for the examination of water and wastewater, 20th ed. Washington, DC: American Public Health Association.

Beun, J.J., van Loosdrecht, M. C. M., and Heijnen, J. J. 2000. Aerobic granulation. *Water Sci Technol* 41: 41–48.

Beun, J. J., van Loosdrecht, M. C. M., and Heijnen, J. J. 2002. Aerobic granulation in a sequencing batch airlift reactor. *Water Res* 36: 702–712.

Beun, J. J., Hendriks, A., Van Loosdrecht, M. C. M., Morgenroth, E., Wilderer, P. A., and Heijnen, J. J. 1999. Aerobic granulation in a sequencing batch reactor. *Water Res* 33: 2283–2290.

Bossier, P. and Verstraete, W. 1996. Triggers for microbial aggregation in activated sludge? *Appl Microbiol Biotechnol* 45: 1–6.

Calleja, G. B. 1984. *Microbial aggregation.* Boca Raton, FL: CRC Press.

Davies, D. G., Parsek, M. R., Pearson, J. P., Iglewski, B. H., Costerton, J. W., and Greenberg, E. P. 1998. The involvement of cell-to-cell signals in the development of a bacterial biofilm. *Science* 280: 295–298.

Giokas, D. L., Daigger, G. T., von Sperling, M., Kim, Y., and Paraskevas, P. A. 2003. Comparison and evaluation of empirical zone settling velocity parameters based on sludge volume index using a unified settling characteristics database. *Water Res* 37: 3821–3836.

Hartel, L. and Pöpel, J. H. 1992. A dynamic secondary clarifier model including processes of sludge thickening. *Water Sci Technol* 25: 267–284.

Hasar, H., Kinaci, C., Unlu, A., Togrul, H., and Ipek, U. 2004. Rheological properties of activated sludge in a sMBR. *Biochem Eng J* 20: 1–6.

Hu, L. L., Wang, J. L., Wen, X. H., and Qian, Y. 2005. The formation and characteristics of aerobic granules in sequencing batch reactor (SBR) by seeding anaerobic granules. *Process Biochem* 40: 5–11.

Hulshoff Pol, L. W., Heijnekamp, K., and Lettinga, G. 1988. The selection pressure as a driving force behind the granulation of anaerobic sludge. In Lettinga, G., Zehnder, A. J. B., Grotenhuis, J. T. C., and Hulshoff Pol, L. W. eds., *Granular anaerobic sludge: Microbiology and technology,* 153–161. Wageningen, the Netherlands: Purdoc.

Jiang, H.-L., Tay, J. H., and Tay, S. T. L. 2002. Aggregation of immobilized activated sludge cells into aerobically grown microbial granules for the aerobic biodegradation of phenol. *Lett Appl Microbiol* 35: 439–445.

Jiang, H.-L., Tay, J. H., Maszenan, A. M., and Tay, S. T. L. 2006. Enhanced phenol biodegradation and aerobic granulation by two coaggregating bacterial strains. *Environ Sci Technol* 40: 6137–6142.

Kosaric, N., Blaszczyk, R., Orphan, L., and Valladares, J. 1990. Characteristics of granules from upflow anaerobic sludge blanket reactors. *Water Res* 24: 14731477.

Li, Y. 2007. Metabolic behaviors of aerobic granules developed in sequencing batch reactor. Ph.D. thesis, Nanyang Technological University, Singapore.

Li, Z. H., Kuba, T., and Kusuda, T. 2006. Selective force and mature phase affect the stability of aerobic granule: An experimental study by applying different removal methods of sludge. *Enzyme Microb Technol* 39: 976–981.

Lin, Y. M., Liu, Y., and Tay, J. H. 2003. Development and characteristics of phosphorus-accumulating microbial granules in sequencing batch reactors. *Appl Microbiol Biotechnol* 62: 430–435.

Liu, Q. S. 2003. Aerobic granulation in sequencing batch reactor. Ph.D. thesis, Nanyang Technological University, Singapore.

Liu, Q. S., Tay, J. H., and Liu, Y. 2003. Substrate concentration-independent aerobic granulation in a sequential aerobic sludge blanket reactor. *Environ Technol* 24: 1235–1242.

Liu, Y. and Tay, J. H. 2002. The essential role of hydrodynamic shear force in the formation of biofilm and granular sludge. *Water Res* 36: 1653–1665.

Liu, Y., Wang, Z. W., and Tay, J. H. 2005. A unified theory for upscaling aerobic granular sludge sequencing batch reactors. *Biotechnol Adv* 23: 335–344.

Liu, Y., Yang, S. F., and Tay, J. H. 2003. Elemental compositions and characteristics of aerobic granules cultivated at different substrate N/C ratios. *Appl Microbiol Biotechnol* 61: 556–561.

Liu, Y., Wang, Z.-W., Liu, Y. Q., Qin, L., and Tay, J. H. 2005a. A generalized model for settling velocity of aerobic granular sludge. *Biotechnol Prog* 21: 621–626.

Liu, Y., Wang, Z.-W., Qin, L., Liu, Y. Q., and Tay, J. H. 2005b. Selection pressure-driven aerobic granulation in a sequencing batch reactor. *Appl Microbiol Biotechnol* 67: 26–32.

Mahoney, E. M., Varangu, L. K., Cairns, W. L., Kosaric, N., and Murray, R. G. E. 1987. The effect of calcium on microbial aggregation during UASB reactor start-up. *Water Sci Technol* 19: 249–260.

Manem, J. and Sanderson, R. 1996. Membrane bioreactor. In Mallevialle, J., Odendaal, P. E., and Wiesner, M. R. eds., *Water treatment membrane processed*. New York: McGraw-Hill, 17.1–17.31.

Metcalf and Eddy. 2003. *Wastewater engineering: Treatment and reuse*, 4th ed. Boston: McGraw-Hill.

McSwain, B. S., Irvine, R. L., and Wilderer, P. A. 2004. The influence of-settling time on the formation of aerobic granules. *Water Sci Technol* 50: 195–202.

Moon, B. H., Seo, G. T., Lee, T. S., Kim, S. S., and Yoon, C. H. 2003. Effects of salt concentration on floc characteristics and pollutants removal efficiencies in treatment of seafood wastewater by SBR. *Water Sci Technol* 47: 65–70.

Morgenroth, E., Sherden, T., van Loosdrecht, M. C. M., Heijnen, J. J., and Wilderer, P. A. 1997. Aerobic granular sludge in a sequencing batch reactor. *Water Res* 31: 3191–3194.

Pan, S. 2003. Inoculation of microbial granular sludge under aerobic conditions. Ph.D. thesis, Nanyang Technological University, Singapore.

Pan, S., Tay, J. H., He, Y. X., and Tay, S. T. L. 2004. The effect of hydraulic retention time on the stability of aerobically grown microbial granules. *Lett Appl Microbiol* 38: 158–163.

Peng, D. C., Bernet, N., Delgenes, J. P., and Moletta, R. 1999. Aerobic granular sludge: A case report. Water Res 33: 890–893.

Pratt, L. A. and Kolter, R. 1998. Genetic analysis of Escherichia coli biofilm formation: Roles of flagella, motility, chemotaxis and type I pili. *Mol Microbiol* 30: 285–293.

Qin, L., Liu, Y., and Tay, J. H. 2004a. Selection pressure is a driving force of aerobic granulation in sequencing batch reactors. *Process Biochem* 39: 579–584.

Qin, L., Liu, Y., and Tay, J. H. 2004b. Effect of settling time on aerobic granulation in sequencing batch reactor. *Biochem Eng J* 21: 47–52.

Renko, E. K. 1998. Modelling hindered batch settling. II. A model for computing solids profile of calcium carbonate slurry. *Water SA* 24: 331–336.

Schmidt, J. E. and Ahring, B. K. 1996. Granular sludge formation in upflow anaerobic sludge blanket (UASB) reactors. *Biotechnol Bioeng* 49: 229–246.

Schwarzenbeck, N., Erley, R., and Wilderer, P. A. 2004. Aerobic granular sludge in an SBR-system treating wastewater rich in particulate matter. *Water Sci Technol* 49: 41–46.

Shapiro, J. A. 1998. Thinking about bacterial populations as multicellular organisms. *Annu Rev Microbiol* 52: 81–104.

Tay, J. H., Liu, Q. S., and Liu, Y. 2001. Microscopic observation of aerobic granulation in sequential aerobic sludge blanket reactor. *J Appl Microbiol* 91: 168–175.

Tay, J. H., Pan, S., He, Y. X., and Tay, S. T. L. 2004. Effect of organic loading rate on aerobic granulation. I. Reactor performance. *J Environ Eng* 130: 1094–1101.

Tokutomi, T. 2004. Operation of a nitrite-type airlift reactor at low DO concentration. *Water Sci Technol* 49: 81–88.

Tsuneda, S., Nagano, T., Hoshino, T., Ejiri, Y., Noda, N., and Hirata, A. 2003. Characterization of nitrifying granules produced in an aerobic upflow fluidized bed reactor. *Water Res* 37: 4965–4973.

Wahlberg, E. J. and Keinath, T. M. 1995. Development of settling flux curves using SVI. An addendum. *Water Environ Res* 67: 872–874.

Wang, Q., Du, G. C., and Chen, J. 2004. Aerobic granular sludge cultivated under the selective pressure as a driving force. *Process Biochem* 39: 557–563.

Wang, Z.-W. 2007. Insights into mechanism of aerobic granulation in a sequencing batch reactor. Ph.D. thesis, Nanyang Technological University, Singapore.

Wang, Z.-W., Liu, Y., and Tay, J. H. 2006. The role of SBR mixed liquor volume exchange ratio in aerobic granulation. *Chemosphere* 62: 767–771.

Yang, S. F., Tay, J. H., and Liu, Y. 2003. A novel granular sludge sequencing batch reactor for removal of organic and nitrogen from wastewater. *J Biotechnol* 106: 77–86.

Zheng, Y. M., Yu, H. Q., and Sheng, G. P. 2005. Physical and chemical characteristics of granular activated sludge from a sequencing batch airlift reactor. *Process Biochem* 40: 645–650.

7 Growth Kinetics of Aerobic Granules

Qi-Shan Liu and Yu Liu

CONTENTS

7.1 INTRODUCTION

In biofilm culture, biofilm thickness has been commonly used to describe the growth behaviors of fixed bacteria at the surface of the biocarrier, and a number of growth models have been developed for biofilm culture. However, these models may not be suitable for the description of the growth of aerobic granules. It has been shown that aerobic granules can grow in a wide range of sizes, from 0.2 to 16.0 mm in mean diameter, as described in chapter 1. Granule size determines the total surface area available for the biodegradation of substrate, and subsequently the substrate surface loading. In biofilm culture, microbial growth kinetics has been reported to be surface loading-dependent (Trinet et al. 1991). In fact, microbial surface growth rate and biodegradation rate of aerobic granules are fairly related to the substrate surface loading, and can be described by the Monod-type equation (Y. Liu et al. 2005). This chapter discusses the growth kinetics of aerobic granules associated with substrate utilization.

7.2   A SIMPLE KINETIC MODEL FOR THE GROWTH OF AEROBIC GRANULES

The growth of aerobic granules after the initial cell-to-cell self-attachment is similar to the growth of biofilm, and can be regarded as the net result of interaction between bacterial growth and detachment (Y. Liu et al. 2003). The balance between the growth and detachment processes in turn will lead to an equilibrium size of aerobic granules (Y. Liu and Tay 2002). Compared with biofilm process, aerobic granulation is a process of cell-to-cell self-immobilization instead of cell attachment to a solid surface. Thus, size evolution of microbial aggregates can be used to describe the growth of aerobic granules. As presented in chapter 1, aerobic granulation is a gradual process from dispersed sludge to mature aerobic granules with a spherical outer shape and a stable size. Under given growth and detachment conditions, the equilibrium size (D_{eq}) of aerobic granules exists when the growth and detachment forces are balanced, that is, the size of aggregate (D) gradually approaches its equilibrium size (D_{eq}). According to Atlas and Bartha (1998), the change rate of population density in terms of size or concentration of a microbial community is a function of the difference between its density at growth equilibrium and that at time t. Thus, the difference between D_{eq} and D represents the growth potential of aerobic granules under given conditions (Yang et al. 2004).

The linear phenomenological equation (LPE) shows that a flux term and a driving force term for transport phenomena are linearly related (De Groof and Mazur 1962). The unqualified success of this linear assumption has been universally recognized as the basis of thermodynamics of transport phenomena (Prigogine 1967; Garfinkle 2002), while the linear relationship between the rate of a microbial process and its driving force had been confirmed (Rutgers, Balk, and Van Dam 1989; Heijnen and van Dijken, 1992). It must be realized that the LPE indeed reveals that the change rate of population density would be a first-order function of the driving force or growth potential. As an analogue to the LPE, Yang et al. (2004) proposed that the growth of aerobic granules in size can be described by the following equation:

$$\frac{dD}{dt} = \mu(D_{eq} - D) \tag{7.1}$$

where μ is the specific growth rate of aggregate by size (day^{-1}). Equation (7.1) can be rearranged to:

$$\frac{dD}{D_{eq} - D} = \mu \ dt \tag{7.2}$$

In general, a newly inoculated culture does not grow immediately over a time, which is often referred to as the lag phase (Gaudy and Gaudy 1980). The lag phase is the time required for bacteria to adapt to new living conditions instead of growth, and is not included in equation 7.1. Thus, only the size of microbial aggregates at the end of the lag phase can be used as the initial value for microbial growth. Integration of equation 7.2 gives:

$$D - D_o = (D_{eq} - D)\, 1 - e^{-(t-t_o)} \qquad (7.3)$$

where t_o is the time at the end of the lag phase, and D_o is the size of microbial aggregates at time t_o. D_{eq}, μ, D_0, and t_0 can be determined experimentally by using the method proposed by Gaudy and Gaudy (1980).

7.2.1 Growth of Aerobic Granules at Different Organic Loading Rates

The formation of aerobic granules was demonstrated in sequencing batch reactors (SBRs) supplied with different organic loading rates, from 1.5 to 9.0 kg COD m^{-3} d^{-1} (see chapter 1). Figure 7.1 shows the evolution of microbial aggregates in terms of mean size at different organic loading rates. The size of microbial aggregates gradually increased up to a stable value, the so-called equilibrium size, during the SBR operation. It can be seen that equation 7.3 can provide a good prediction to the growth data of aerobic granules obtained at different organic loading rates, indicated by a correlation coefficient greater than 0.95 (figure 7.1). The effects of organic loading rate on the equilibrium size (D_{eq}) of aerobic granules and the size-dependent specific

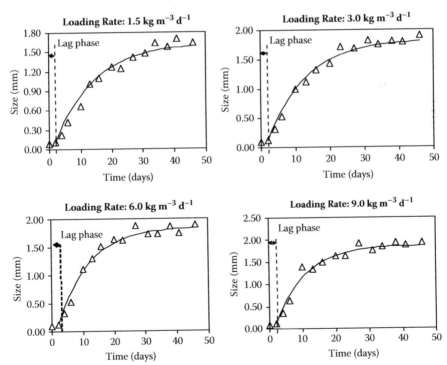

FIGURE 7.1 Size evolution of microbial aggregates cultivated at different organic loading rates. The prediction given by equation 7.3 is shown by a solid line. (Data from Yang, S. F., Liu, Q. S., and Liu, Y. 2004. *Lett Appl Microbiol* 38: 106–112.)

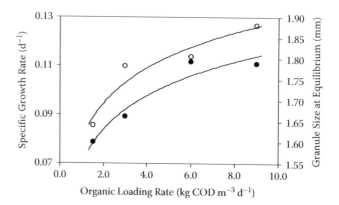

FIGURE 7.2   Effect of organic loading rate on size of microbial aggregate at equilibrium (○) and specific growth rate by size (●). (Data from Yang, S. F., Liu, Q. S., and Liu, Y. 2004. *Lett Appl Microbiol* 38: 106–112.)

growth rate (μ) are presented in figure 7.2. It was found that both the size of the microbial aggregate at equilibrium (D_{eq}) and the size-dependent specific growth rate (μ) tended to increase with the increase of organic loading rate in the range studied. A similar phenomenon was also observed by Moy et al. (2002). Obviously, the relationship observed between the growth rate of aerobic granules and the organic loading rate is subject to the best-known Monod equation, that is, a high substrate loading results in a high microbial growth rate. The development of bigger aerobic granules at the higher organic loading rate is simply due to its loading-associated growth rate. In the biofilm process, biofilm thickness was also found to be proportionally related to the applied organic loading rate (Tijhuis et al. 1996; Kwok et al. 1998).

7.2.2   Growth of Aerobic Granules at Different Shear Forces

In a column SBR, hydrodynamic shear force is mainly created by aeration that can be quantified by superficial upflow air velocity (see chapter 2). The effect of shear force in terms of superficial upflow air velocity on the growth of aerobic granules is illustrated in figure 7.3. It can be seen that the prediction by equation 7.3 is in good agreement with the experimental data obtained at different shear forces. Both the size of the microbial aggregate at equilibrium and the size-dependent specific growth rate show decreasing trends as the shear force increases (figure 7.4).

It is known that high shear force would lead to more collision among particles, and friction between particle and liquid, leading to a high detachment force. This may in part explain why smaller aerobic granules were developed at higher shear force. A similar phenomenon was also observed in the biofilm culture where thinner biofilm was cultivated at higher shear force (van Loosdrecht et al. 1995; Gjaltema, van Loosdrecht, and Heijnen 1997; Y. Liu and Tay 2001; Horn, Reiff, and Morgenroth 2003). Y. Liu et al. (2003) proposed that the growth kinetics of biofilm is highly dependent on the ratio of growth force normalized to detachment force. At a given organic loading rate, a microbial community can regulate its metabolic pathways in response to changes in external shear force, for example more

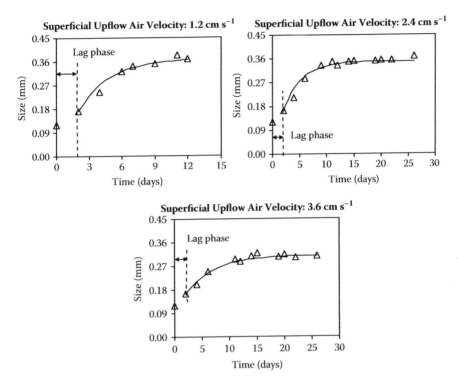

FIGURE 7.3 Size evolution of microbial particles at different shear forces. The prediction given by equation 7.3 is shown by a solid line. (Data from Yang, S. F., Liu, Q. S., and Liu, Y. 2004. *Lett Appl Microbiol* 38: 106–112.)

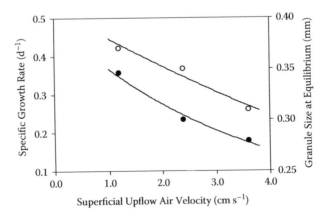

FIGURE 7.4 Effect of shear force on size of microbial aggregate at equilibrium (○) and specific growth rate by size (●). (Data from Yang, S. F., Liu, Q. S., and Liu, Y. 2004. *Lett Appl Microbiol* 38: 106–112.)

extracellular polysaccharides would be produced (see chapter 2). This is the reason behind a reduced equilibrium size and growth rate with the increase of shear force. In fact, it has been demonstrated that suspended bacteria can respond to hydrodynamic shear by altering their growth rate, cell density, and metabolism (Meijer et al. 1993; Chen and Huang 2000; Q. S. Liu et al., 2005).

7.2.3 Growth of Aerobic Granules at Different Substrate N/COD Ratios

Aerobic granules can form in a wide range of different substrate N/COD ratios for nutrient and carbon removal (see chapter 1). The growth of aerobic granules at the N/COD ratios of 0.05 to 0.3 is shown in figure 7.5. It can be seen that the prediction of equation 7.3 fitted the experimental data very well, indicated by a correlation coefficient greater than 0.97. The relationships between the size of microbial aggregate at equilibrium, the size-dependent specific growth rate, and the substrate N/COD ratio are presented in figure 7.6. Both the size of the microbial aggregate at equilibrium and the size-dependent specific growth rate were found to decrease with the increase of substrate N/COD ratio. This seems to imply that the substrate N/COD

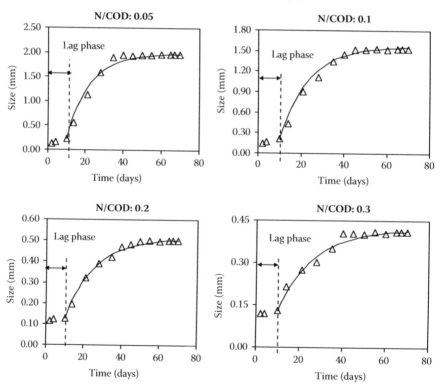

FIGURE 7.5 Size evolution of microbial particles at different substrate N/COD ratios. The prediction given by equation 7.3 is shown in a solid line. (Data from Yang, S. F., Liu, Q. S., and Liu, Y. 2004. *Lett Appl Microbiol* 38: 106–112.)

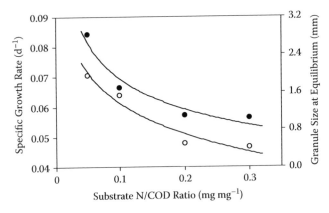

FIGURE 7.6 Effect of substrate N/COD ratio on size of microbial aggregate at equilibrium (○) and specific growth rate by size (●). (Data from Yang, S. F., Liu, Q. S., and Liu, Y. 2004. *Lett Appl Microbiol* 38: 106–112.)

ratio might select microbial populations in aerobic granules, that is, high substrate N/COD ratio will promote the growth of nitrifying populations (Yang, Tai, and Liu 2004, 2005). It is well known that a nitrifying population grows much slower than heterotrophs do. Consequently, an enriched nitrifying population in aerobic granules developed at high substrate N/COD ratio would be responsible for the overall low growth rate of granular sludge and smaller size, as shown in figure 7.6.

7.3 EFFECT OF SURFACE LOADING ON KINETIC BEHAVIOR OF AEROBIC GRANULES

7.3.1 Effect of Surface Loading on Growth Rate

Y. Liu et al. (2005) studied the effect of surface loading rate on the growth of aerobic granules, and found that the specific surface area of aerobic granules is inversely correlated to the mean diameter of the aerobic granules, that is, bigger granules have a smaller specific surface area (figure 7.7). According to the specific surface area of aerobic granules, the substrate surface loading of aerobic granules can be calculated based on the volumetric organic loading rate applied. Figure 7.8 further exhibits the effect of substrate surface loading on the surface growth rate of aerobic granules. It appears that a higher surface loading results in faster growth of aerobic granules, and the relationship between the surface growth rate of aerobic granules and the substrate surface loading is subject to the Monod-type equation:

$$\mu_S = \mu_{S,max} \frac{L_S}{L_S + K_S} \tag{7.4}$$

where μ_S and $\mu_{S,max}$ are, respectively, the surface growth rate and the maximum surface growth rate of aerobic granules (g biomass m^{-2} h^{-1}) and L_s is the surface loading (g COD m^{-2}), while K_s is the Monod constant. Equation 7.4 can satisfactorily describe

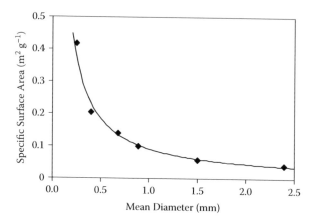

FIGURE 7.7 Specific surface area versus the mean diameter of aerobic granules. (From Liu, Y. et al. 2005. *Appl Microbiol Biotechnol* 67: 484–488. With permission.)

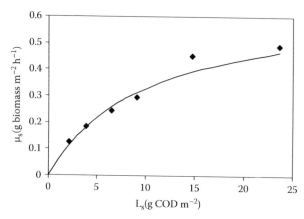

FIGURE 7.8 Effect of the substrate surface loading (L_s) on the surface growth rate (μ_s) of aerobic granules. The prediction given by equation 7.4 is shown by a solid curve. $\mu_{s,max} = 0.62$ g biomass m^{-2} h^{-1}; $K_s = 9.6$ g COD m^{-2}; and correlation coefficient = 0.994. (From Liu, Y. et al. 2005. *Appl Microbiol Biotechnol* 67: 484–488. With permission.)

the experimental data, indicated by a correlation coefficient of 0.99 (figure 7.8). In addition, figure 7.9 shows the effect of the substrate surface loading on the surface oxygen utilization rate (SOUR) of aerobic granules. A trend similar to μ_s is observed in figure 7.9. It seems that the microbial activity of aerobic granules increases with the increase of substrate surface loading rate.

7.3.2 EFFECT OF SURFACE LOADING ON SUBSTRATE BIODEGRADATION RATE

The surface COD removal rate (q_s) by aerobic granules versus the substrate surface loading is presented in figure 7.10, showing that an increased substrate surface loading leads to a higher surface COD removal rate until a maximum value is reached. Analogous to equation 7.4, q_s versus L_s can be described by a Monod-type equation:

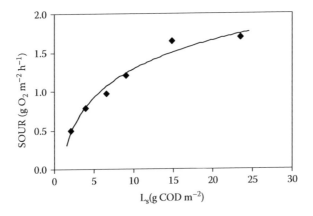

FIGURE 7.9   Effect of substrate surface loading (L_s) on the surface oxygen utilization rate (SOUR) of aerobic granules. (From Liu, Y. et al. 2005. *Appl Microbiol Biotechnol* 67: 484–488. With permission.)

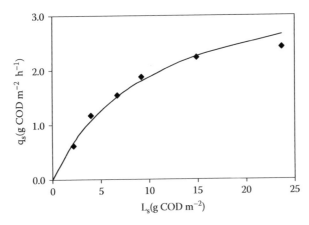

FIGURE 7.10   Effect of the substrate surface loading (L_s) on the substrate surface removal rate (q_s) by aerobic granules. The prediction given by equation 7.5 is shown by a solid curve. $q_{s,max} = 4.67$ g COD m^{-2} h^{-1}; $K_s = 14.2$ g COD m^{-2}; and correlation coefficient $= 0.991$. (From Liu, Y. et al. 2005. *Appl Microbiol Biotechnol* 67: 484–488. With permission.).

$$q_S = q_{S,max} \frac{L_S}{L_S + K_S} \tag{7.5}$$

where $q_{s,max}$ is the maximum substrate surface removal rate by aerobic granules (g COD m^{-2} h^{-1}). It is obvious that the equation 7.5 prediction is in good agreement with the experimental data (figure 7.10). It is known that the kinetic behavior of a microbial culture is associated with the interaction between anabolism and catabolism, and catabolism is coupled to anabolism (Lehninger 1975). This implies that substrate oxidation is tied up with oxygen reduction during the aerobic culture of microorganisms. Figure 7.11 shows the close correlation of q_s to SOUR, which reveals that 1.0 g substrate-COD oxidized by aerobic granules requires 0.68 g oxygen.

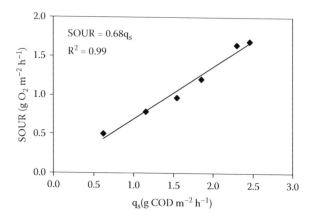

FIGURE 7.11 Correlation of SOUR to q_s. (From Liu, Y. et al. 2005. *Appl Microbiol Biotechnol* 67: 484–488. With permission.).

7.3.3 RELATIONSHIP OF SURFACE GROWTH RATE TO SUBSTRATE BIODEGRADATION RATE

It has been recognized that aerobic granules can be differentiated from suspended activated sludge by their size, spherical shape, excellent settleability, and highly organized microbial structure (Y. Liu and Tay 2002). Figure 7.7 shows that the specific surface area of aerobic granules is closely related to their mean diameter, while figures 7.8 to 7.10 clearly indicate that the surface growth rate and the substrate surface biodegradation rate of aerobic granules in terms of μ_S, q_s, and SOUR increase with the substrate surface loading, that is, the kinetic behavior of aerobic granules is dependent on the substrate surface loading. According to Tempest and Neijssel (1978), the Pirt maintenance equation can be linearized as follows:

$$q_S = m_S + \frac{1}{Y_G}\mu_S \qquad (7.6)$$

where m_s is the Pirt maintenance coefficient and Y_G is the theoretical maximum growth yield. Figure 7.12 shows the linear relationship of q_s to μ_S with a m_s value of 0.24 g COD m^{-2} h^{-1} and a Y_G value of 0.2 g biomass g^{-1} COD. At the lowest substrate surface loading of 2.2 g COD m^{-2}, about 40% of the input substrate is consumed through the maintenance metabolism, while only 10% of input substrate goes into the maintenance at the highest substrate surface loading (24 g COD m^{-2}). In fact, these are in good agreement with the Pirt maintenance theory, stating that more substrate will be used for maintenance purposes at lower substrate availability (Pirt 1965). Compared with conventional activated sludge with a typical growth yield of 0.4 to 0.6 g biomass g^{-1} COD (Droste 1997), the theoretical maximum growth yield of aerobic granules is low. In fact, there is evidence showing that the productivity of aerobic granules fell into a range of 0.1 to 0.2 g biomass g^{-1} COD (Pan 2003).

As discussed earlier, the rate of substrate utilization is well expressed as a Monod equation, and can be used to describe the relationship between the bacterial growth

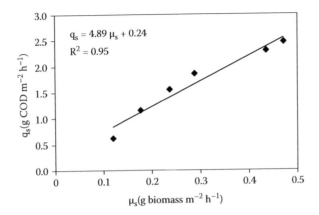

FIGURE 7.12 Surface growth rate (μ_s) versus substrate surface biodegradation rate (q_s). (From Liu, Y. et al. 2005. *Appl Microbiol Biotechnol* 67: 484–488. With permission.)

TABLE 7.1
Kinetic Comparison of Aerobic Granules, Activated Sludge, and Anaerobic Granules

	Activated Sludge	Anaerobic Granules from UASB	Aerobic Granules from SBR
Start-up period	Several weeks	3 months	Several days
MLSS (g L^{-1})[a]	1–2	15–25	8
OLR (g COD L^{-1} d^{-1})[b]	0.5–1	10	4
Effluent COD (mg L^{-1})	<40	>100	<30
Maximum specific substrate utilization rate (d^{-1})	2–10	0.9–3	23.65
Half-velocity coefficient (mg L^{-1})	15–70	100–250	3367.05
Growth yield coefficient (mg MLSS mg^{-1} COD)	0.25–0.4	0.04–0.10	0.1927–0.2022
Bacteria decay rate coefficient (d^{-1})	0.04–0.075	0.02–0.04	0.00845–0.0135

[a] MLSS, mixed liquor suspended solids.
[b] OLR, organic loading rate.
Source: Data from L. Liu et al. 2005. *Enzyme Microb Technol* 36: 307–313.

rate and the concentration of growth-limiting substrate. L. Liu et al. (2005) determined the Monod kinetics coefficients for aerobic granules cultivated from glucose substrate in an SBR operated at a mean cell residence time of 10 days (table 7.1). It can be seen that by comparing with a conventional activated sludge system and an upflow anaerobic sludge blanket (UASB) reactor, the aerobic granular sludge system had a shorter start-up period, high substrate utilization rate, less surplus sludge production, and low effluent COD. These results further demonstrated the excellence of aerobic granules for wastewater treatment.

Lubken, Schwarzenbeck, and Wilderer (2004) applied Activated Sludge Model No. 3 (ASM3) to describe an aerobic granular sludge SBR system. ASM3 is a model developed for the activated sludge process to simulate oxygen consumption, sludge production, nitrification, and denitrification. However, it was shown that the model could also be used for the simulation of nutrient removal with aerobic granular sludge after adjusting some biological parameters. It was shown that this model can provide a good control as well as the design basis for aerobic granular sludge reactors.

Figure 7.11 shows that the unit oxygen requirement of aerobic granules is 0.68 g O_2 g^{-1} COD oxidized. In a conventional activated sludge process, the unit oxygen uptake may vary over a wide range of 0.21 to 0.54 g O_2 g^{-1} COD (Burkhead and McKinney 1969). For comparison, it is assumed that aerobic granules would have a similar empirical formula to that of activated sludge, that is, $C_5H_7NO_2$. Based on the values of the unit oxygen uptake by aerobic granules and the growth yield of aerobic granules as given earlier, it is possible to generate an oxidative assimilation equation of aerobic granules grown on acetate, that is:

$$C_2H_4O_2 + 1.36O_2 + 0.1NH_3 \rightarrow 0.1C_5H_7NO_2 + 1.5CO_2 + 1.52H_2O \qquad (7.7)$$

It appears from equation 7.7 that 75% of the input acetate carbon is channeled to carbon dioxide-carbon during aerobic granular culture. The respirometric tests with aerobic granules showed that about 74% of acetate-carbon is converted to carbon dioxide, which is fairly consistent with equation 7.7. When more substrate-carbon goes to carbon dioxide, less sludge is produced, that is, the input substrate can be finally respired to carbon dioxide and water, resulting in a lower biomass production (equation 7.7). The higher carbon dioxide production in aerobic granular culture offers a plausible explanation for the lower growth yield of aerobic granules.

Basically, metabolism is the sum of biochemical transformation, which includes interrelated catabolic and anabolic reactions; and the behavior of a microbial culture is determined by catabolism and anabolism (Lehninger 1975). As discussed earlier, aerobic granules have a low growth yield and high carbon dioxide production. This may imply that the energy generated from catabolism cannot be mainly used for the growth of aerobic granules; and aerobic granules seem to expend energy on functions that are not readily growth-associated. One major difference between aerobic granules and activated sludge is the highly organized three-dimensional structure of the aerobic granules (Tay, Liu, and Liu 2001). Compared with loose and nonorganized activated sludge, it is likely that, for aerobic granules, part of the energy generated from the oxidation of organic substrate is used to maintain the highly organized three-dimensional microbial structure and integrity of the aerobic granule, which in turn leads to high carbon dioxide production and the subsequent low growth yield of aerobic granules. In order to interpret the low growth yield of a biofilm, Y. Liu and Tay (2001) proposed a hypothesis showing that the biofilm community would have to regulate its metabolic pathway so as to maintain its structural integrity and stability through consuming nongrowth-associated energy.

According to Burkhead and McKinney (1969), the oxidative assimilation equation of activated sludge grown on acetate can be roughly determined, that is:

$$C_2H_4O_2 + 0.95O_2 + 0.21NH_3 \rightarrow 0.21C_5H_7NO_2 + 0.95CO_2 + 1.58H_2O \qquad (7.8)$$

The above equation shows that 1 mol acetate oxidized by activated sludge requires only 0.95 mol oxygen, while to oxidize 1 mol acetate by aerobic granules, 1.36 mol oxygen must be provided (equation 7.7). This implies that the oxygen requirement in aerobic granular culture is almost 1.4-fold higher than that in the activated sludge process. In the aerobic oxidation process, ATP is generated by oxidative phosphory-lation, during which process electrons are transported through the electron transport system from an electron donor (substrate) to a final electron acceptor (oxygen). More-over, a clear linkage of oxygen reduction to proton translocation has been shown (Babcock and Wikstrom 1992; Wolfe 1993). Therefore, the high unit oxygen uptake by aerobic granules and low growth yield of aerobic granules seem to indicate an enhanced catabolic activity over anabolism, that is, more energy is generated, but less biomass is produced. It appears from equations 7.7 and 7.8 that the anabolism of activated sludge is highly coupled to its catabolism. However, a significant dis-crepancy between the energy production by catabolism and the energy utilization by anabolism occurs during aerobic granular culture. Equations 7.7 and 7.8 also show that, for 1 mol acetate oxidized, 1.5 mol carbon dioxide are produced in aerobic granular culture, while only 0.95 mol carbon dioxide is generated in activated sludge culture. In fact, abnormally high carbon dioxide production is a good indication of energy uncoupling in aerobic systems (Russell and Cook 1995). In a study on the energy metabolism of *Saccharomyces cerevisiae*, Lagunas (1976) found that as much as 60% of the energy generated from catabolism was spent in functions other than net biosynthesis, while it is reported that the provision of support matrices within microbial structure could result in high maintenance energy (Mayhew and Stephenson 1997). Therefore, it is believed that the energy metabolism of aerobic granules is dissociated and, to some extent, granule structure-related. However, further study is required to demonstrate the structure-energy metabolism relation-ship of aerobic granules from both theoretical and experimental aspects.

7.4 SUBSTRATE CONCENTRATION-ASSOCIATED KINETIC BEHAVIORS OF AEROBIC GRANULES

L. Liu et al. (2005) correlated the kinetic behaviors of aerobic granules to substrate concentration in terms of milligrams COD per liter, and some key kinetic constants were also determined. In general, the rate of substrate utilization is a function of the biomass concentration as well as available substrate concentration for microbial growth. In a study of aerobic granulation in SBRs, L. Liu et al. (2005) applied the well-known Monod equation to describe the relationship between substrate utiliza-tion rate and substrate concentration. The maximum specific substrate utilization rate was estimated as 23.6 mg COD mg MLSS^{-1} day^{-1}, while an extremely high Monod constant of 3367 mg L^{-1} was obtained. This indicates that the affinity of aerobic granules to substrates is low, that is, the biodegradation rate depends closely on the mass transfer of molecules from the bulk solution to aerobic granules, as discussed in chapter 8.

According to Metcalf and Eddy (2003), the substrate utilization rate (U_s) can be related to sludge retention time (θ) in a way such that:

TABLE 7.2

Kinetic Constants Determined from Aerobic Granular Sludge SBR

Time (days)	1	2	3	4	5	6
S_0 (mg COD L^{-1})	559.7	560.8	564.0	560.3	561.5	558.9
S_e (mg COD L^{-1})	27.5	30.7	27.0	24.1	27.2	25.6
X (g MLSS L^{-1})	7.56	7.59	7.57	7.6	7.61	7.6
Y (mg MLSS mg^{-1} COD)	0.20	0.195	0.20	0.20	0.193	0.20
U_s (mg COD mg^{-1} MLSS d^{-1})	0.56	0.560	0.56	0.56	0.56	0.56
K_d (d^{-1})	0.014	0.0093	0.013	0.012	0.0085	0.012

Source: Data from L. Liu et al. 2005. *Enzyme Microb Technol* 36: 307–313.

$$\frac{1}{\theta} = YU_S - K_d \tag{7.9}$$

where Y is the observed growth yield and K_d is a decay rate coefficient. On the other hand, U_s is subject to a Monod-type equation:

$$U_S = \frac{kS_e}{K_S + S_e} = \frac{S_0 - S_e}{Xt} \tag{7.10}$$

where k is the maximum value of U_s, K_s is the Monod constant, X is biomass concentration, t is time, S_0 and S_e are initial and effluent substrate concentrations, respectively. According to equation 7.9, the observed growth yield (Y) and the microbial decay rate coefficient (K_d) can be determined (table 7.2). It was found that U_s was much lower than its maximum value of 23.6 mg COD mg MLSS^{-1} day^{-1}. This may imply that the aerobic granular sludge SBR has a greater potential to handle wastewater with higher COD concentration or organic loading rate (OLR) than was applied in the study (L. Liu et al. 2005).

7.5 A GENERAL MODEL FOR AEROBIC GRANULAR SLUDGE SBR

Su and Yu (2006) developed a general model for aerobic granular sludge SBR, comprising reactor hydrodynamics, oxygen transfer, diffusion within granules, and biological reactions. In this approach, aerobic granules were classified into various fractions according to their sizes, and each granule was composed of a number of specific slices. The model development is based mainly on the following assumptions: (1) the SBR is operated at steady state; aerobic granules are spherical in shape and have a constant size distribution in one cycle of operation; (2) the liquid phase is completely mixed in the SBR, and suspended flocs are integrated with tiny granules; (3) only radial diffusion is taken into account and it is subject to Fick's law; (4) the kinetic constants and density of aerobic granules with different sizes are constant; (5) the effective diffusivity of a substance is constant; and (6) no biological reaction occurs in filling, settling, and decanting periods of the SBR operation. It is well known that an SBR is a discontinuous biosystem, that is, there are dynamic changes in both biomass and substrate concentrations within each operation cycle. This seems to imply that the first assumption given above might not be valid. Because of different kinetic behaviors and physical properties of aerobic granules over suspended flocs, integration of suspended flocs with tiny granules sounds unreasonable. In addition, density of aerobic granules will vary with granule size, and cannot be simply attributed to a constant regardless of size.

7.5.1 DESCRIPTION OF SUBSTRATE UTILIZATION

Su and Yu (2006) thought that an SBR can be regarded as a series of continuous stirred tank reactor (CSTRs) in time sequence. In each CSTR, the hydraulic resident time is Δt. For a CSTR at time t, the influent substrate concentration and effluent substrate concentration are $S^i_{(t-\Delta t)}$ and $S^i_{(t)}$, respectively. A mass balance on substrate gives equation 7.11.

$$S^i_{(t)} = S^i_{(t-\Delta t)} + k^i_{(t)} \times \Delta t \tag{7.11}$$

$$S^i_{(0)} = S^i_o \text{ for } t = 0$$

S_o is the initial substrate concentration at the beginning of each cycle. In order to estimate the reaction rate $k^i_{(t)}$, $S^i_{(t)}$ is considered to be equal to $S^i_{(t-\Delta t)}$ if Δt was short enough. The overall reaction rate of granule i in the bulk liquid is the sum of the reaction rates of all slices in all granules:

$$k^i(t) = \sum_{m=1}^{M} \left(\left(\sum_{n=1}^{N} k^i_{m,n} f_{V,m,n} \right) f_{V,m} \right) \tag{7.12}$$

where N is the number of slices for a granule, and M is the number of granule size fractions, $f_{V,m}$ and $f_{V,m,n}$ are the volume fractions of the granules belonging to the mth size fraction and those of the nth slice. $k^i_{m,n}$ is the reaction rate of the nth slice of granules with the mth size fraction.

7.5.2 DESCRIPTION OF OXYGEN TRANSFER

The rates of gas-liquid oxygen transfer are assumed to be proportional to the difference in the oxygen concentration between gas and liquid interfaces, and the proportionality factor is given by the volumetric oxygen transfer coefficient $k_L\alpha$ (Nicolella, van Loosdrecht, and Heijnen 1998). If oxygen transferred from the gas phase is equal to that diffused into granules at the granule surface, then,

$$D_e\alpha \frac{\partial S}{\partial r}\bigg|_{r=R} = J_{sur}\alpha = k_L\alpha(S_{gas} - S_{sur}) \tag{7.13}$$

where S_{gas} is the oxygen concentration in the gas phase, S_{sur} is the oxygen concentration on the granule surface, equal to that in bulk liquid when the liquid-solid oxygen transfer resistance is ignored. D_s is the diffusion coefficient of oxygen in water. S is the oxygen concentration at the radius of r. J_{sur} is oxygen flux on the granule surface, and α is the gas-liquid interfacial coefficient.

The volumetric oxygen transfer coefficient ($k_L\alpha$) has been found to be proportional to the gas holdup, which is affected by operating conditions and sludge characteristics, such as solid fraction and superficial gas velocity (U_g) (Nicolella, van Loosdrecht, and Heijnen 1998):

$$k_L\alpha = C_1\Phi^{C_2}(U_g \times 10^2)^{C_3} = C_4\varepsilon \tag{7.14}$$

where ε is gas holdup, C_1, C_2, C_3, and C_4 are constants (Su and Yu 2005). Thus, oxygen concentration on the granule surface can be determined and used in the calculation of the oxygen profiles within granules.

7.5.3 Description of Diffusion of Substance

Aerobic granules can be classified into various size fractions. In various-sized aerobic granules, the substance concentration profiles and reaction rates are different. In the approach by Su and Yu (2006), granule size is classified into M fractions, and the radius of granules in the mth size fraction is expressed as follows:

$$R_m = R_{min} + \frac{1}{2}\left((2m-1)\frac{R_{max} - R_{min}}{M}\right), \quad m = 1, 2, 1, M \tag{7.15}$$

where R_{min} and R_{max} are the minimum and maximum radius of the granules, respectively.

Based on the normal distribution of granule sizes, the frequency of granules for the mth size fraction ($f_{num,m}$) can be calculated from equation (7.16) (Su and Yu 2006):

$$f_{num,m} = \frac{A}{R_{var}\sqrt{2\pi}}e^{-(R_m - R_{mean})^2/2R_{var}^2} \tag{7.16}$$

where R_m is radius of the mth fraction calculated from equation 7.15, R_{mean} is the mathematical expectations, R_{var} is variance, and A is a constant. If the suspended solids (SS) concentration and wet density (ρ_w) of granules are known, the volume fraction of the granules belonging to the mth fraction is given by:

$$f_{V,m} = \frac{f_{num,m}R_m^3}{\sum_{m=1}^{M} f_{num,m}R_m^3} \tag{7.17}$$

Su and Yu (2006) applied equation 7.18 to describe changes in concentration of the substances (O_2, S_S, NH_4^+, NO_3^-) involved in the biological reactions. Equation 7.18 is obtained from a mass balance of substance i for a slice of granule in the mth size fraction:

$$\frac{\partial S_m^{\ i}}{\partial t} = \frac{\partial^2 S_m^{\ i}}{\partial r^2} + \frac{2}{r}\frac{\partial S_m^{\ i}}{\partial r} \pm \frac{k_m^{\ i}}{D_e^{\ i}} \tag{7.18}$$

with the boundary conditions of:

$$S_m^{\ i} = S_{sur}^{\ i}, \ at \ r = R_m$$

$$\frac{\partial S_m^{\ i}}{\partial r} = 0, \ at \ r = \delta_m^{\ i}$$

where $\delta_m^{\ i}$ is the penetration depth of component i into the mth size fraction granules, in which the gradient of component concentration vanishes by symmetry; r is the distance of the slice from the granule center. $S_m^{\ i}$ is the concentration of component i at the distance of r, and $S_{sur}^{\ i}$ is the concentration of component i at the surface of granule. R_m is the average radius of aerobic granules of one slice, which is the same as in equation 7.16. Equation 7.18 is the central equation in the approach by Su and Yu (2006), but in fact it is similar to that proposed by Li and Liu (2005).

If the SBR is operated at steady state, and can be regarded as a series of CSTRs, the left-hand side of equation 7.18 becomes zero. In order to determine $\delta_m^{\ i}$, which is essential for solving equation 7.18 numerically, Nicolella, van Loosdrecht, and Heijnen (1998) proposed that the substance transition can be characterized with a biological penetration rate ($\beta_m^{\ i}$):

$$\beta_m^{\ i} = \sqrt{\frac{2D_e^{\ i}S_{sur}^{\ i}}{k_i R_m^{\ 2}}} \tag{7.19}$$

where $D_e^{\ i}$ is the diffusion coefficient of component i inside aerobic granules, k_i is the reaction rate of component i in aerobic granules, $S_{sur}^{\ i}$ is the concentration of component i at the surface of a granule whose average radius is R_m.

For $\beta_m^{\ i} < 1$, the ith substance is partially penetrated in a granule with radius of mth size fraction (R_m). Thus,

$$\delta_m^{\ i} = \left(\frac{2D_e^{\ i}S_{sur}^{\ i}}{k_i}\right)^{1/2} \tag{7.20}$$

For $\beta_m^{\ i} > 1$, the ith substance is completely penetrated, that is, $\delta_m^{\ i} = 0$. The granule of the mth size fraction is taken as N slices. The concentrations of substances within each slice was assumed to be uniform over the entire cross section of the slice (Su and Yu 2006). In this case, the concentration of substance in each slice of granule can be calculated by equation 7.18, thus concentrations and their gradients of substance i in the nth slice of the mth size-fraction granule are given by $S_{m,n}^{\ i}$ and $\partial S_{m,n}^{\ i}/\partial r$, respectively.

7.5.4 DESCRIPTION OF BIOLOGICAL REACTIONS

In consideration of the differences between aerobic granules and sludge flocs, Su and Yu (2006) slightly modified the growth rate in order to accommodate specific features of aerobic granular sludge SBRs. As discussed in chapter 8, oxygen diffusion is often a limiting factor in aerobic granules. Su and Yu (2006) thought that the competition for oxygen would favor the growth of heterotrophic bacteria over slow-growing nitrifying bacteria, and subsequently this would result in a limitation of nitrification. According to Su and Yu (2006), in order to obtain a lower specific growth rate at a higher substrate concentration, the maximum specific growth rate ($\mu_{max,A}$) of autotrophic bacteria would be corrected by replacing $\mu_{max,A}$ with $\mu_{max,A}(t)$:

$$\mu_{max,A}(t) = \mu_{max,A}(t)e^{-P_1 S_S(t)} \tag{7.21}$$

where P_1 is a constant. For the modified maximum growth rate, Su and Yu (2006) further proposed that the parameter values can be calibrated by the following objective function:

$$Objective\ function = \sum \frac{(y_{measured} - y_{simulated})^2}{y_{measured}^2} \tag{7.22}$$

where $y_{measured}$ and $y_{simulated}$ are the measured and simulated values of parameters, respectively. The study by Su and Yu (2006) showed that the proposed model system could provide a pretty good simulation of the performance of aerobic granular sludge SBRs.

7.6 CONCLUSIONS

The kinetic growth model developed from the linear phenomenological equation can describe the growth of aerobic granules under various conditions. The growth of aerobic granules in terms of size and size-dependent growth rate is inversely related to the shear force, but positively related to the organic loading rate, while substrate N/COD ratio affects the growth kinetics of aerobic granules through change in the microbial population. The effect of substrate surface loading rate on the microbial surface growth rate and biodegradation rate can be described by the Monod-type equation. The operation and performance of aerobic granular sludge SBRs can be reasonably simulated by a combined diffusion-growth model.

REFERENCES

Atlas, R. M. and Bartha, R. 1998. *Microbial ecology*: Fundamentals and applications, 4th ed. Menlo Park, CA: Benjamin/Cummings.

Babcock, G. T. and Wikstrom, M. 1992. Oxygen activation and the conservation of energy in cell respiration. *Nature* 356: 301–309.

Burkhead, C. E. and McKinney, R. E. 1969. Energy concepts of aerobic microbial metabolism. *ASCE J Sanitary Eng Division* 95: 253–268.

Chen, S. Y. and Huang, S. Y. 2000. Shear stress effects on cell growth and L-DOPA production by suspension culture of *Stizolobium hassjoo* cells in an agitated bioreactor. *Bioprocess Eng* 22: 5–12.

De Groof, S. R. and Mazur, P. 1962. *Nonequilibrium thermodynamics.* Amsterdam: North-Holland.

Droste R. L. 1997. *Theory and practice of water and wastewater treatment.* New York: John Wiley.

Garfinkle, M. 2002. A thermodynamic-probabilistic analysis of diverse homogeneous stoichiometric chemical reactions. *J Phys Chem A* 106: 490–497.

Gaudy, A. F. and Gaudy, E. T. 1980. *Microbiology for environmental scientists and engineers.* New York: McGraw-Hill.

Gjaltema, A., van Loosdrecht, M. C. M., and Heijnen, J. J. 1997. Abrasion of suspended biofilm pellets in airlift reactors: Effect of particle size. *Biotechnol Bioeng* 55: 206–215.

Guiot, S. R., Pauss, A., and Costerton, J. W. 1992. A structured model of the anaerobic granules consortium. *Water Sci Technol* 25: 1–10.

Heijnen, J. J. and van Dijken, J. P. 1992. In search of a thermodynamic description of biomass yields for the chemotropic growth of microorganisms. *Biotechnol Bioeng* 39: 833–858.

Horn, H., Reiff, H., and Morgenroth, E. 2003. Simulation of growth and detachment in biofilm systems under defined hydrodynamic conditions. *Biotechnol Bioeng* 81: 607–617.

Kwok, W. K., Picioreanu, C., Ong, S. L., van Loosdrecht, M. C. M., Ng, W. J., and Heijnen, J. J. 1998. Influence of biomass production and detachment forces on biofilm structures in a biofilm airlift suspension reactor. *Biotechnol Bioeng* 58: 400–407.

Lagunas, R. 1976. Energy metabolism of *Saccharomyces cerevisiae* discrepancy between ATP balance and known metabolic functions. *Biochim Biophys Acta* 440: 661–674.

Lehninger, A. L. 1975. Biochemistry: *The molecular basis of cell structure and function,* 2nd ed. New York: Worth Publishers.

Li, Y. and Liu, Y. 2005. Diffusion of substrate and dissolved oxygen in aerobic granule. *Biochem Eng J* 24: 45–52.

Liu, L., Wang, Z., Yao, J., Sun, X., and Cai, W. (2005) Investigation on the properties and kinetics of glucose-fed aerobic granular sludge. *Enzyme Microb Technol* 36: 307–313.

Liu, Q. S., Liu, Y., Tay, J. H., and Show, K. Y. 2005. Responses of sludge flocs to shear strength. *Process Biochem* 40: 3213–3217.

Liu, Y, and Tay, J. H. 2001. Metabolic response of biofilm to shear stress in fixed-film culture. *J Appl Microbiol* 90: 337–342.

Liu, Y. and Tay, J. H. 2002. The essential role of hydrodynamic shear force in the formation of biofilm and granular sludge. *Water Res* 36: 1653–1665.

Liu, Y., Lin, Y. M., Yang, S. F., and Tay, J. H. 2003. A balanced model for biofilms developed at different growth and detachment forces. *Process Biochem* 38: 1761–1765.

Liu, Y., Liu, Y. Q., Wang, Z.-W., Yang, S. F., and Tay, J. H. 2005. Influence of substrate surface loading on the kinetic behaviour of aerobic granules. *Appl Microbiol Biotechnol* 67: 484–488.

Lubken, M. W. M., Schwarzenbeck, N., and Wilderer, P. A. 2004. Sequencing batch reactor technologies. In Proceedings of SBR3 Conference, February 22–26, 2004, Queensland, Australia.

Mayhew, M. and Stephenson, T. 1997. Low biomass yield activated sludge: A review. *Environ Technol* 18: 883–892.

Meijer, J. J., Tenhoopen, H. J. G., Luyben, K., and Libbenga, K. R. 1993. Effects of hydrodynamic stress on cultured plant-cells: A literature survey. *Enzyme Microb Technol* 15: 234–238.

Metcalf and Eddy, 2003. *Wastewater engineering: Treatment and reuse*, 4th ed., revised by George Tchobanoglous, Franklin L. Burton, and H. David Stensel. Boston: McGraw-Hill.

Moy, B. Y. P., Tay, J. H., Toh, S. K., Liu, Y., and Tay, S. T. L. 2002. High organic loading influences the physical characteristics of aerobic sludge granules. *Lett Appl Microbiol* 34: 407–412.

Nicolella, C., van Loosdrecht, M. C. M., and Heijnen, J. J. 1998. Mass transfer and reaction in a biofilm airlift suspension reactor. *Chem Eng Sci* 53: 2743–2753.

Pan, S. (2003) Inoculation of microbial granular sludge under aerobic conditions. Ph.D. thesis, Nanyang Technological University, Singapore.

Pirt, S. J. 1965. The maintenance energy of bacteria in growing cultures. *Proc. Royal Soc. London Ser B* 163: 224–231.

Prigogine, I. 1967. *Introduction to the thermodynamics of irreversible processes.* New York: Wiley-Interscience.

Russell, J. B. and Cook, G. M. 1995. Energetics of bacterial growth: Balance of anabolic and catabolic reactions. *Microbiol Rev* 59: 48–62.

Rutgers, M., Balk, P. A., and Vandam, K. 1989. Effect of concentration of substrates and products on the growth of *Klebsiella pneumoniae* in chemostat cultures. *Biochim Biophys Acta* 977: 142–149.

Sponza, D. T. 2001. Anaerobic granule formation and tetrachloroethylene (TCE) removal in an upflow anaerobic sludge blanket (UASB) reactor. *Enzyme Microb Technol* 29: 417–427.

Su, K. Z. and Yu, H. Q. 2005. Gas holdup and oxygen transfer in an aerobic granule-based sequencing batch reactor. *Biochem Eng J* 25: 201–207.

Su, K. Z. and Yu, H. Q. 2006. A generalized model for aerobic granule-based sequencing batch reactor. I. Model development. *Environ Sci Technol* 40: 4703–4708.

Tay, J. H., Liu, Q. S., and Liu, Y. 2001. Microscopic observation of aerobic granulation in sequential aerobic sludge blanket reactor. *J Appl Microbiol* 91: 168–175.

Tempest, D. W. and Neijssel, O. M. 1978. Eco-physiological aspects of microbial growth in aerobic nutrient-limited environments. *Adv Microb Ecol* 2: 105–153.

Tijhuis, L., Hijman, B., van Loosdrecht, M. C. M., and Heijnen, J. J. 1996. Influence of detachment, substrate loading and reactor scale on the formation of biofilms in airlift reactors. *Appl Microbiol Biotechnol* 45: 7–17.

Trinet, F., Heim, R., Amar, D., Chang, H. T., and Rittmann, B. E. 1991. Study of biofilm and fluidization of bioparticles in a three-phase liquid-fluidized-bed reactor. *Water Sci Technol* 23: 1347–1354.

van Loosdrecht, M. C. M., Eikelboom, D., Gjaltema, A., Mulder, A., Tijhuis, L., and Heijnen, J. J. 1995. Biofilm structures. *Water Sci Technol* 32: 35–43.

Wolfe, S. L. 1993. *Molecular and cellular biology.* Belmont, CA: Wadsworth.

Yang, S. F., Tay, J. H., and Liu, Y. 2004. Respirometric activities of heterotrophic and nitrifying populations in aerobic granules developed at different substrate N/COD ratios. *Curr Microbiol* 49: 42–46.

Yang, S. F., Tay, J. H., and Liu, Y. 2005. Effect of substrate nitrogen/chemical oxygen demand ratio on the formation of aerobic granules. *J Environ Eng* 131: 86–92.

Yang, S. F., Liu, Q. S., Tay, J. H., and Liu, Y. 2004. Growth kinetics of aerobic granules developed in sequencing batch reactors. *Lett Appl Microbiol* 38: 106–112.

8 Diffusion of Substrate and Oxygen in Aerobic Granules

Yong Li, Zhi-Wu Wang, and Yu Liu

CONTENTS

8.1 INTRODUCTION

Up-to-date, intensive research has been dedicated to the effect of various operating parameters on aerobic granulation in sequencing batch reactors (SBRs) (chapters 1 to 7). However, very limited information is currently available about the diffusion behaviors of substances inside aerobic granules. Tay et al. (2002) found that a model dye was only able to penetrate 800 μm beneath the surface of aerobic granules, while Jang et al. (2003) reported a penetration depth of 700 μm for dissolved oxygen from the surface of aerobic granules. Meanwhile, oxygen diffusion limitation in nitrifying aerobic granules was detected by microelectrode (Wilen, Gapes, and Keller 2004). Moreover, the unbalanced microbial growth inside aerobic granules has been supposed to be due to the mass diffusion limitation in aerobic granules, which would further lead to a heterogeneous internal structure (see chapter 11). These indicate that without a proper control of the mass diffusion, the structural stability of aerobic granules might not be sustainable.

In view of the importance of substrate and oxygen in microbial culture, this chapter attempts to offer insights into the diffusion behaviors of substrate and oxygen in aerobic granules. The factors that determine their respective diffusion in aerobic granules are also discussed, with special focus on the reactor operation as well as aerobic granule characteristics.

8.2 SIZE-DEPENDENT KINETIC BEHAVIORS OF AEROBIC GRANULES

Mass diffusion limitation often occurs in attached microbial communities, such as biofilms, and it suppresses microorganisms from fully accessing the substrate and oxygen in bulk solution. As a result, the microbial activity is lowered by deficiency of the energy source. To inspect the existence of mass diffusion limitation in aerobic granules, Y. Q. Liu, Liu, and Tay (2005) determined the specific COD removal rates and specific growth rates of aerobic granules with different sizes. It appears from figure 8.1 that the specific COD removal rate decreased markedly with the increase in the size of aerobic granules, indicating that the microbial activity inside aerobic granules is inhibited as the granule size increases. Moreover, the specific growth rate of aerobic granules was inversely dependent on the granule size (figure 8.2). A plot of the specific growth rate against the specific substrate utilization rate further reveals that the slow substrate utilization by large-sized aerobic granules results in a low specific growth rate (figure 8.3). Such a relationship between the specific growth and substrate utilization rates is consistent with the prediction by microbial growth theory (Metcalf and Eddy 2003). Consequently, the kinetic behaviors of aerobic granules depicted by the specific growth and substrate utilization rates are size dependent.

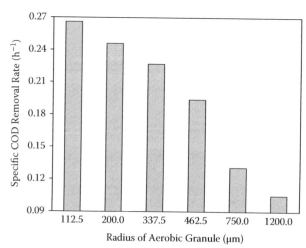

FIGURE 8.1 Granule size-dependent specific chemical oxygen demand (COD) removal rate. (Data from Liu, Y. Q., Liu, Y., and Tay, J. H., 2005. *Lett Appl Microbiol* 40: 312–315.)

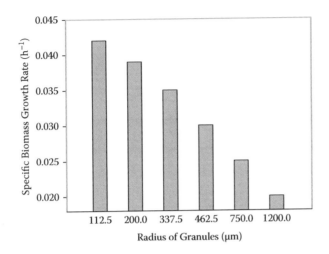

FIGURE 8.2   Granule size-dependent specific biomass growth rate of aerobic granules. (Data from Liu, Y. Q., Liu, Y., and Tay, J. H. 2005. *Lett Appl Microbiol* 40: 312–315.)

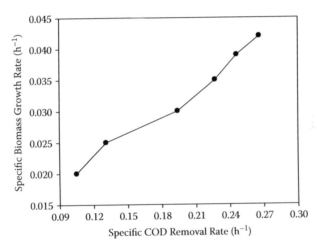

FIGURE 8.3   Specific growth rate of aerobic granules versus specific substrate utilization rate; data from figures 8.1 and 8.2.

8.3   DESCRIPTION OF DIFFUSION RESISTANCE IN AEROBIC GRANULES

It appears from figures 8.1 and 8.2 that the observed kinetic behaviors of aerobic granules with different sizes is ultimately the result of diffusion limitation of substrate or oxygen in aerobic granules. In order to look into this, Y. Q. Liu, Liu, and Tay (2005) introduced the concept of an effectiveness factor (η), which can be calculated as follows:

$$\eta = \frac{\text{rate with diffusion limitation}}{\text{rate without diffusion limitation}} \tag{8.1}$$

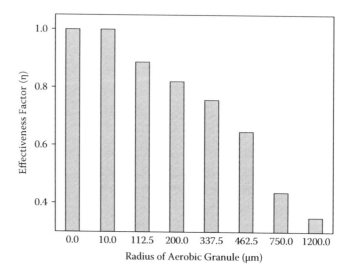

FIGURE 8.4 Effectiveness factor versus (η) the radius of aerobic granule. (Data from Liu, Y. Q., Liu, Y., and Tay, J. H. 2005. *Lett Appl Microbiol* 40: 312–315.)

In a case where diffusion limitation is negligible, η approaches unity, while η is close to zero if the diffusion limitation becomes significant as compared to the reaction. Figure 8.4 shows that the η tended to drop quickly with the increase of the granule radius. This would result from the presence of mass diffusion limitation in large aerobic granules.

The Thiele modulus is a measurement of the ratio of granule surface reaction rate to the mass diffusion rate. If the Monod kinetics for microbial reaction is applied, Y. Q. Liu, Liu, and Tay (2005) proposed the following modified Thiele modulus (φ) by introducing dimensionless concentrations:

$$\phi = R(\frac{X_o \mu_m}{Y_{X/S} S_o D_{es}})^{1/2} \tag{8.2}$$

The Thiele modulus combines the individual effect of specific growth rate μm, granule radius R, initial biomass X_0, substrate concentrations S_0, and diffusivity of substrate D_{es} in granules. Obviously, a high ϕ value means high surface reaction rate and low diffusion rate, and vice versa. A quasi-linear relationship of the Thiele modulus to the granule sizes is shown in figure 8.5. This seems to indicate that the mass diffusion dominates the surface reaction and becomes limiting in large aerobic granules. Under similar cultivation conditions of granules, Tay et al. (2003) also reported that small granules with radius of approximately 300 μm consisted entirely of live biomass. As can be seen in figure 8.5, a very low Thiele modulus was found in aerobic granules with a radius of 300 μm, that is, the microbial reaction is dominant over diffusion resistance in granules, and maintained the live cells throughout the entire aerobic granules. In contrast, Tay et al. (2002) detected an anaerobic layer at a depth of 800 to 900 μm from the surface of aerobic granules by the fluorescence

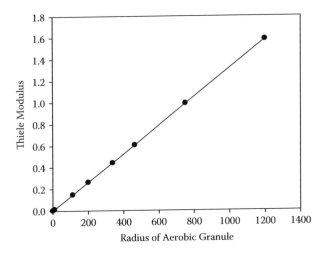

FIGURE 8.5   Thiele modulus versus radius of aerobic granule. (Data from Liu, Y. Q., Liu, Y., and Tay, J. H. 2005. *Lett Appl Microbiol* 40: 312–315.)

in situ hybridization (FISH) method. This suggests that the oxygen diffusion became the rate limiting factor over microbial reaction, which is evidenced by a large Thiele modulus, as shown in figure 8.5.

Picioreanu, van Loosdrecht, and Heijnen (1998) reported that a porous, mushroom-like, loose biofilm structure was observed at a low mass transfer rate, while the biofilm structure was compact and smooth at increased mass transfer rates. In general, the radius of aerobic granules is bigger than the thickness of biofilm, thus the effect of mass transfer resistance on the structure of aerobic granules is remarkable. It appears from equation 8.2 that mass transfer resistance in aerobic granules is closely related to the radius of the aerobic granules, the maximum specific growth rate, initial biomass, and substrate concentrations and diffusivity of the substrate. So far, it has been reported that selection of slow-growing bacteria and increase of shear rate favors the formation of compact and stable aerobic granules (Tay, Liu, and Liu 2001; Y. Liu, Yang, and Tay 2004). Obviously, slow-growing bacteria have small specific growth rates, leading to small-sized granules. According to equation 8.2, a small value of the Thiele modulus can be expected in this case, that is, the mass transfer resistance in this kind of aerobic granule would be lowered. Similarly, small aerobic granules can be cultivated at relatively high shear force, and this in turn results in a low value of the Thiele modulus (equation 8.2). In fact, the Thiele modulus and the effectiveness factor have been widely applied in biofilm research for decades. These two parameters also can provide useful information for quantitatively understating the mass transfer resistance in aerobic granules.

8.4   SIMULATION OF MASS TRANSFER IN AEROBIC GRANULES

Mass transfer limitation has been observed in aerobic granules. The modeling of the substrate diffusion in biofilms has been well studied, for example, a one-dimensional model for biofilms has been proposed by Wanner and Gujer (1986), and a three-dimensional model reflecting the heterogeneous structures of biofilms was also established by Picioreanu, van Loosdrecht, and Heijnen (1998). Based on the study of biofilm, Li and Liu (2005) investigated the description of diffusion process in aerobic granules.

8.4.1 MODEL DEVELOPMENT

Mature aerobic granules have an equilibrium or stable size when growth and detachment forces are balanced (Y. Liu and Tay 2002). In the development of the one-dimensional model for aerobic granules, Li and Liu (2005) made the following assumptions:

1. An aerobic granule is isotropic in physical, chemical, and biological properties, such as density and diffusion coefficient
2. An aerobic granule is ideally spherical
3. No nitrification and anaerobic degradation happen in the process
4. Aerobic granules respond to the change of bulk substrate concentration so quickly that the response time can be ignored

According to Bailey and Ollis (1986), the mass balance equations between the two layers whose radiuses are, respectively, r and $r + dr$ can be expressed as follows:

$$D_s\left(\frac{d^2s}{dr^2} + \frac{2}{r}\frac{ds}{dr}\right) = v \tag{8.3}$$

in which v is the substrate conversion rate, s is the substrate concentration, and D_s is the diffusion coefficient. In the approach by Li and Liu (2005), the substrate conversion rate was given by the Monod-type equation:

$$v = \frac{\rho_x}{Y_{x/s}}\mu_{max}\frac{s}{K_s + s} \tag{8.4}$$

in which ρ_x is the biomass density, $Y_{x/s}$ is the biomass yield, and μ and μ_{max} are the specific growth rate and the maximum specific growth rate, respectively. Substituting equation 8.4 into equation 8.3 leads to the following expression:

$$D_s\left(\frac{d^2s}{dr^2} + \frac{2}{r}\frac{ds}{dr}\right) = \frac{\mu_{max}s}{K_s + s}\frac{\rho_x}{Y_{x/s}} \tag{8.5}$$

It was assumed that the derivative at the center of the granule is zero and the substrate concentration at the surface of the granule equals the bulk solution (Li and Liu 2005), that is:

$$\left.\frac{ds}{dr}\right|_{r=0} = 0 \tag{8.6}$$

$$s\big|_{r=R} = S_{bulk} \tag{8.7}$$

Perez, Picioreanu, and van Loosdrecht (2005) applied equation 8.5 to the study of spherical biofilms, and analytically solved this equation by assuming that the growth rate is zero-order or first-order. As noted by Li and Liu (2005), such a simplified treatment of equation 8.5 leads to the inaccuracy of the prediction, and also limits the application of this equation to a very narrow range. As equation 8.5 is a non-homogenous equation, a numerical method to solve it completely was developed without any assumption on it (Li and Liu 2005). This method is based on the finite difference method (FDM) (Hoffman 2001). The radius is thus divided into n grids, that is:

$$D_s \left(\frac{s_{i+1} + s_{i-1} - 2s_i}{\Delta r^2} + \frac{s_{i+1} - s_{i-1}}{r\Delta r} \right) = \frac{\mu_{max} s_i}{K_s + s_i} \frac{\rho_x}{Y_{x/s}} \tag{8.8}$$

This numerical scheme is applied to all situations without simplifying assumptions and therefore the accuracy is increased. The program was written in Matlab™ language and run under Matlab 7.0 which allows an easy visualization of the simulated data (Li and Liu 2005). It should be emphasized that equation 8.8 can also be applied to oxygen if the set of parameters for the substrate is replaced with the set of parameters for dissolved oxygen.

After the substrate concentration is determined, the substrate utilization rate (v_1) of a single aerobic granule can be calculated as:

$$v_1 = \frac{1}{Y_{x/s}} \int_0^R \rho_x \mu(s) 4r^2 dr \tag{8.9}$$

Summing up the substrate utilization rate of all the aerobic granules gives the total substrate utilization rate v_{all}:

$$v_{all} = \sum_{i=1}^m \frac{1}{Y_{x/s}} \int_0^{R_i} \rho_x \mu(s) 4r^2 dr \tag{8.10}$$

in which m is the number of aerobic granules in the reactor and R_i is the radius of the granule being calculated (Li and Liu 2005). According to Li and Liu (2005), the average radius R of aerobic granules can be expressed as follows:

$$\bar{R} = \frac{1}{m} \sum_{i=1}^m R_i \tag{8.11}$$

Substituting equation 8.11 into equation 8.10 yields:

$$V_{all} = \frac{m}{Y_{x/s}} \int_0^{\bar{R}} \rho_x \mu(s) 4r^2 dr \qquad (8.12)$$

After deformation, equation 8.12 becomes:

$$V_{all} = \frac{4\pi \bar{R} \rho_x}{3} \frac{m}{Y_{x/s}} \int_0^{\bar{R}} \frac{\mu(s) 4r^2}{4/3\pi R^3} dr \qquad (8.13)$$

in which $4\pi \bar{R}^3 \rho_x m$ is equal to the total biomass in the reactor. Thus, equation 8.13 can be simplified to:

$$V_{all} = \frac{XV}{Y_{x/s}} \int_0^{\bar{R}} \frac{\mu(s) 4r^2}{4/3\pi R^3} dr \qquad (8.14)$$

in which V is the reactor volume. At time dt, the change of substrate concentration in the reactor can be described as:

$$dS_{bulk} = \frac{V_{all}}{V} dt = \frac{X}{Y_{x/s}} \int_0^{\bar{R}} \frac{\mu(s) 4r^2}{4/3\pi R^3} drdt \qquad (8.15)$$

In a time period from T_0 to T, the change in the substrate concentration is given by equation 8.16:

$$\Delta S_{bulk} = \frac{X}{Y_{x/s}} \int_{T_0}^{T} \left(\int_0^{\bar{R}} \frac{\mu(s) 4r^2}{4/3\pi R^3} dr \right) dt \qquad (8.16)$$

At an initial substrate concentration, $S_{bulk\,0}$, the bulk substrate concentration at any time t can be calculated as:

$$S_{bulk} = S_{bulk0} - \Delta S_{bulk}$$

$$= S_{bulk0} - \frac{X}{Y_{x/s}} \int_{T_0}^{T} \left(\int_0^{\bar{R}} \frac{\mu(s) 4r^2}{4/3\pi R^3} dr \right) dt \qquad (8.17)$$

As discussed earlier, equation 8.17 cannot be solved analytically, and the method based on the finite difference principle is thus applied to solve this equation (Li and Liu 2005). At each time step, the state is considered pseudo-static, which means the bulk substrate concentration is constant at each time step. Then change of the bulk substrate concentration is a process of mapping, that is, the substrate concentration is determined by the previous time step. This model is valid for different substrates so long as the parameters are replaced with the specific substrate parameters. In this study, equation 8.8 was applied to organic substrate and dissolved oxygen under various operation conditions.

8.4.2 SUBSTRATE PROFILE IN AEROBIC GRANULES WITH DIFFERENT RADIUSES

Li and Liu (2005) simulated the substrate profiles in bioparticles with a mean size of 0.1 to 1.0 mm, and the initial acetate concentration and granule biomass concentration were kept at 465 mg COD L^{-1} and 6250 mg L^{-1} volatile solids, respectively, while the initial DO concentration was controlled at 8.3 mg L^{-1}. Model predictions show that the acetate in terms of COD is able to penetrate to the center of all sizes of aerobic granules, and the microbial utilization of COD was able to proceed throughout aerobic granules with the radius of 0.1 to 0.4 (figure 8.6a and b). In contrast, the COD profiles in figure 8.6c to e show the platforms beneath the depths of 0.23, 0.11, and 0.07 mm from the surface of aerobic granules with radiuses of 0.5, 0.75, and 1.0 mm, respectively. As assumed, there should be no autotrophic bacteria in aerobic granule, thus these COD platforms in figure 8.6c to e seem to imply that bacteria at those platform depths of the aerobic granule should have ceased normal metabolic activity. Furthermore, the noticeable level of COD present at those granule centers points to another possibility accounting for the lowered microbial activity, that is, dissolved oxygen could be a limiting factor at those depths (figure 8.6c to e).

8.4.3 OXYGEN PROFILES IN AEROBIC GRANULES WITH DIFFERENT RADIUSES

Oxygen profiles in aerobic granules with radiuses of 0.1 to 1.0 mm are simulated (figure 8.7). Similar to the COD profiles in figure 8.6, oxygen can diffuse into the entire aerobic granules with radius of 0.1 and 0.4 mm (figure 8.7a and b). For aerobic granules with a radius bigger than 0.5 mm, prominent oxygen diffusion limitation turns out. Oxygen is only able to penetrate to 0.27, 0.64, and 0.1 mm from the surface of aerobic granules with radiuses of 0.5, 0.75, and 1.0 mm, and the remaining depth in the aerobic granule is deficient in dissolved oxygen (DO) (figure 8.7c, d, and e). Zero DO depths inside the aerobic granules are in good agreement with those COD platforms (figure 8.6c to e), and microbial activity at the depth of those COD platforms was seriously limited by the availability of DO.

The DO-reachable depths shown in figure 8.7 appear to be inversely related to the size of the aerobic granule. This suggests that larger aerobic granules will be subjected to an even more severe diffusion limitation of DO and further cause a drop in microbial activity. It can thus be concluded that, in large aerobic granules, acetate or organic substrate would not be a limiting factor, and the whole microbial process would be dominated by the availability of DO inside the aerobic granule.

8.4.4 DIFFUSION PROFILES OF SUBSTRATE IN AEROBIC GRANULES AT DIFFERENT BULK SUBSTRATE CONCENTRATIONS

The results presented in figures 8.6 and 8.7 were obtained at a fixed bulk COD concentration of 465 mg L^{-1}, the effect of bulk COD concentration on the mass diffusion in aerobic granule is not taken into account. In order to clarify this point, the COD

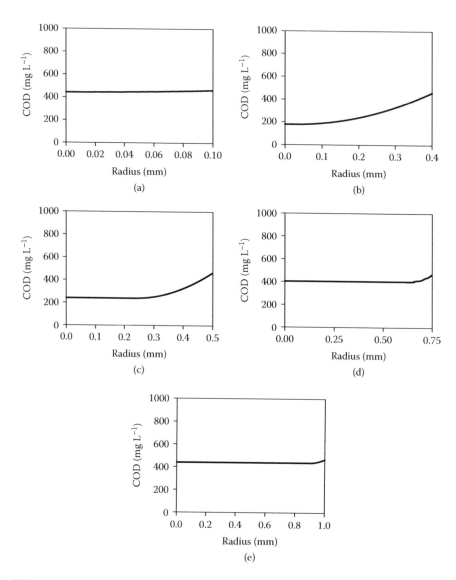

FIGURE 8.6 Substrate profiles in aerobic granules with radiuses of 0.10 (a), 0.40 (b), 0.50 (c), 0.75 (d), and 1.00 mm (e). (Data from Li, Y. and Liu, Y. 2005. *Biochem Eng J* 27: 45–52.)

profiles in an aerobic granule with radius of 0.5 mm were simulated at different bulk COD concentrations of 100, 200, and 300 mg L^{-1}, respectively (Li and Liu 2005). Figure 8.8 shows soluble COD can penetrate through the entire aerobic granule at a bulk COD concentration of 300 mg L^{-1} or above, while COD becomes a limiting factor and sharply drops to nil at depths of 0.25 and 0.1 mm at the bulk COD concentrations of 100 and 200 mg L^{-1}, respectively. This suggests that COD availability may also be a limiting factor for microbial growth at low concentrations, and this is strongly dependent on the level of external substrate concentration in bulk solution.

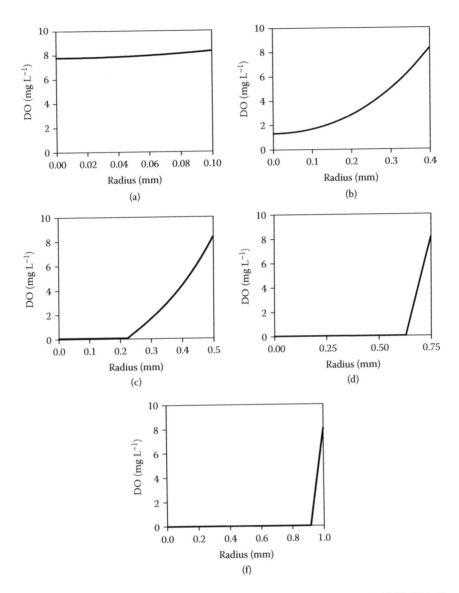

FIGURE 8.7 Oxygen profiles in aerobic granules with radiuses of 0.10 (a), 0.40 (b), 0.50 (c), 0.75 (d), and 1.00 mm (e). (Data from Li, Y. and Liu, Y. 2005. *Biochem Eng J* 27: 45–52.)

8.4.5 Diffusion Profiles of Dissolved Oxygen in Aerobic Granules at Different Substrate Concentrations

Oxygen profiles at different acetate concentrations in aerobic granules with radius of 0.5 mm are shown in figure 8.9. It can be seen that oxygen can penetrate to the center of aerobic granules at low bulk COD concentrations of 100 and 200 mg COD L^{-1}. However, it can only diffuse to the depth of 0.38 mm from the surface of aerobic granules at the bulk COD of 300 mg L^{-1}. This means that either DO or COD can be

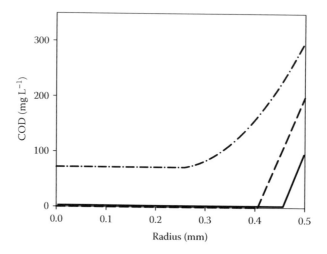

FIGURE 8.8 Substrate profiles in aerobic granules with a radius of 0.5 mm at bulk COD of 100 (—), 200 (– – –), and 300 mg L⁻¹ (— · —). (Data from Li, Y. and Liu, Y. 2005. *Biochem Eng J* 27: 45–52.).

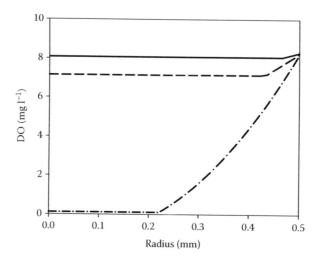

FIGURE 8.9 Substrate profiles in aerobic granules with a radius of 0.5 mm at bulk COD of 100 (—), 200 (– – –), and 300 (— · —) mg L⁻¹. (Data from Li, Y. and Liu, Y. 2005. *Biochem Eng J* 27: 45–52.)

a limiting factor for microbial growth in aerobic granules, which strongly depends on the relative level of bulk COD concentration to DO, that is, a high bulk COD would cause DO limitation in aerobic granules, and COD limitation would become dominant at low bulk COD concentrations.

Most previous research on aerobic granulation investigated the profiles of DO and substrate separately (Tay et al. 2002; Jang et al. 2003; Wilen, Gapes, and Keller 2004). However, the fact that oxygen profile inside an aerobic granule is dynamic and is closely related with substrate profile and metabolic activity has been ignored (figures 8.8 and 8.9). In another study on activated sludge by Beun, Heijnen, and van Loosdrecht (2001), the oxygen profile in activated sludge floc was considered dynamic, but the interrelation between oxygen and substrate profiles was also ignored. Beun et al. (1999) concluded that oxygen penetration depth in a sequential batch airlift reactor was smaller than the acetate penetration depth. However, this point would be valid only when the substrate concentration is high, whereas it is invalid in the case where substrate is the limiting factor, as shown in figure 8.8. In addition, in modeling biofilm, substrate was regarded as the limiting substance without taking account of the oxygen (Picioreanu, van Loosdrecht, and Heijnen 1998; Laspidou 2003). As shown in figures 8.8 and 8.9, diffusion of substances in aerobic granules is a dynamic process, and is interrelated with one another. The approach presented here may offer a more reasonable tool for the study of diffusion phenomena in aerobic granules.

8.4.6 Prediction of Bulk Substrate Concentration in an Aerobic Granules Reactor

According to equation 8.16, the substrate conversion rate is proportional to the biomass concentration (X) and is inversely related to the yield coefficient at a given granule radius. Four sets of experiments were also conducted using bioparticles with radius of 0.1 to 1.0 mm, and figure 8.10 shows the bulk substrate concentration profiles in the reactors containing bioparticles with different sizes. It can be seen in figure 8.10 that the simulation is in good agreement with the experimental data. When the aerobic granule size was as small as 0.1 mm in radius, diffusion is not a limiting factor, indicated by a sharp COD drop to zero within 16 minutes (figure 8.10a). In this case, equation 8.14 reduces to a formulation similar to the activated sludge model (ASM) (Henze et al. 1987). In fact, a satisfactory agreement between equation 8.14 and the ASM was observed when the granule size was smaller than or close to the size of biofloc (Li and Liu 2005). It appears from figure 8.10 that under the same conditions of operation, the substrate can be removed rapidly in the reactor with small aerobic granules, for example, the substrate removal rate by 0.5-mm granules is three times higher than that by 1.0-mm granules (figure 8.10b and d). This implies that: (1) the reactor with small aerobic granules is more efficient, and has a higher treatment capacity than the reactor with large granules; and (2) the size of the aerobic granules needs to be properly controlled in order to maximize their metabolic activity. According to Li and Liu (2005), the inflection points in figure 8.10a to d indeed represent a turning point from oxygen limitation to substrate limitation. Under the condition of oxygen limitation, the substrate removal rate is determined mainly by the availability of dissolved oxygen, that is, the substrate removal is independent of the bulk substrate concentration, and thereby a pseudo-zero-order reaction kinetics with respect to substrate is observed in figure 8.10a to d. This is consistent with other experimental observations (Q. S. Liu 2003; Yang et al. 2004). However, under the substrate limitation condition, figure 8.10a to d shows that the substrate removal tends to decrease with the decrease in the substrate concentration, that is, it is a function of the substrate concentration. It is apparent that

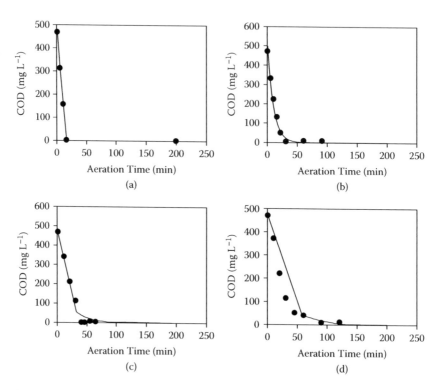

FIGURE 8.10 COD removal by aerobic granules with radiuses of 0.1 (a), 0.5 (b), 0.75 (c), and 1.0 mm (d) during the cycle operation of SBR. ●: Experimental point; model prediction. (From Li, Y. and Liu, Y. 2005. *Biochem Eng J* 27: 45–52. With permission.)

the performance of an aerobic granular sludge SBR is mainly controlled by the availability of dissolved oxygen.

As discussed earlier, the substrate removal kinetics by aerobic granules is size dependent (figures 8.1 and 8.4). In order to investigate the interaction of diffusion and reaction in aerobic granules, the effectiveness factor (η) was calculated and further compared with the η determined from experiments with different-sized aerobic granules (figure 8.11). It can be seen in figure 8.11 that both theoretical and experimental values are in good agreement. At a radius smaller than 0.5 mm, no significant change in η is observed. However, the effectiveness factor decreases quickly with the further increase of radius of the aerobic granules, which indicates that the mass transfer limitation begins to play an important role in the overall reaction of aerobic granules. This is in good agreement with the results presented in figures 8.5 and 8.6.

8.5 CONCLUSIONS

Mass diffusion limitation in aerobic granules result in a significant drop in microbial activity and growth rate. It was found that the mass diffusion overtakes the microbial reaction, and becomes the process-rate limiting factor for large aerobic granules. The model simulation also supports the size-associated diffusion limitation in aerobic

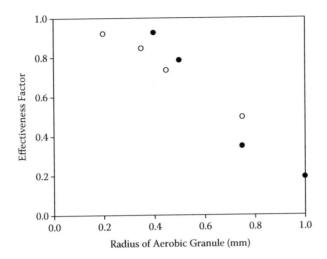

FIGURE 8.11 Effectiveness factor versus radius of aerobic granules, theoretical values (●) and experimental data (○). (From Li, Y. and Liu, Y. 2005. *Biochem Eng J* 27: 45–52. With permission.)

granules. The simulation results showed that diffusion of substrate and oxygen in aerobic granules is an interrelated dynamic process and is interrelated. Smaller aerobic granules exhibited higher metabolic activity in terms of the substrate removal rate, while in an SBR dominated by aerobic granules with a size larger than 0.5 mm, the dissolved oxygen is the bottleneck that limits the substrate utilization. It is evident that the simulation model presented can provide an effective and useful tool for predicting and optimizing the performance of an aerobic granular sludge SBR.

SYMBOLS

D_s Acetate diffusion coefficient
D_{fo} Oxygen diffusion coefficient
$\mu(s)$ Growth rate at substrate concentration s
μ_{max} Maximum growth rate
K_s Half rate constant
$Y_{x/o}$ Growth yield of oxygen
$Y_{x/s}$ Growth yield of substrate
S_{bulk} Bulk substrate concentration
ρ_x Biomass density
R The radius of a granule
r One-dimensional coordinate
Δr The thickness of one layer
$\bar{R}$ Average radius of all granules
s_i Substrate concentration at no. i layer
s Substrate concentration at a point of a granule
ΔS_{bulk} The change of bulk substrate concentration

$S_{bulk\ 0}$	Initial bulk substrate concentration
v	Substrate conversion rate
v_1	Substrate conversion rate of a granule
V	Volume of the reactor
v_{all}	Substrate conversion rate of all granules
X	Biomass concentration
m	Number of aerobic granules in a reactor

REFERENCES

Bailey, J. E. and Ollis, D. F. 1986. *Biochemical engineering and fundamentals*, 2nd ed. New York: McGraw-Hill.

Beun, J. J., Heijnen, J. J., and van Loosdrecht, M. C. M. 2001. N removal in a granular sludge sequencing batch airlift reactor. *Biotechnol Bioeng* 75: 82–92.

Beun, J. J., Hendriks, A., van Loosdrecht, M. C. M., Morgenroth, E., Wilderer, P. A., and Heijnen, J. J. 1999. Aerobic granulation in a sequencing batch reactor. *Water Res* 33: 2283–2290.

Henze, M., Grady, C. P. L., Gujer, W., Marais, G. R., and Matsuo, T. 1987. *Activated sludge model no. 1*. Scientific and Technical Report No. 1. London: International Association on Water Pollution Research and Control.

Hoffman, J. D. 2001. *Numerical methods for engineers and scientists*, 2nd ed. New York: Marcel Dekker.

Jang, A., Yoon, Y. H., Kim, I. S., Kim, K. S., and Bishop, P. L. 2003. Characterization and evaluation of aerobic granules in a sequencing batch reactor. *J Biotechnol* 105: 71–82.

Laspidou, C. S. 2003. Modeling heterogeneous biofilms including active biomass, inert biomass and extracellular polymeric substances. Ph.D. thesis, Northwestern University, Evanston, IL.

Li, Y. and Liu, Y. 2005. Diffusion of substrate and oxygen in aerobic granule. *Biochem Eng J* 27: 45–52.

Liu, Q. S. 2003. Aerobic granulation in a sequencing batch reactor. Ph.D. thesis, Nanyang Technological University, Singapore.

Liu, Y. and Tay, J. H. 2002. The essential role of hydrodynamic shear force in the formation of biofilm and granular sludge. *Water Res* 36: 1653–1665.

Liu, Y., Yang, S. F., and Tay, J. H. 2004. Improved stability of aerobic granules by selecting slow-growing nitrifying bacteria. *J Biotechnol* 108: 161–169.

Liu, Y. Q., Liu, Y., and Tay, J. H. 2005. Relationship between size and mass transfer resistance in aerobic granules. *Lett Appl Microbiol* 40: 312–315.

Metcalf and Eddy, 2003. *Wastewater engineering: Treatment and reuse*, 4th ed., revised by George Tchobanoglous, Franklin L. Burton, and H. David Stensel. Boston: McGraw-Hill.

Perez, J., Picioreanu, C., and van Loosdrecht, M. 2005. Modeling biofilm and floc diffusion processes based on analytical solution of reaction-diffusion equations. *Water Res* 39: 1311–1323.

Picioreanu, C., van Loosdrecht, M. C. M., and Heijnen, J. J. 1998. Mathematical modeling of biofilm structure with a hybrid differential-discrete cellular automaton approach. *Biotechnol Bioeng* 58: 101–116.

Tay, J. H., Liu, Q. S., and Liu, Y. 2001. The effects of shear force on the formation, structure and metabolism of aerobic granules. *Appl Microbiol Biotechnol* 57: 227–233.

Tay, J. H., Ivanov, V., Pan, S., and Tay, S. T. L. 2002. Specific layers in aerobically grown microbial granules. *Lett Appl Microbiol* 34: 254–257.

Tay, J. H., Tay, S. T. L., Ivanov, V., Pan, S., Jiang, H. L., and Liu, Q. S. 2003. Biomass and porosity profiles in microbial granules used for aerobic wastewater treatment. *Lett Appl Microbiol* 36: 297–301.

Wanner, O. and Gujer, W. 1986. Multispecies biofilm model. *Biotechnol Bioeng* 28: 314–328.

Wilen, B. M., Gapes, D., and Keller, J. 2004. Determination of external and internal mass transfer limitation in nitrifying microbial aggregates. *Biotechnol Bioeng* 86: 445–457.

Yang, S. F., Liu, Q. S., Tay, J. H., and Liu, Y. 2004. Growth kinetics of aerobic granules developed in sequencing batch reactors. *Lett Appl Microbiol* 38: 106–112.

9 The Essential Role of Cell Surface Hydrophobicity in Aerobic Granulation

Yu Liu and Zhi-Wu Wang

CONTENTS

9.1 INTRODUCTION

Aerobic granulation is a process of cell-to-cell self-immobilization that results in a form of regular shape. In view of mass transfer and utilization of substrate, bacteria indeed would prefer a dispersed rather than aggregated state. There should be triggering forces that can bring bacteria together and further make them aggregate. It appears from the preceding chapters that cell hydrophobicity induced by culture conditions can serve as a triggering force for aerobic granulation. In fact, it has been well known that the physicochemical properties of the cell surface have profound effects on the formation of biofilms and both anaerobic and aerobic granules (Bossier and Verstraete 1996; Zita and Hermansson 1997; Kos et al. 2003;

149

Liu et al. 2004b). When bacteria became more hydrophobic, increased cell-to-cell adhesion was observed, that is, cell surface hydrophobicity may contribute to the ability of cells to aggregate (Kjelleberg, Humphrey, and Marshall 1983; Del Re et al. 2000; Kos et al. 2003; Liu et al. 2004b). This chapter looks at the role of cell surface hydrophobicity in the formation of aerobic granular sludge in a sequencing batch reactor (SBR).

9.2 CELL SURFACE HYDROPHOBICITY

9.2.1 WHAT IS HYDROPHOBICITY?

Hydrophobicity attraction is the strongest binding force occurring between particles or polymers immersed in water. The attraction between two apolar surfaces, or between one apolar and one polar surface, in water, is traditionally called the hydrophobic effect. Hydrophobic surfaces do not repel water but instead attract water (Hildebrand 1979). Because of water hydrogen bonds, water molecules often present in the form of water clusters (figure 9.1), and the size of these clusters tends to decrease with increase of temperature. The classical macroscopic scale interactions between apolar and/or polar surfaces, immersed in a liquid, have been often described by the well-known DLVO theory, which shows apolar Lifshitz–van der Waals (LW) attraction and electrical double layer (EL) repulsion as a function of distance. It can be shown that hydrophobic interaction becomes the main driving force which represents nearly all the macro-scale interactions in water in terms of

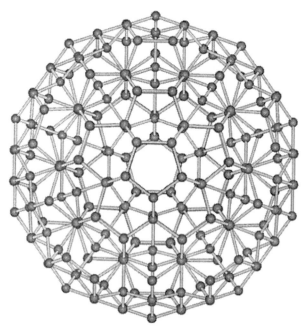

FIGURE 9.1 Illustration of water molecules cluster. (From Chaplin, M. F. 2000. *Biophys Chem* 83: 211–221. With permission.)

attraction or repulsion (Bergendahl et al 2002). Hydrophilic repulsion occurs only when polar molecules, particles, or cells attract water molecules more strongly than the acid-base (AB) cohesive attraction between water molecules.

9.2.2 Cell Surface Property-Associated Hydrophobicity

Most biological surfaces have a low γ^+ in the order of 0.1 mJ m^{-2}. The cell surface is composed mainly of proteins, polysaccharides, and phospholipids. The combination characteristics of these substances in turn determine the overall cell surface hydrophobicity.

9.2.2.1 Surface Properties of Amino Acids

According to Parker, Guo, and Hodges (1986), the order of amino acid side chains beginning with the most hydrophobic can probably be summarized as follows: Trp, Phe, Leu, Ile, Met, Val, Tyr, Cys, Ala, Pro, His, Arg, Thr, Lys, Gly, Glu, Ser, Asx, Glu, Asp, where the amino acids to the right of Thr are more hydrophilic. It is evident that an amino acid with a larger hydrophobic side chain is more hydrophobic than those with a small hydrophobic side chain. This seems to indicate that the surface property of amino acids can significantly influence the cell surface hydrophobicity.

9.2.2.2 Surface Properties of Proteins

Proteins are made up of hydrophobic and/or hydrophilic amino acids. For water-soluble protein, the majority of its hydrophilic amino acids presents at the water interface, whereas the more hydrophobic amino acids are located inside the three-dimensional framework of the macromolecule. However, once protein has made contact with a hydrophobic surface, it can orient its most hydrophobic sites to the hydrophobic interface (Lee et al. 1973; van Oss 1994a). This seems to indicate that some proteins can shift between hydrophobicity and hydrophilicity, depending on actual conditions.

9.2.2.3 Surface Properties of Polysaccharides

In contrast with protein that comprises hydrophilic and/or hydrophobic amino acids, polysaccharides are made up of different sugars that are hydrophilic and soluble in water (van Oss 1995). Obviously, the high solubility of polysaccharides in water means a low hydrophobicity. As polymeric substances, these sugars may become more hydrophilic or more hydrophobic, depending on the structure of the polymer molecule. It has been reported that the amount of extracellular polymers affects the contribution of electrostatic interaction to cell attachment onto a solid surface (Tsuneda et al. 2003). Furthermore, in a study of hydrophobic and hydrophilic properties of activated sludge, it was found that a significant portion of extracellular polymers are hydrophobic (Jorand et al. 1998). Likely, extracellular polymer-induced cell surface hydrophobic changes may be fundamental in microbial aggregation.

9.2.2.4 Surface Properties of Phospholipids

The general structure of biological membranes is a phospholipids bilayer. Phospholipids contain both highly hydrophobic (fatty acid) and relatively hydrophilic (glycerol) moieties and can exist in many different chemical forms as a result of variation in the nature of the fatty acids or phosphate-containing groups attached to the glycerol backbone (Madigan, Martinko, and Parker 2003). As phospholipids aggregate in an aqueous solution, they tend to form a bilayer structure spontaneously with the fatty acids in a hydrophobic environment, and the hydrophilic portions remain exposed to the aqueous external environment. Saturated alkyl chains of phospholipids can attract each other strongly in water, with a hydrophobic energy of attraction of -102 mJ m^{-2} in all cases (van Oss 1994b). The major proteins of the cell membrane generally have very hydrophobic external surfaces in the regions of the protein that span the membrane and have hydrophilic surfaces exposed on both the inside and the outside of the cell. The overall structure of the cytoplasmic membrane is stabilized by hydrogen bonds and hydrophobic interactions (Madigan, Martinko, and Parker 2003).

9.2.3 DETERMINATION OF CELL SURFACE HYDROPHOBICITY

There are a number of methods available to characterize the cell surface hydrophobicity, including contact angle measurement, bacterial adherence to hydrocarbons, hydrophobic interaction chromatography, salting out aggregation, adhesion to solid surfaces, and binding of fatty acids to bacterial cells (Rosenberg and Kjelleberg 1986; Mozes and Rouxhet 1987). Contact angle measurement is the traditional and widely used method, and it involves measuring the contact angle of a sessile drop with a flat bacteria fixed filter or a lawn of the bacteria on an agar plate (Mozes and Rouxhet 1987). As the cell surface moisture decreases with evaporation, the contact angle increases over time. The stationary-phase contact angle is often used to characterize the cell surface hydrophobicity (Absolom et al. 1983). According to the water contact angle, cell surface hydrophobicity may be roughly classified into three categories: a hydrophobic surface with a contact angle greater than 90°, a medium hydrophobic surface with a contact angle between 50° and 60°, and a hydrophilic surface with a contact angle below 40° (Mozes and Rouxhet 1987).

It should be noted that γ^{LW}, γ^+, γ^- can be determined by contact angle measurements with at least three different liquids (of which two must be polar) by the Young equation (van Oss, Chaudhury, and Good 1987, 1988):

$$(1 + \cos\theta)\gamma_L = 2\left(\sqrt{\gamma_m^{LW}\gamma_L^{LW}} + \sqrt{\gamma_m^+\gamma_L^-} + \sqrt{\gamma_m^-\gamma_L^+}\right)$$

Becausee the contact angle method requires specific equipment, microbial adhesion to solvents (MATS) has been developed to characterize microbial cell surfaces (Bellon Fontaine, Rault, and van Oss 1996). This method is based on the comparison between microbial cell affinity to a monopolar solvent and a polar solvent. Table 9.1 lists the commonly used the monopolar solvents in the MATS method. Acidic solvent serves as electron acceptor, and basic solvent as electron donor.

9.3 THE ROLE OF CELL SURFACE HYDROPHOBICITY IN AEROBIC GRANULATION

An aerobic granule can form through cell-to-cell self-adhesion, and its formation is a multiple-step process, and both physicochemical and biological forces are involved.

TABLE 9.1
Surface Tensions of Typical Organic Solvents

Liquid	Formula	γ_l^{LW} (mJ m^{-2})	γ_l^+ (mJ m^{-2})	γ_l^- (mJ m^{-2})
Decane	$C_{10}H_{22}$	23.9	0	0
Ethyl acetate	$C_4H_{10}O$	23.9	0	19.4
Hexadecane	$C_{16}H_{34}$	27.7	0	0
Chloroform	$CHCl_3$	27.2	3.8	0

Source: Data from Bellon Fontaine, M/ N., Rault, J., and van Oss, J.
1996. *Colloid Surface B* 7: 47–53.

It has been suggested that microbial adhesion can be defined in terms of the energy involved in the formation of the adhesive junction. When a bacterium approaches another bacterium, the hydrophobic interaction between them is a crucial force. Wilschut and Hoekstra (1984) proposed a local dehydration model, and suggested that under the physiological conditions, the strong repulsive hydration interaction was the main force to keep the cells apart.

So far, aerobic granules have been developed with various substrates, and aerobic granulation by heterotrophic, nitrifying, denitrifying, and phosphorus-accumulating bacteria has been reported. This implies that aerobic granulation is not strictly restricted to some specific substrate and microbial species, and it can be regarded as a process in which individual cells aggregate together through cell-to-cell hydrophobic interaction and binding. It is believed that cell hydrophobicity is one of the most important affinity forces in microbial aggregation. Hydrophobicity and hydrophilicity are usually used to describe a molecule or a structure having the feature of being rejected from an aqueous medium (i.e., hydrophobicity), or being positively attracted (i.e., hydrophilicity). Hydration interaction becomes significant at surface separations of 2 to 5 nm or less, depending on the nature of bacterial surfaces. In terms of process thermodynamics, microbial aggregation is driven by decreases of free energy, that is, increasing cell surface hydrophobicity results in a corresponding decrease in the Gibbs energy of the surface, which in turn promotes cell-to-cell interaction and further serves as an inducing force for cells to aggregate out of hydrophilic liquid phase. Local dehydration of the surfaces that are a short distance apart has been identified as the prerequisite for bacterial adhesion (Tay, Xu, and Teo 2000).

Concrete evidence shows that the formation of aerobic granules under different culture conditions is correlated very closely to an increase in cell surface hydrophobicity, as discussed in the preceding chapters. Li, Kuba, and Kusuda (2006) reported that the surface negative charge of bacteria decreased from 0.203 to 0.023 meq g VSS^{-1} along with aerobic granulation in an SBR, while such a decrease in the density of cell surface negative charge was accompanied by an increase in the relative cell surface hydrophobicity from 28.8% to 60.3% (figure 9.2). As discussed earlier, reduced density of cell surface charge results in weakened repulsive force of bacterium to bacterium, that is, the decreased surface negative charge promotes cell to cell aggregation, ultimately

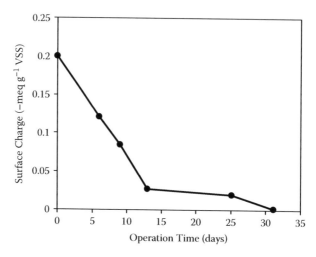

FIGURE 9.2 Change in the density of cell surface negative charge along with aerobic granulation in an SBR. (Data from Li, Z. H., Kuba, T., and Kusuda, T. 2006. *Enzyme Microb Technol* 38: 670–674.)

leading to aerobic granulation. In fact, an inverse proportional correlation between cell surface hydrophobicity and surface negative charge has been established for activated sludge microorganisms (Liao et al. 2001). This seems to imply that high cell surface hydrophobicity favors aerobic granulation.

The cell surface hydrophobicity of acetate-fed aerobic granules was found to be nearly two times higher than that of suspended seed sludge (Tay, Liu, and Liu 2002), while Yang, Tay, and Liu (2004) reported that nitrifying bacteria exposed to high free ammonia concentration could not form granules, and a low cell surface hydrophobicity of the nitrifying biomass was detected. As discussed in the preceding chapters, cell surface hydrophobicity is very sensitive to the shear force and hydraulic selection pressure present in an SBR; however, the effect of the organic loading rate in the range of 1.5 to 9.0 kg COD m^{-3} d^{-1} on the cell surface hydrophobicity was not significant. Zheng, Yu, and Sheng (2005) also found that there was a significant difference in cell surface hydrophobicity before and after the formation of aerobic granular sludge, for example, the mean contact angle values were 35.0° and 46.3° for seed sludge and granular sludge, respectively. This suggests that the formation of aerobic granular sludge is associated with an increase in the cell surface hydrophobicity, whereas the specific gravity of sludge increased with the increase of the cell surface hydrophobicity along with aerobic granulation.

Toh et al. (2003) investigated the cell surface hydrophobicity of aerobic granules of various sizes, and in their study cell surface hydrophobicity was expressed as the specific surface hydrophobicity determined by measuring phenanthrene adsorption according to the procedure proposed by Kim, Stabnikova, and Ivanov (2000). In this method, a 2-ml sample was added to 4 ml of 60% (w/v) phenanthrene solution, while in the control test, 2 ml of deionized water is used to replace the sample. All mixtures were incubated without shaking in the dark for 30 minutes. The incubated mixtures were then filtered, and the filtrates were subsequently used for the determination of dry biomass, while the supernatants were assayed for phenanthrene concentration, using a luminescence spectrometer. The specific surface hydrophobicity (H_s) in milligrams phenanthrene per gram volatile solids (VS) can be calculated as follows:

$$H_s = \frac{V(F_c - F_e)}{X}$$

in which F_c and F_e are the phenanthrene concentration in the control and the sample, respectively, V is the volume from which the concentration was measured, and X is the total dry biomass used in the hydrophobicity test. It was found that the specific cell surface hydrophobicity tended to increase from 2.46 to 5.92 mg phenanthrene g^{-1} VS as the granule size increased from <1 mm to >4 mm in diameter (Toh et al. 2003). This was also confirmed by confocal laser scanning microscopy (CLSM) examination showing that when a granule grew larger, the biomass density of aerobic granules increased in the surface layer and thus the accumulative hydrophobicity on these bacterial cell surfaces could generate a higher hydrophobicity on the exterior face of the granule (Toh et al. 2003).

Changes in cell surface hydrophobicity result from bacterial responses to certain stressful culture conditions (Bossier and Verstraete 1996; Mattarelli et al. 1999). It is most likely that the cell surface hydrophobicity induced by stressful conditions would strengthen cell-to-cell interaction, leading to a stronger microbial self-attachment, which in turn provides a protective shell for cells exposed to the unfavorable environments.

To date, aerobic granulation phenomena have been observed only in SBRs, while no successful example of aerobic granulation has been reported in continuous culture. Compared to a continuous culture, the unique feature of an SBR is its cycle operation. As a result of the cycle operation, microorganisms are subject to a periodic fasting and feasting, that is, there is a periodic starvation phase during the cycle operation of an SBR (see chapter 14). It has been shown that the starvation phase has a profound impact on the surface properties of bacteria (Kjelleberg, Humphrey, and Marshall 1983; Hantula and Bamford 1991; Bossier and Verstraete 1996). Some studies showed that starvation conditions could induce cell surface hydrophobicity that in turn facilitates microbial adhesion and aggregation (Chesa, Irvine, and Manning 1985; Bossier and Verstraete 1996). Through controlling feasting and fasting cycles by operating activated sludge systems in a plug flow or by feeding the sludge intermittently, Chesa, Irvine, and Manning (1985) found that such an operation strategy yielded better-settling sludge with high cell surface hydrophobicity. It is most likely that microorganisms can change their surface properties when faced with starvation, and such changes can contribute to their ability to aggregate. Therefore, the periodic starvation cycle would induce the cell surface hydrophobicity, and then the induced cell surface hydrophobicity can help initiate cell-to-cell self-aggregation.

It should be pointed out that the effect of starvation on cell surface hydrophobicity is still debatable, as discussed in chapter 14. The negative effect of starvation on changes in cell surface hydrophobicity has been reported, for example, upon transfer from a rich growth medium to starvation conditions, cell surface hydrophobicity dropped sharply but recovered its initial value within 24 to 48 hours (Castellanos,

Ascencio, and Bashan 2000). On the other hand, Sanin, Sanin, and Bryers (2003) reported that cell surface hydrophobicities stayed more or less constant during carbon starvation conditions, whereas there was a significant decrease in hydrophobicity when all three cultures were starved for nitrogen. Castellanos, Ascencio, and Bashan (2000) also noted that starvation was not a major factor in inducing changes in the cell surface that led to the primary phase of attachment of *Azospirillum* to surfaces.

9.4   FACTORS INFLUENCING CELL SURFACE HYDROPHOBICITY

Microbial cells favor a dispersed rather than aggregated state under normal culture conditions. Aerobic granulation is the result of cell response to stressful environments, which lead to changes in the surface characteristics of bacteria (see chapter 2). The high hydrophobicity of microorganisms is usually associated with the presence of specific cell wall proteins (Singleton, Masuoka, and Hazen 2001; Kos et al. 2003). As discussed earlier, extracellular polymeric substances produced by bacteria mainly consist of proteins and polysaccharides. Proteins are polymers of amino acids covalently bonded by peptide bonds. Amino acid has a hydrophilic carboxylic acid group (-COOH) and a hydrophobic or hydrophilic side chain. The side chains of amino acids have twenty different structures whose hydrophobicity varies markedly (Parker, Guo, and Hodges 1986). If the carboxylic acid group (-COOH) of the amino acid is connected with an amino group ($-NH_2$) of another amino acid, the connected polymer becomes a polar molecule, either monopolar (hydrophilic) or amphipathic (one hydrophilic and one hydrophobic) (Parker, Guo, and Hodges 1986). Cell wall proteins may work in two ways: (1) exposed hydrophobic proteins directly bind to extracellular matrix proteins; or (2) alternatively, cell surface hydrophobicity may mediate attachment by facilitating and maintaining specific receptor-ligand interactions (Singleton, Masuoka, and Hazen 2001). Obviously, a sound understanding of the factors that may influence cell surface hydrophobicity is important for developing the strategy for a fast aerobic granulation.

Extracellular polysaccharides have been considered to play an important role in both the formation and stability of biofilms and anaerobic and aerobic granules by mediating both cohesion and adhesion of cells (Schmidt and Ahring 1994; Tay, Liu, and Liu 2001b, 2001a; Liu and Tay 2002; Qin, Liu, and Tay 2004). Polymers can bridge physically or electrostatically to form a three-dimensional structure, which favors attachment of bacterial cells (Ross 1984). In a pilot-scale upflow anaerobic sludge blanket (UASB) reactor, Quarmby and Forster (1995) found that anaerobic granules tended to become weaker as the surface negative charge of cells increased. At the usual pH value, suspended bacteria are negatively charged and electrostatic repulsion exists between cells. It has been suggested that extracellular polymers can change the surface negative charge of bacteria, and thereby bridge two neighboring bacterial cells to each other as well as other inert particulate matters, and settle out as floccus aggregates (Shen, Kosaric, and Blaszczyk 1993; Schmidt and Ahring 1994). This seems to indicate that the formation and stability of immobilized cell communities have a strong association with extracellular polysaccharides.

The high cell surface hydrophobicity is usually associated with the presence of fibrillar structures on the cell surface and specific cell wall proteins. Fibrils may

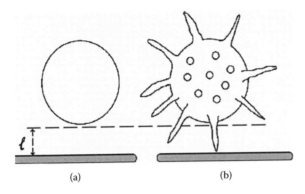

FIGURE 9.3   Illustration of the different accessibilities of spheres with smooth (A) and rough (B) surfaces to a flat plate surface. (From van Oss, C. J. 2003. *J Mol Recognit* 16: 177–190. With permission.)

attach to the surface of receptors by piercing through the energy barrier between cells into a strong Lewis acid-base (AB) force interaction distance (van Oss 2003). Figure 9.3 schematically presents differences in accessibility of round spherical bodies and a flat plate. For cell-to-cell interaction, the flat plate shown in figure 9.3 can be displaced by another cell. The smooth hydrophilic spherical cell cannot make contact with a smooth flat hydrophilic surface because their mutual specific, macroscopic-scale repulsion prevents a closer approach. However, a similar spherical cell with long thin spiky fibrils can easily penetrate the microscopic-scale repulsion field, leading to a macroscopic-scale specific contact. In figure 9.3, the dotted line indicates the limit of closest approach for a smooth hydrophilic cell with a relatively large radius of curvature.

Starvation may induce changes in cell surface hydrophobicity. Kjelleberg and Hermansson (1984) observed a large increase in surface roughness throughout the starvation period for all studied strains that showed marked changes in physicochemical characteristics. As discussed earlier, fibrillar surface structure can help to overcome the intercellular energy barrier. It is likely that increased cell surface roughness might have the same function as cell surface hydrophobicity in microbial aggregation.

Culture temperature may also influence cell surface hydrophobicity. Thermodynamically, water will become a much stronger Lewis acid at higher temperature (van Oss 1993, 1994b), for example, at 20°C, $\gamma_w^+ = \gamma_w^- = 25.5$ mJ m^{-2}, whereas at 38°C, $\gamma_w^+ = 32.4$ and $\gamma_w^- = 18.5$ mJ m^{-2}. Usually, γ_m^+ values of most biological surfaces are extremely low, whereas the γ_m^- value can be high or low, depending one whether it is hydrophilic or hydrophobic. Thus, a cell surface may be designated as essentially electron donor monopolar (van Oss, Chaudhury, and Good 1987; van Oss 1994b). In this case, for electron donor monopolar compounds, the increase of γ_w^+ with temperature will markedly increase the value of the term

$$\sqrt{\gamma_m^- \gamma_w^+}$$

This is in line with the finding by Blanco et al. (1997) that the majority of the forty-two strains of *Candida albicans* studied were hydrophobic at 22°C, but hydrophilic at 37°C, and the hydrophobic cells showed a consistent adherence capacity that was absent from the hydrophilic strains.

The growth rate of microorganisms is another factor that can influence cell surface hydrophobicity. In a study of the impact of brewing yeast cell age on fermentation performance, Powell, Quain, and Smart (2003) reported that the flocculation potential of cells and cell surface hydrophobicity increased in conjunction with cell age, whereas in a selection of xenobiotic-degrading microorganisms, a similar trend was found by Asconcabrea and Lebeault (1995), that is, the cell surface hydrophobicity tended to increase with the growth rate in terms of dilution rate. van Loosdrecht et al. (1987) also found that for a certain species, at high growth rates, bacterial cells would become more hydrophobic. Meanwhile, Malmqvist (1983) observed an increase in cell surface hydrophobicity during exponential growth. Expression of cell surface hydrophobicity is influenced by growth conditions, and is often expressed after growth under nutrient-poor conditions, or starvation (Ljungh and Wadstrom 1995).

It has been shown that the change in cell surface hydrophobicity can result from bacterial stress responses to certain culture conditions, such as low pH, high temperature, and hyperosmotic stress (Danniels, Hanson, and Phillips 1994; Bossier and Verstraete 1996; Correa, Rivas, and Barneix 1999; Mattarelli et al. 1999). Blanco et al. (1997) reported that the majority of the forty-two strains of *Candida albicans* studied were hydrophobic at 22°C, but hydrophilic at 37°C. As presented in chapter 1, cell surface hydrophobicities of aerobic granules grown on glucose and acetate showed no significant difference, whereas cell surface hydrophobicity was found to increase with increase in hydrodynamic shear force (c2). It appears from chapter 1 that aerobic granules cultivated at different organic loading rates of 1.5 to 9.0 kg COD m^{-3} d^{-1} exhibited comparable cell surface hydrophobicity of 78% to 86%; however, cell surface hydrophobicity was significantly improved as the cycle time was shortened (chapter 6). In the preceding chapters, it can be seen that all selection pressures may improve cell hydrophobicity, including settling time, discharge time, and exchange ratio of the SBR. This means that the cell surface hydrophobicity induced by culture conditions strengthen cell-to-cell interaction, leading to a stronger microbial structure.

Sun, Yang, and Li (2007) investigated the effect of carbon source on aerobic granulation. It appears from figure 9.4 that during the period of the reactor start-up, the type of carbon source influences the surface property of sludge in terms of zeta potential, which is often used to quantify the density of surface charges, for example, sludge grown on peptone shows the lowest surface charge density among all four organic carbon sources studied. As a result, the peptone-fed aerobic granules had the highest biomass density over granules grown on acetate, glucose, and fecula (Sun, Yang, and Li 2007). As discussed earlier, the density of the surface charge is inversely related to the cell surface hydrophobicity, that is, a lower charge density means a higher cell surface hydrophobicity. In addition, figure 9.4 also implies that the property of feed may influence cell surface hydrophobicity during aerobic granulation.

When exposed to toxic or inhibitory substrates, microorganisms are able to regulate their surface properties, especially cell surface hydrophobicity. In a study on

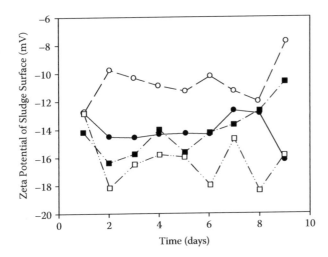

FIGURE 9.4   Change in zeta potential of sludge in the course of aerobic granulation in SBRs fed with acetate (●), glucose (○), peptone (■), and fecula (□). (Data from Sun, F. Y., Yang, C. Y., and Li, J. Y. 2007. Beijing Jiaotong Daxue Xuebao [J Beijing Jiaotong University] 31: 106–110.)

substrate-dependent autoaggregation of *Pseudomonas putida* CP1 during the degradation of monochlorophenols and phenol, Farrell and Quilty (2002) found that cells grown on the higher concentrations of monochlorophenol were more hydrophobic than those grown on phenol and lower concentrations of monochlorophenol, whereas when *Pseudomonas putida* was exposed to toxic alcohols, the bacterium changed degrees of cell surface hydrophobicity and adapted to the alcohols (Tsubata, Tezuka, and Kurane 1997). Jiang, Tay, and Tay (2004) studied the toxic effect of phenol on aerobic granules, and found that the surface hydrophobicity of cultivated sludge significantly decreased from 60% to 40% as phenol loading was increased from 1 to 2.5 kg phenol $m^{-3}\,d^{-1}$. This may imply that increased hydrophobicity and the resultant autoaggregation of bacteria is a microbial response to the toxicity of substrates.

Bulk liquid surface tension may also alter cell surface hydrophobicity. Thaveesri et al. (1994; 1995) reported that anaerobic granules grown in protein-rich media exhibited lower hydrophobicity than carbohydrate, which in turn slowed down the anaerobic granulation. On the contrary, van Loosdrecht et al. (1987) found that the effect of the growth substrate on cell surface hydrophobicity was not significant. Different observations could result from the fact that all substrates used by van Loosdrecht (1987) were carbohydrates and no protein-rich media were included. Thaveesri et al. (1995) proposed a model to interpret the influence of growth substrate on the surface hydrophobicity of anaerobic granules, and they thought that if γ_{lv} is high, microorganisms with low surface energy (low γ_{mv} or hydrophobic bacteria) can aggregate in order to obtain minimal free energy. Nevertheless, when γ_{lv} is low, bacteria with high surface energy (high γ_{mv} or hydrophilic bacteria) can form aggregates easily. Since there is no high-concentration chemical solvent present in municipal wastewater, the first case (high γ_{lv}) is common in municipal wastewater treatment practice, whereas the second case becomes true in treating chemical solvent-containing industrial wastewater. In both cases, microbial aggregation seems to be related to the relative hydrophobicity of microorganisms to the aqueous phase in which they grow. Protein-rich substrate can lower the surface tension of reactor liquid, and consequently lead to the formation of anaerobic granules with low hydrophobicity because low hydrophobicity of anaerobic granules could decrease the free energy at the interface between aggregates and liquid.

Cell surface hydrophobicity is also related to gene expression of microorganisms. The analytical study of biofilm has demonstrated that adhesion launches the expression of a set of genes that ends with the typical biofilm phenotype, particularly with an enhanced resistance to antimicrobial agents (Stickler 1999; Davey and O'Toole 2000; Watnick and Kolter 2000; Goldberg 2002). Likewise, the gene expression for enzyme is also different under free-living and immobilized conditions (Hata et al. 1998; Vallim et al. 1998; Akao et al. 2002; Asther et al. 2002). For example, Vallim et al. (1998) found that *Phanerochaete chrysosporium* showed a differential gene expression for cellobiohydrolase when it was cultured under immobilized conditions. Therefore, it is possible that the new gene expression gives rise to changes in the cell surface properties that in turn helps the cells adapt to the new living environment.

In conclusion, changes in cell surface hydrophobicity can result from bacterial stressful responses to certain culture conditions. The cell surface hydrophobicity induced by culture conditions strengthens cell-to-cell interaction, leading to a stronger microbial structure that provides a protective shell for cells exposed to the unfavorable environments. Therefore, it is necessary to identify the key engineered parameter that can induce cell surface hydrophobicity during biogranulation.

9.5 SELECTION PRESSURE-INDUCED CELL SURFACE HYDROPHOBICITY

Aerobic granulation is a gradual process evolving from dispersed seed sludge to mature and stable aerobic granules with spherical outer shape. It is thought that cell surface hydrophobicity may trigger and initiate aerobic granulation. As the seed sludge used to inoculate bioreactors often has a very low cell surface hydrophobicity, high cell surface hydrophobicity reported in aerobic granulation would not be an extant property of the seed sludge, and it would be induced during aerobic granulation. Many factors as discussed earlier may induce cell surface hydrophobicity, but most of them cannot be manipulated in terms of process operation. Thus, it is necessary to identify the key engineered parameters that can induce cell hydrophobicity during aerobic granulation.

Hydraulic selection pressure has been proved to be a decisive parameter in the formation of biogranules. Absence of anaerobic granulation in UASB reactors was observed at very weak hydraulic selection pressure in terms of liquid upflow velocity, while it has been demonstrated that aerobic granulation is a process driven by selection pressure in SBRs. Under strong selection pressures, only the good-settling and heavy sludge particles are retained in the reactor, while the light and poor-settling sludge is washed out. Compared to seed sludge with a cell surface hydrophobicity of 20%, when microorganisms are subject to high selection pressure in terms of settling time, the cell surface hydrophobicity gradually increases to 70% during aerobic granulation. The hydraulic selection pressure seems to induce changes in cell surface hydrophobicity, and a strong selection pressure results in a more hydrophobic cell surface (Qin, Liu, and Tay 2004). Similarly, it was found that anaerobic granular sludge in UASB reactors was more hydrophobic than the nongranular sludge that washed out (Mahoney et al. 1987). Under the strong selection pressure, microorganisms have to adapt their surface properties in order to avoid being washed out from the reactors through microbial self-aggregation. In this sense, biogranulation would be an effective defensive strategy of microbial communities against external selection pressure, and thus one may expect to manipulate the formation and characteristics of biogranules by controlling selection pressure (chapter 6).

9.6 THERMODYNAMIC INTERPRETATION OF CELL SURFACE HYDROPHOBICITY

Similar to a chemical process, microbial aggregation is a function of changes in free energy at interfaces. Bacterial attachment to a solid surface can be described by the following:

$$\text{Microorganism} + \text{surface} \rightarrow \text{Microorganism} - \text{surface} \quad \Delta G_{adh}^{o'} \quad (9.1)$$

in which microorganism–surface means microorganism attached onto a solid surface, and $\Delta G_{adh}^{o'}$ is the effective free energy change of microorganism-to-solid surface attachment, which changes with the proceeding of microbial attachment on the surface. For bacterial attachment onto a solid surface, $\Delta G_{adh}^{o'}$ can be expressed as:

$$\Delta G_{adh}^{o'} = \gamma_{sm} - \gamma_{sl} - \gamma_{ml} \quad (9.2)$$

in which γ_{sm}, γ_{sl}, and γ_{ml} are the solid-bacteria, solid-liquid, and bacteria-liquid interfacial free energies, respectively. The surface free energies are related to contact angles (θ) according to the Young equation:

$$\gamma_{lv} \cos \theta = \gamma_{sv} - \gamma_{sl} \quad (9.3)$$

in which γ_{lv} and γ_{sv} are the liquid-vapor and solid-vapor interfacial free energies, respectively. It should be realized that equation 9.3 cannot be solved for the interfacial free energies (γ_{sv} and γ_{sl}) by measuring the contact angle and liquid surface tension (γ_{lv}) without additional assumptions (Bos, van der Mei, and Busscher 1999). In a study of anaerobic granulation, Thaveesri et al. (1995) used the equation developed by Neumann et al. (1974) as a supplementary equation to equation 9.3:

$$\gamma_{sl} = \frac{(\sqrt{r_{sv}} - \sqrt{\gamma_{lv}})^2}{1 - 0.015\sqrt{\gamma_{sv}\gamma_{lv}}} \quad (9.4)$$

Compared to microbial attachment to a solid surface, microbial aggregation is a microorganism-to-microorganism self-immobilization process, which can be depicted as follows:

$$\text{Microorganism} + \text{Microorganism} \rightarrow \text{Microorganism} - \text{Microorganism} \quad \Delta G^{o'}_{agg} \quad (9.5)$$

in which microorganism–microorganism represents microbial aggregation, and $\Delta G^{o'}_{agg}$ is the effective free energy change of microorganism-to-microorganism aggregation, which changes with the proceeding of microbial aggregation. It has been proposed that equation 9.2 can also be applied to the microbial aggregation process by replacing the solid by an identical microorganism (Bos, van der Mei, and Busscher 1999). Thus, $\Delta G^{o'}_{agg}$ of microbial aggregation can be written as follows:

$$\Delta G^{o'}_{agg} = \gamma_{mm} - 2\gamma_{ml} \quad (9.6)$$

in which γ_{mm} is the microorganism-microorganism interfacial free energy. In a study of anaerobic granulation, Thaveesri et al. (1995) postulated that when adhesion of two identical bacteria is considered, γ_{mm} is equal to zero. Hence, equation 9.6 reduces to:

$$\Delta G'_{agg} = -2\gamma_{ml} \quad (9.7)$$

Substituting equation 9.4 into equation 9.7 by replacing the subscript s by m yields:

$$\Delta G'_{agg} = -2\frac{(\sqrt{\gamma_{mv}} - \sqrt{\gamma_{lv}})^2}{1 - 0.015\sqrt{\gamma_{mv}\gamma_{lv}}} \quad (9.8)$$

According to equation 9.8, Thaveesri et al. (1995) proposed that if γ_{lv} is high, low-energy surface types of microorganisms (low γ_{mv} or hydrophobic bacteria) can aggregate in order to obtain minimal free energy, while when γ_{lv} is low, high-energy surface types of bacteria (high γ_{mv} or hydrophilic bacteria) can aggregate better. As there is no high-concentration chemical solvent present in municipal wastewater, the first case (high γ_{lv}) is most commonly encountered in municipal wastewater treatment practice, whereas the second case applies to treating chemical solvent-containing industrial wastewater. No matter what the case is, microbial aggregation seems to be related to relative hydrophobicity of microorganisms to the aqueous phase in which they survive. It should be realized that equation 9.8 is not correlated to a characteristic parameter of microbial aggregation; it only offers a qualitative interpretation of surface thermodynamics of microbial aggregation. Moreover, in the real microbial aggregation process, there are no completely identical bacteria, and the simplicity assumed by Thaveesri et al. (1995) such that γ_{mm} is zero is still debatable. Cells favor a dispersed rather than aggregated state under normal culture conditions, hence γ_{mm} should exist during cell-to-cell interaction.

As Bos, van der Mei, and Busscher (1999) noted, it is impossible to determine experimentally the interfacial free energies in equation 9.2. Evidence shows that the γ_{mm} and γ_{ml} terms in equation 9.4 are related to both cell surface charge and hydrophobicity (Zita and Hermansson 1997; Kos et al. 2003), while cell surface hydrophobicity is inversely correlated to the quantity of surface charge of microorganisms (Liao et al. 2001). There is strong evidence that the microbial autoaggregation process is correlated to cell surface hydrophobicity (Del Re et al. 2000; Kos et al. 2003; Liu et al. 2003). According to extended DLVO theory by Van Oss, Good and Chaudhury (1986), cell surface hydrophobicity represents an attractive force, while cell surface hydrophilicity reflects repulsion between cells. This indicates that both γ_{mm} and γ_{ml} in equation 9.4 are functions of the interaction between cell surface hydrophobicity and cell surface hydrophilicity, that is, with the increase of cell surface hydrophobicity or the decrease of cell surface hydrophilicity, repulsive forces between cells becomes weaker and weaker. According to Liu et al.

(2004a), $\Delta G_{agg}^{o'}$ can be expressed as:

$$\Delta G_{agg}^{o'} = \Delta G_{agg}^{o} - aRT \ln H_{o/w}$$ (9.9)

in which ΔG_{agg}^{o} is change of the standard free energy of the microbial aggregation process, a is a positive coefficient, and $H_{o/w}$ is relative cell hydrophobicity, which is defined as follows:

$$H_{o/w} = \frac{\text{Cell hydrophobicity}}{\text{Cell hydrophilicty}}$$ (9.10)

It has been recognized that the formation of biofilms and microbial aggregates is a multiple-step process, to which physicochemical and biological forces make significant contributions (Liu and Tay 2002). Some parameters other than cell surface hydrophobicity, such as cell surface charge and extracellular polymers, may also influence or contribute to microbial attachment and aggregation. In general, the surface of microorganisms is negatively charged under the usual pH conditions, and bacterial surface charge might affect bacterial attachment (Rouxhet and Mozes 1990). There is evidence that microbial attachment on a solid surface can be improved or enhanced by reducing electrical repulsion between cell and solid surfaces (Changui et al. 1987; Rouxhet and Mozes 1990; Masui, Takata, and Kominami 2002), while Strand, Varum, and Ostgaard (2003) reported that charge neutralization was not the main microbial flocculation mechanism. As pointed out earlier, microbial aggregation is cell-to-cell interaction that is different from bacterial adhesion on a solid surface. When bacteria that carry charges with the same sign approach each other, there should be an electrical repulsive force between bacterial surfaces. Zita and Hermansson (1997) studied the correlations of cell surface hydrophobicity and charge to adhesion of *Escherichia coli* strains to activated sludge flocs, and found that there was a strong correlation between the cell surface hydrophobicity of the *E. coli* strains and adhesion to the sludge flocs, while for positive cell surface charges the correlation was weaker than for surface hydrophobicity, and negative cell surface charges showed no correlation to adhesion. In order to enhance initial interaction between microorganisms, Jiang et al. (2003) added calcium ions to neutralize the negatively charged bacterial surface, but the results were not as satisfactory as expected.

Therefore, it seems that electrical interaction is not a triggering force of aggregation of microorganisms. In addition, extracellular polymers play a role in microbial attachment and aggregation. Tsuneda et al. (2003) reported that if the amount of extracellular polymers is relatively small, cell adhesion onto solid surfaces is inhibited by electrostatic interaction, and if it is relatively large, cell adhesion is enhanced by polymeric interaction. Jorand et al. (1998) studied hydrophobic and hydrophilic properties of activated sludge extracellular polymers, and found that a significant proportion of extracellular polymers fraction was hydrophobic. Such results support the hypothesis that hydrophobic extracellular polymers are involved in the organization of microbial flocs, biofilm, and aggregates, that is, hydrophobic interaction may be fundamental in microbial aggregation, as discussed earlier.

Similar to any chemical process, the overall change of free energy of the microbial aggregation process (ΔG_{agg})inc reases with the increase of process resistance, and decreases with the increase of the driving force of microbial aggregation process. Liu et al. (2004a) put forward a hypothesis that the overall change of free energy of the microbial aggregation process should be formulated as a function of the driving force and resistance of the process, such that:

$$\Delta G_{agg} = \Delta G_{agg}^{o'} + bRT \ln \frac{\text{Resistance}}{\text{Driving force}} \qquad (9.11)$$

in which b is a positive coefficient. Equation 9.11 is similar to those developed for biological systems (Roels 1983). It has been demonstrated that aerobic granulation is a gradual process from dispersed sludge to stable granules, and both granule size and density finally approach respective equilibrium value (Tay, Liu, and Liu 2001c). Thus, the driving force of microbial aggregation is the potential of the process towards aggregation that can be described by the difference between the current state of microbial aggregates and the balanced state that microbial aggregates can thermodynamically achieve. It is obvious that increasing the cell surface hydrophobicity can simultaneously cause a decrease in the excess energy of the microbial surface, which in turn enhances cell-to-cell interaction, leading to a more compact structure of microbial aggregates (Liu et al. 2003). In the environmental engineering field, the density of microbial aggregates is often used to describe how compact and strong the microbial interaction is. The observed density of microbial aggregates is the result of balanced cell-to-cell interaction, which is a characteristic parameter representing the state of microbial aggregates at time t. Figure 9.5 shows an example of the evolution of the density (ρ) of microbial aggregates against time observed in the SBR run at a substrate N/COD ratio of 5/100. It appears that ρ is gradually close to its equilibrium.

Liu et al. (2004a) further proposed that the driving force of microbial aggregation can be defined as the difference between the density at time t (ρ) and the density of microbial aggregates at equilibrium state (ρ_{eq}). With the increase of ρ, the aggregation process tends to reach its equilibrium. As a result, the driving force ($\rho_{eq} - \rho$) decreases and the resistance would increase accordingly. This shows that ρ indeed would reflect the magnitude of aggregation resistance. Hence, equation 9.11 can be translated to

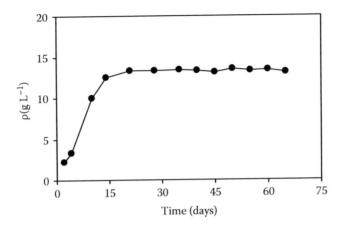

FIGURE 9.5 The density (ρ) of aggregates against operation time observed in the SBR operated at substrate N/COD ratio of 5/100. (From Liu, Y. et al. 2004a. *Appl Microbiol Biotechnol* 64: 410–415. With permission.)

$$\Delta G_{agg} = \Delta G^{o'}_{agg} + bRT \ln \frac{\rho}{\rho_{eq} - \rho} \tag{9.12}$$

When $\rho = 0.5\rho_{eq}$, $\Delta G^{o'}_{agg}$ is equal to ΔG_{agg}. This implies that $\Delta G^{o'}_{agg}$ can be defined as the overall free energy change at $\rho = 0.5\rho_{eq}$, that is, the driving force of microbial aggregation is equal to the resistance force. As ρ approaches ρ_{eq}, ΔG_{agg} goes to infinity and further aggregation becomes energetically impossible. In fact, this is in agreement with previous research (Bos, van der Mei, and Busscher 1999). Substitution of equation 9.9 into equation 9.12 yields:

$$\Delta G_{agg} = \Delta G^{o}_{agg} - aRT \ln H_{olw} + bRT \ln \frac{\rho}{\rho_{eq} - \rho} \tag{9.13}$$

Solving equation 9.13 for ρ gives:

$$\rho = \rho_{eq} \frac{(H_{olw})^{a/b}}{\left(e^{(\frac{\Delta G^{o}_{agg} - \Delta G_{agg}}{RT})} \right)^{1/b} + (H_{olw})^{a/b}} \tag{9.14}$$

Equation 9.14 can be rearranged as:

$$\rho = \rho_{eq} \frac{(H_{olw})^{n}}{K_{agg} + (H_{olw})^{n}} \tag{9.15}$$

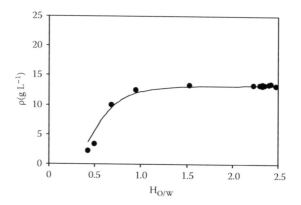

FIGURE 9.6 ρ versus $H_{o/w}$ observed in aerobic granulation at substrate N/COD ratio of 5/100. The prediction given by equation 9.15 is shown by a solid curve with a correlation coefficient of 0.97. $\rho_{eq} = 13.3$ g L^{-1}; $K_{agg} = 0.086$; and $n = 4.0$, (From Liu, Y. et al. 2004a. *Appl Microbiol Biotechnol* 64: 410–415. With permission.)

in which

$$K_{agg} = \left[e^{(\frac{\Delta G^o_{agg} - \Delta G_{agg}}{RT})} \right]^m \qquad (9.16)$$

and n and m equal a/b and $1/b$, respectively. A simple least-square method can be used to evaluate the constants involved in equation 9.15. If ρ_{eq} is known, K_{agg} and n can also be easily estimated from a linear plot of $\ln[\rho/(\rho_{eq} - \rho)]$ versus $\ln H_{o/w}$.

$$\ln \frac{\rho}{\rho_{eq} - \rho} = n \ln H_{o/w} - \ln K_{agg} \qquad (9.17)$$

Values of Ka_{gg} and n can be determined from the slope and intercept of equation 9.17.

Liu et al. (2004a) determined cell surface hydrophobicity in the course of the formation of aerobic granules in SBRs operated at different substrate N/COD ratios of 5/100 to 30/100 by weight. Thus, $H_{o/w}$ can be calculated according to equation 9.10:

$$H_{o/w} = \frac{\text{Cell hydrophobicity (\%)}}{100\% - \text{cell hydrophobicity (\%)}} \qquad (9.18)$$

The value of $H_{o/w}$ is in between infinite (absolutely hydrophobic) and 0 (absolutely hydrophilic). The density (ρ) of microbial aggregates versus $H_{o/w}$ is shown in figures 9.6 to 9.9 for different substrate N/COD ratios. It can be seen that the prediction given by equation 9.15 is in good agreement with the experimental data obtained, indicated by a correlation coefficient greater than 0.92. For microbial aggregation at substrate N/COD ratios of 5/100 to 30/100, the respective value of n

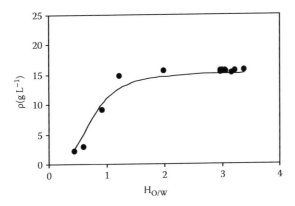

FIGURE 9.7 ρ versus $H_{o/w}$ observed in aerobic granulation at substrate N/COD ratio of 10/100. The prediction given by equation 9.15 is shown by a solid curve with a correlation coefficient of 0.97. $\rho_{eq} = 15.3$ g L^{-1}; $K_{agg} = 0.39$; and $n = 3.2$. (From Liu, Y. et al. 2004a. *Appl Microbiol Biotechnol* 64: 410–415. With permission.)

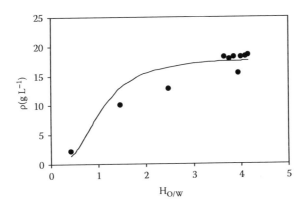

FIGURE 9.8 ρ versus $H_{o/w}$ observed in aerobic granulation at substrate N/COD ratio of 20/100. The prediction given by equation 9.15 is shown by a solid curve with a correlation coefficient of 0.93. $\rho_{eq} = 17.9$ g L^{-1}; $K_{agg} = 1.11$; and $n = 2.8$. (From Liu, Y. et al. 2004a. *Appl Microbiol Biotechnol* 64: 410–415. With permission.)

estimated is 4.0, 3.2, 2.8, and 1.7. This implies that cell surface hydrophobicity has a more pronounced effect on aerobic granulation at the low substrate N/COD ratio, which is further confirmed by the values of K_{agg}, which increases with the increase of the substrate N/COD ratio. On the other hand, the density of aerobic granules at equilibrium was found to increase with the increase of the substrate N/COD ratio. This implies that a more compact microbial structure of aerobic granules can be obtained at high substrate N/COD ratio. Figure 9.10 shows the relationship between the relative activity of the nitrifying population over the heterotrophic population and the specific growth rate (μ_{agg}) of aerobic granules by size, and figure 9.11 further exhibits the effect of μ_{agg} on n and ρ_{eq}.

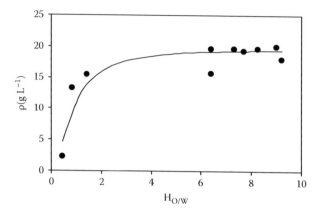

FIGURE 9.9 ρ versus $H_{o/w}$ observed in aerobic granulation at substrate N/COD ratio of 30/100. The prediction given by equation 9.15 is shown by a solid curve with a correlation coefficient of 0.92. $\rho_{eq} = 19.8$ g L^{-1}; $K_{agg} = 0.77$; and $n = 1.7$. (From Liu, Y. et al. 2004a. *Appl Microbiol Biotechnol* 64: 410–415. With permission.)

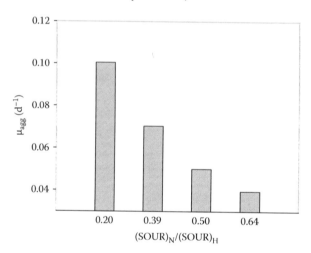

FIGURE 9.10 Effect of $(SOUR)_N/(SOUR)_H$ on μ_{agg} of aerobic granules. (Data from Liu, Y. et al. 2004a. *Appl Microbiol Biotechnol* 64: 410–415. With permission.)

In terms of chemistry, ΔG_{agg} is supposed to correlate to the mass quotient of the microbial aggregation process (Q_{agg}) through the following equation:

$$\Delta G_{agg} = \Delta G^o_{agg} + RT \ln Q_{agg} \tag{9.19}$$

Substitution of equation 9.15 into equation 9.19 leads to:

$$K_{agg} = \left(\frac{1}{Q_{agg}}\right)^m \tag{9.20}$$

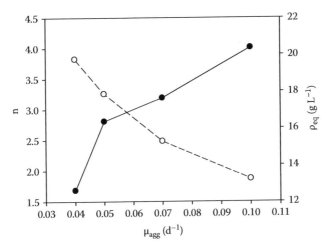

FIGURE 9.11   Effect of μ_{agg} on n (●) and ρ_{eq} (○). (Data from Liu, Y. et al. 2004a. *Appl Microbiol Biotechnol* 64: 410–415.)

Equation 9.20 reveals that K_{agg} is inversely related to the process mass quotient. When K_{agg} is small or Q_{agg} is large, microorganisms proceed toward aggregation; on the other hand, when K_{agg} is large or Q_{agg} is small, microbial interaction tends away from the aggregation. This provides a plausible explanation for why ρ approaches its maximum much more slowly when K_{agg} is large, whereas when K_{agg} is small, ρ increases steeply and breaks to the right sharply, as shown in figures 9.6 to 9.9. Obviously, equation 9.20 offers a theoretical interpretation for the physical meaning of K_{agg}. If ΔG_{agg} is close to zero, Q_{agg} in equations 9.19 and 9.20 can be replaced by the equilibrium constant (K_{agg}^{eq}) of the microbial aggregation process. Consequently, the magnitude of the K_{agg} value may represent the equilibrium position of a microbial aggregation process.

Equation 9.15 can fit the experimental data very well. The values of n estimated by equation 9.15 are closely related to the substrate N/COD ratios, that is, a low substrate N/COD ratio results in a high n value. In terms of reaction kinetics, the magnitude of n describes how fast microorganism-to-microorganism hydrophobic interaction is under given culture conditions. In fact, Chen and Strevett (2003) reported that various substrate N/COD ratios influence microbial surface thermodynamics reflected in cell surface hydrophobicity, while evidence also indicates that cell surface hydrophobicity and hydrophobic interaction are closely related to microbial activity (Liao et al. 2001; Tay, Liu, and Liu2001c; Liu et al. 2003). Liu et al. (2004a) determined the overall specific growth rate (μ_{agg}) of microbial aggregates by size and the activity distribution of heterotrophic and nitrifying populations in stable aerobic granules in terms of $(SOUR)_N/(SOUR)_H$. It was found that μ_{agg} tended to decease with the increase of substrate N/COD ratio. Such a changing trend of μ_{agg} with the substrate N/COD ratio indeed is within expectation because a high substrate N/COD ratio results in a sustainable enrichment of slow-growing nitrifying population with low growth rate in aerobic granules (figure 9.10). Figure 9.11 further shows the relationships among μ_{agg}, n, and ρ_{eq}. It appears that n is positively correlated to μ_{agg}, that is, the microbial aggregation process is faster at high μ_{agg} than that at low μ_{agg}. Meanwhile, the lowered growth rate of aerobic granules results in a higher ρ_{eq}, that is, a more compact and stronger microbial structure. In fact, high growth rate encourages the outgrowth of aerobic granules, leading to a loose structure accompanied with a low density. In a study of biofilms, the strength of biofilms was found to be negatively related to the growth rate of microorganisms (Tijhuis, van Loosdrecht, and Heijnen 1995), while Kwok et al. (1998) reported that the biofilm density decreased as the growth rate increased. Similarly, in the anaerobic granulation process, a high biomass growth rate also led to a reduced strength of anaerobic granules (Quarmby and Forster 1995).

It seems that microorganism-to-microorganism interaction with a $H_d/_w$ value greater than 1 favors cell self-aggregation towards a tight and compact microbial structure. Hydrophobic cells may attach not only on the surface, but also can aggregate to form microbial granules, whereas hydrophilic cells did not (Olofsson, Zita, and Hermanson 1998; Sharma and Rao 2003). Ryoo and Choi (1999) attempted to develop the fungal pellets from the surface thermodynamic balance between fungal cell and liquid media. Aerobic granulation is a dynamic process evolving from dispersed sludge to mature and stable aerobic granules. Equation 9.15 sheds light on a sound thermodynamic understanding of cell surface hydrophobicity in aerobic granulation, and it would be applicable for biofilm and anaerobic granulation.

9.7 ENHANCED AEROBIC GRANULATION BY HIGHLY HYDROPHOBIC MICROBIAL SEED

It is apparent that cell surface hydrophobicity can serve as an important inducing force for aerobic granulation. This in turn implies that enrichment cultures of microorganisms with high cell surface hydrophobicity or self-aggregation ability can be used to accelerate or enhance aerobic granulation. For this purpose, a four-step procedure to select highly hydrophobic bacteria has been developed (Wang 2004; Ivanov et al. 2005).

Step 1: precultivated aerobic granules are disintegrated by a beater for 5 minutes and are further filtered through a 35-μm cell strainer cap in order to remove particles bigger than 35 μm.

Step 2: bacteria collected at the bottom of the test tube after 5 minutes of settling time are transferred into a liquid medium.

Step 3: the transferred bacteria are cultivated in the liquid medium for 48 hours, and then microbial aggregates formed are removed after 5 minutes of sedimentation time.

Step 4: those bacteria at the air–liquid interface are finally harvested as seed for enhanced aerobic granulation.

The selected bacteria had an average aggregation index of 60% to 80% compared to the activated sludge with a typical aggregation index of 5% to 20% (Ivanov et al. 2005).

Using the selected highly hydrophobic bacteria as a seed, Ivanov et al. (2005) showed that aerobic granulation was shortened from 8 days in the culture of

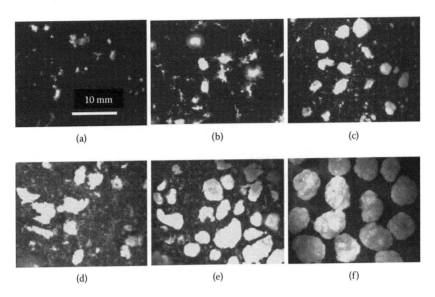

FIGURE 9.12   Aerobic granulation after 5 days (a, d), 10 days (b, e), and 20 days (c, f) observed in two SBRs seeded with conventional activated sludge flocs (a to c) and selected highly hydrophobic bacteria (d to f), respectively. (From Ivanov, V. et al. 2005. *Water Sci Technol* 52: 13–19. With permission.)

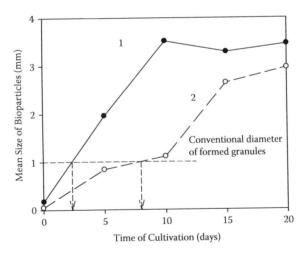

FIGURE 9.13   Evolution of bioparticle size during aerobic granulation with selected highly hydrophobic bacteria (1) and conventional activated sludge flocs (2). (Data from Ivanov, V. et al. 2005. *Water Sci Technol* 52: 13–19.)

conventional activated sludge to 2 days in the enrichment culture (figure 9.12 and figure 9.13). These offer a new operation strategy by which the time required for aerobic granulation can be shortened markedly.

According to the interfacial chemistry, microorganisms situated at the water–air interface are highly hydrophobic. Following the same procedure as presented earlier,

FIGURE 9.14 Selection of highly hydrophobic cells at the water–air interface. (From Wang, Z.-W. 2004. Unpublished report. Nanyang Technological University, Singapore.)

TABLE 9.2

Characteristics of Conventional Activated Sludge and Selected Highly Hydrophobic Microbial Seed

	Activated Sludge	Selected Highly Hydrophobic Seed
Total soluble solids (g L^{-1})	2.43	0.33
Sludge volume index (mL g^{-1})	226	76.9
Particle size (μm)	42	102.9
Hydrophobicity (%)	32.6	80.1

Source: Wang, Z.-W. 2004. Unpublished report. Nanyang Technologiccal University, Singapore.

Wang (2004) artificially selected highly hydrophobic cells at the water–air interface using an inoculation loop as shown in figure 9.14. Two SBRs, namely R1 and R2, were then inoculated with conventional activated sludge with low cell surface hydrophobicity and selected highly hydrophobic cells as seeds, respectively.

Table 9.2 summarizes the main characteristics of conventional activated sludge and selected highly hydrophobic seed inoculated to R1 and R2. Obviously, the selected seed was much more hydrophobic than the conventional activated sludge. Figure 9.15 shows changes in the sludge volume index (SVI) in the course of R1 and R2 operations. The SVI in R2 remained at a very low level of 60 mL g^{-1}, while a sharp increase in the SVI, followed by a significant decrease, was observed in R1. It is clear that after aerobic granulation from day 10 onwards, the SVI observed in R1 decreased significantly, indicating an improved sludge settleability.

The evolution of sludge morphology in the course of operation of R1 and R2 was tracked by image analysis technique. As can be seen in figure 9.16, smooth, round aerobic granules appeared in R2 inoculated by the selected highly hydrophobic microbial seed after 10 days of operation, while the same phenomenon was observed

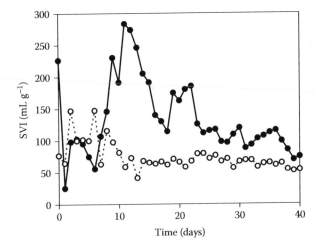

FIGURE 9.15 Changes in sludge volume index (SVI) in the course of operation observed in R1 seeded by conventional activated sludge (●) and R2 (○) inoculated by selected highly hydrophobic microbial seed. (Data from Wang, Z.-W. 2004. Unpublished report. Nanyang Technological University, Singapore.)

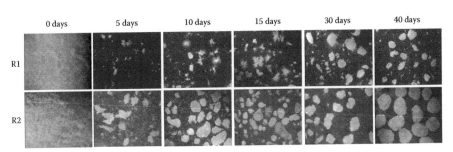

FIGURE 9.16 Evolution of sludge morphology in the course of aerobic granulation in R1 and R2 seeded with conventional activated sludge and highly hydrophobic cells, respectively. (From Wang, Z.-W. 2004. Unpublished report. Nanyang Technological University, Singapore.)

in R1 seeded by conventional activated sludge 20 days later. These observations seem to indicate that aerobic granulation is shortened by 20 days if the selected highly hydrophobic microbial seed is used. Meanwhile, figure 9.17 clearly shows that aerobic granules developed from the selected highly hydrophobic microbial seed look more compact and denser in structure than those cultivated from conventional activated sludge. Moreover, the mechanical shearing tests also reveal that aerobic granules developed from the selected highly hydrophobic microbial seed are much more resistant to external mechanical stirring strength than those cultivated from conventional activated sludge (figure 9.18).

Changes in biomass concentration in R1 and R2 were followed throughout the granulation process (figure 9.19). Although the initial biomass concentration seeded to R1 was much higher than that inoculated to R2, after only 1 day of operation the biomass concentration in R1 and R2 became comparable. This was due mainly to

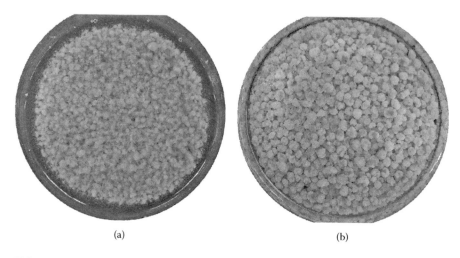

FIGURE 9.17 Morphologies of aerobic granules harvested 40 days after cultivation from the inoculums of conventional activated sludge (A) and high hydrophobic cells (B), respectively. (From Wang, Z.-W. 2004. Unpublished report. Nanyang Technological University, Singapore.)

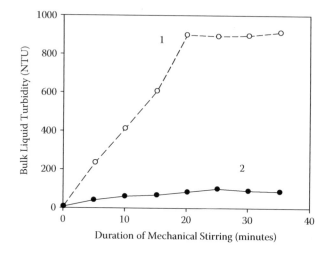

FIGURE 9.18 Responses of aerobic granules cultivated from high hydrophobicity bacteria (1) and conventional activated sludge (2) to mechanical stirring. (Data from Wang, Z.-W. 2004. Unpublished report. Nanyang Technological University, Singapore.)

the washout of poorly settling bioflocs present in R1 by strong selection pressure. Because of rapid aerobic granulation of the selected highly hydrophobic seed in R2 (figure 9.16), more and more biomass was aggregated and was subsequently accumulated until a steady state was reached in R2. On the contrary, slow aerobic granulation in R1 resulted in a slow build-up of biomass concentration. This once again confirms that highly hydrophobic seed accelerates the aerobic granulation process and further improves the property of aerobic granules. The performances of two kinds of aerobic granules developed in R1 and R2 were also evaluated in terms of

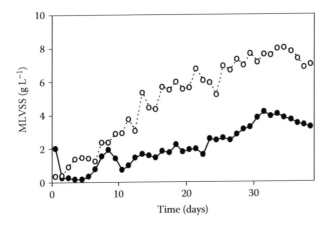

FIGURE 9.19   Profiles of biomass concentration in R1 (●) and R2 (○) seeded with conventional activated sludge and highly hydrophobic cells, respectively, in the course of aerobic granulation. (Data from Wang, Z.-W. 2004. Unpublished report. Nanyang Technological University, Singapore.)

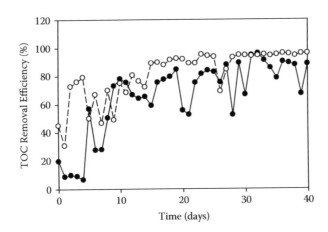

FIGURE 9.20   Total organic carbon (TOC) removal efficiencies of R1 (●) and R2 (○) seeded with conventional activated sludge and highly hydrophobic cells, respectively, in the course of aerobic granulation. (Data from Wang, Z.-W. 2004. Unpublished report. Nanyang Technological University, Singapore.)

the removal efficiency of total organic carbon (TOC). No significant difference was observed in the two reactors, as shown in figure 9.20.

Cell surface hydrophobicity has been proved to be a triggering force that enhances cell-to-cell aggregation. The inoculation with selected highly hydrophobic microbial seed can significantly shorten the aerobic granulation process in SBRs and further improve the stability of aerobic granules. It is expected that selection and inoculation of special microbial seed would be a feasible way to accelerate the start-up of a full-scale aerobic granular sludge SBR.

9.8   CONCLUSIONS

This chapter shows that cell surface hydrophobicity is an essential triggering force of cell-to-cell aggregation that is a crucial initial step towards microbial granulation, and can further strengthen the aggregated microbial structure. High cell surface hydrophobicity associated with aerobic granulation can be induced by various culture conditions, and enrichment and selection of highly hydrophobic bacteria greatly facilitates aerobic granulation. As a result, the time required for aerobic granulation in SBRs can be shortened significantly.

REFERENCES

Absolom, D. R., Lamberti, F. V., Policova, Z., Zingg, W., van Oss, C. J., and Neumann, A. W. 1983. Surface thermodynamics of bacterial adhesion. *Appl Environ Microbiol* 46: 90–97.

Akao, T., Gomi, K., Goto, K., Okazaki, N., and Akita, O. 2002. Subtractive cloning of cDNA from Aspergillus oryzae differentially regulated between solid-state culture and liquid (submerged) culture. *Curr Genet* 41: 275–281.

Asconcabrera, M. A. and Lebeault, J. M. 1995. Cell hydrophobicity influencing the activity stability of xenobiotic-degrading microorganisms in a continuous biphasic aqueous-organic system. *J Ferment Bioeng* 80: 270–275.

Asther, M., Haon, M., Roussos, S., Record, E., Delattre, M., Lesage-Meessen, L., and Labat, M. 2002. Feruloyl esterase from Aspergillus niger: A comparison of the production in solid state and submerged fermentation. *Process Biochem* 38: 685–691.

Bellon Fontaine, M. N., Rault, J., and van Oss, C. J. 1996. Microbial adhesion to solvents: A novel method to determine the electron-donor/electron-acceptor or Lewis acid-base properties of microbial cells. *Colloid Surface* B 7: 47–53.

Bergendahl, J., Grasso, D., Strevett, K., Butkus, M., and Subramanian, K. 2002. A review of non-DLVO interactions in environmental colloidal systems. *Rev Environ Sci Biotechnol* 1: 17–38.

Blanco, M. T., Blanco, J., Sanchez Benito, R., Perez Giraldo, C., Moran, F. J., Hurtado, C., and Gomez Garcia, A. C. 1997. Incubation temperatures affect adherence to plastic of Candida albicans by changing the cellular surface hydrophobicity. *Microbios* 89: 23–28.

Bos, R., van der Mei, C. H, and Busscher, H. J. 1999. Physico-chemistry of initial microbial adhesive interactions: Its mechanisms and methods for study. *FEMS Microbiol Rev* 23: 179–229.

Bossier, P. and Verstraete, W. 1996. Triggers for microbial aggregation in activated sludge? *Appl Microbiol Biotechnol* 45: 1–6.

Castellanos, T., Ascencio, F., and Bashan, Y. 2000. Starvation-induced changes in the cell surface of Azospirillum lipoferum. *FEMS Microbiol Ecol* 33: 1–9.

Changui, C., Doren, A., Stone, W. E. E., Mozes, N., and Rouxhet, P. G. 1987. Surface properties of polycarbonate and promotion of yeast cells adhesion. *J Chim Phys* 84: 275–281.

Chaplin, M. F. 2000. A proposal for the structuring of water. *Biophys Chem* 83: 211–221.

Chen, G. and Strevett, K. A. 2003. Impact of carbon and nitrogen conditions on E. coli surface thermodynamics. *Colloids and Surfaces B: Biointerfaces* 28: 135–146.

Chesa, S. C., Irvine, R. L., and Manning, J. F. 1985. Feast/famine growth environments and activated sludge population selection. *Biotechnol Bioeng* 27: 562–569.

Correa, O. S., Rivas, E. A., and Barneix, A. J. 1999. Cellular envelopes and tolerance to acid pH in Mesorhizobium loti. *Curr Microbiol* 38: 329–334.

Danniels, L., Hanson, R. S., and Phillips, J. A. 1994. Chemical analysis. In Gerhardt, P. (ed.), *Methods for general and molecular bacteriology*. Washington, DC: American Society of Microbiology, 512–554.

Davey, M. E. and O'Toole, G. A. 2000. Microbial biofilms: From ecology to molecular genetics. *Microbiol Mol Biol Rev* 64: 847.

Del Re, B., Sgorbati, B., Miglioli, M., and Palenzona, D. 2000. Adhesion, autoaggregation and hydrophobicity of 13 strains of Bifidobacterium longum. *Lett Appl Microbiol* 31: 438–442.

Farrell, A. and Quilty, B. 2002. Substrate-dependent autoaggregation of Pseudomonas putida CP1 during the degradation of mono-chlorophenols and phenol. *J Ind Microbiol Biotechnol* 28: 316–324.

Goldberg, J. 2002. Biofilms and antibiotic resistance: A genetic linkage. *Trends Microbiol* 10: 264.

Hantula, J. and Bamford, D. H. 1991. The efficiency of the protein dependent flocculation of Flavobacterium sp. is sensitive to the composition of growth medium. *Appl Microbiol Biotechnol* 36: 100–104.

Hata, Y., Ishida, H., Ichikawa, E., Kawato, A., Suginami, K., and Imayasu, S. 1998. Nucleotide sequence of an alternative glucoamylase-encoding gene (glaB) expressed in solid-state culture of Aspergillus oryzae. *Gene* 207: 127–134.

Hildebrand, J. H. 1979. Is there a "hydrophobic effect"? *PNAS* 76: 194.

Ivanov, V., Tay, S. T. L., Liu, Q. S., Wang, X. H., Wang, Z.-W., and Tay, J. H. 2005. Formation and structure of granulated microbial aggregates used in aerobic wastewater treatment. *Water Sci Technol* 52: 13–19.

Jiang, H. L., Tay, J. H., and Tay, S. T. L. 2004. Changes in structure, activity and metabolism of aerobic granules as a microbial response to high phenol loading. *Appl Microbiol Biotechnol* 63: 602–608.

Jiang, H. L., Tay, J. H., Liu, Y., and Tay, S. T. L. 2003. Ca2+ augmentation for enhancement of aerobically grown microbial granules in sludge blanket reactors. *Biotechnol Lett* 25: 95–99.

Jorand, F., Boue-Bigne, F., Block, J. C., and Urbain, V. 1998. Hydrophobic/hydrophilic properties of activated sludge exopolymeric substances. *Water Sci Technol* 37: 307–315.

Kim, I. S., Stabnikova, E. V., and Ivanov, V. N. 2000. Hydrophobic interactions within biofilms of nitrifying and denitrifying bacteria in biofilters. *Bioprocess Eng* 22: 285–290.

Kjelleberg, S. and Hermansson, M. 1984. Starvation-induced effects on bacterial surface characteristics. *Appl Environ Microbiol* 48: 497–503.

Kjelleberg, S., Humphrey, B. A., and Marshall, K. C. 1983. Initial phases of starvation and activity of bacteria at surfaces. *Appl Environ Microbiol* 46: 978–984.

Kos, B., Suskovic, J., Vukovic, S., Simpraga, M., Frece, J., and Matosic, S. 2003. Adhesion and aggregation ability of probiotic strain Lactobacillus acidophilus M92. *J Appl Microbiol* 94: 981–987.

Kwok, W. K., Picioreanu, C., Ong, S. L., van Loosdrecht, M. C. M., Ng, W. J., and Heijnen, J. J. 1998. Influence of biomass production and detachment forces on biofilm structures in a biofilm airlift suspension reactor. *Biotechnol Bioeng* 58: 400–407.

Li, Z. H., Kuba, T., and Kusuda, T. 2006. The influence of starvation phase on the properties and the development of aerobic granules. *Enzyme Microb Technol* 38: 670–674.

Liao, B. Q., Allen, D. G., Droppo, I. G., Leppard, G. G., and Liss, S. N. 2001. Surface properties of sludge and their role in bioflocculation and settleability. *Water Res* 35: 339–350.

Liu, Y. and Tay, J. H. 2002. The essential role of hydrodynamic shear force in the formation of biofilm and granular sludge. *Water Res* 36: 1653–1665.

Liu, Y., Yang, S. F., Liu, Q. S., and Tay, J. H. 2003. The role of cell hydrophobicity in the formation of aerobic granules. *Curr Microbiol* 46: 270–274.

Liu, Y., Yang, S. F., Qin, L., and Tay, J. H. 2004a. A thermodynamic interpretation of cell hydrophobicity in aerobic granulation. *Appl Microbiol Biotechnol* 64: 410–415.

Liu, Y., Yang, S. F., Tay, J. H., Liu, Q. S., Qin, L., and Li, Y. 2004b. Cell hydrophobicity is a triggering force of biogranulation. *Enzyme Microb Technol* 34: 371–379.

Ljungh, A. and and Wadstrom, T. 1995. Growth conditions influence expression of cell surface hydrophobicity of staphylococci and other wound infection pathogens. *Microbiol Immunol* 39: 753–757.

Madigan, M. T., Martinko, J. M., and Parker, J. 2003. Brock biology of microorganisms. Upper Saddle River, NJ: Prentic Hall/Pearson Education.

Mahoney, E. M., Varangu, L. K., Cairns, W. L., Kosaric, N., and Murray, R. G. E. 1987. The effect of calcium on microbial aggregation during UASB reactor start-up. *Water Sci Technol* 19: 249–260.

Malmqvist, T. 1983. Bacterial hydrophobicity measured as partition of palmitic acid between the two immiscible phases of cell surface and buffer. *Acta Pathol Microbiol Immunol Scand* B 91: 69–73.

Masui, M., Takata, H., and Kominami, T. 2002. Cell adhesion and the negative cell surface charges in embryonic cells of the starfish Asterina pectinifera. *Electrophoresis* 23: 2087–2095.

Mattarelli, P., Biavati, B., Pesenti, M., and Crociani, F. 1999. Effect of growth temperature on the biosynthesis of cell wall proteins from Bifidobacterium globosum. *Res Microbiol* 150: 117–127.

Mozes, N. and Rouxhet, P. G. 1987. Methods for measuring hydrophobicity of micro-organisms. *J Microbiol Meth* 6: 99–112.

Neumann, A., Good, R., Hope, C., and Sejpal, M. 1974. An equation of state approach to determine surface tension of low-energy solids from contact angles. *J Colloid Interf Sci* 49: 291–304.

Olofsson, A. C., Zita, A., and Hermansson, M. 1998. Floc stability and adhesion of green-fluorescent-protein-marked bacteria to flocs in activated sludge. *Microbiol-UK* 144: 519–528.

Parker, J. M. R., Guo, D., and Hodges, R. S. 1986. New hydrophilicity scale derived from high-performance liquid chromatography retention data: Correlation of predicted surface residues with antigenicity and x-ray derived accessible sites. *Biochemistry* 25: 5425.

Powell, C. D., Quain, D. E., and Smart, K. A. 2003. The impact of brewing yeast cell age on fermentation performance, attenuation and flocculation. *FEMS Yeast Res* 3: 149–157.

Qin, L., Liu, Y., and Tay, J. H. 2004. Selection pressure is a driving force of aerobic granulation in sequencing batch reactors. *Process Biochem* 39: 579–584.

Quarmby, J. and Forster, C. F. 1995. An examination of the structure of UASB granules. *Water Res* 29: 2449–2454.

Roels, J. A. 1983. *Energetics and kinetics in biotechnology.* Amsterdam: Elsevier Biomedical Press.

Rosenberg, M. and Kjelleberg, S. 1986. Role in bacterial adhesion: Advances in microbial ecology. In Marshall, K. C. (ed.), *Hydrophobic interactions*. New York: Plenum Press, 353–393.

Ross, W. 1984. The phenomenon of sludge pelletization in the treatment of maize processing waste. *Water SA* 10: 197–204.

Rouxhet, P. G. and Mozes, N. 1990. Physical chemistry of the interface between attached micro-organisms and their support. *Water Sci Technol* 22: 1–16.

Ryoo, D. H. and Choi, C. S. 1999. Surface thermodynamics of pellet formation in Aspergillus niger. *Biotechnol Lett* 21: 97–100.

Sanin, S. L., Sanin, F. D., and Bryers, J. D. 2003. Effect of starvation on the adhesive properties of xenobiotic degrading bacteria. *Process Biochem* 38: 909–914.

Schmidt, J. E. and Ahring, B. K. 1994. Extracellular polymers in granular sludge from different upflow anaerobic sludge blanket (UASB) reactors. *Applied Microbiol Biotechnology* 42: 457–462.

Sharma, P. K. and Rao, K. H. 2003. Adhesion of Paenibacillus polymyxa on chalcopyrite and pyrite: Surface thermodynamics and extended DLVO theory. *Colloid Surface B* 29: 21–38.

Shen, C. F., Kosaric, N., and Blaszczyk, R. 1993. The effect of selected heavy metals (Ni, Co and Fe) on anaerobic granules and their extracellular polymeric substance (EPS). *Water Res* 27: 25–33.

Singleton, D. R., Masuoka, J., and Hazen, K. C. 2001. Cloning and analysis of a Candida albicans gene that affects cell surface hydrophobicity. *J Bacteriol* 183: 3582–3588.

Stickler, D. 1999. Biofilms. *Curr Op Microbiol* 2: 270–275.

Strand, S. P., Varum, K. M., and Ostgaard, K. 2003. Interactions between chitosans and bacterial suspensions: Adsorption and flocculation. *Colloid Surface B* 27: 71–81.

Sun, F. Y., Yang, C. Y., and Li, J. Y. 2007. Effect of carbon substrates on the morphological characteristics of aerobic granule in SBR. *Beijing Jiaotong Daxue Xuebao* [Journal of Beijing Jiaotong University] 31: 106–110.

Tay, J. H., Liu, Q. S., and Liu, Y. 2001a. The effects of shear force on the formation, structure and metabolism of aerobic granules. *Appl Microbiol Biotechnol* 57: 227–233.

Tay, J. H., Liu, Q. S., and Liu, Y. 2001b. The role of cellular polysaccharides in the formation and stability of aerobic granules. *Lett Appl Microbiol* 33: 222–226.

Tay, J. H., Liu, Q. S., and Liu, Y. 2001c. Microscopic observation of aerobic granulation in sequential aerobic sludge blanket reactor. *J Appl Microbiol* 91: 168–175.

Tay, J. H., Liu, Q. S., and Liu, Y. 2002. Characteristics of aerobic granules grown on glucose and acetate in sequential aerobic sludge blanket reactors. *Environ Technol* 23: 931–936.

Tay, J. H., Xu, H. L., and Teo, K. C. 2000. Molecular mechanism of granulation. I. H+ translocation-dehydration theory. *J Environ Eng* 126, 403–410.

Thaveesri, J., Gernaey, K., Kaonga, B., Boucneau, G., and Verstraete, W. 1994. Organic and ammonium nitrogen and oxygen in relation to granular sludge growth lab-scale UASB reactors. Proceedings of the IAWQ 7th International Symposium on Anaerobic Digestion, Jan 23–28, 1994. *Water Sci Technol* 30: 43–53.

Thaveesri, J., Daffonchio, D., Liessens, B., Vandermeren, P., and Verstraete, W. 1995. Granulation and sludge bed stability in upflow anaerobic sludge bed reactors in relation to surface thermodynamics. *Appl Environ Microbiol* 61: 3681–3686.

Tijhuis, L., van Loosdrecht, M. C. M., and Heijnen, J. J. 1995. Dynamics of biofilm detachment in biofilm airlift suspension reactors. *Biotechnol Bioeng* 45: 481–487.

Toh, S. K., Tay, J. H., Moy, B. Y. P., Ivanov, V., and Tay, S. T. L. 2003. Size-effect on the physical characteristics of the aerobic granule in a SBR. *Appl Microbiol Biotechnol* 60: 687–695.

Tsubata, T., Tezuka, T., and Kurane, R. 1997. Change of cell membrane hydrophobicity in a bacterium tolerant to toxic alcohols. *Can J Microbiol* 43: 295–299.

Tsuneda, S., Aikawa, H., Hayashi, H., Yuasa, A., and Hirata, A. 2003. Extracellular polymeric substances responsible for bacterial adhesion onto solid surface. *FEMS Microbiol Lett* 223: 287–292.

Vallim, M. A., Janse, B. J. H., Gaskell, J., Pizzirani-Kleiner, A. A., and Cullen, D. 1998. Phanerochaete chrysosporium cellobiohydrolase and cellobiose dehydrogenase transcripts in wood. *Appl Environ Microbiol* 64: 1924–1928.

van Loosdrecht, M. C., Lyklema, J., Norde, W., Schraa, G., and Zehnder, A. J. 1987. Electrophoretic mobility and hydrophobicity as a measure to predict the initial steps of bacterial adhesion. *Appl Environ Microbiol* 53: 1898–1901.

van Oss, C. J. 1993. Acid-base interfacial interactions in aqueous media. Colloids Surfaces A: *Physicochem Eng Aspects* 78: 1.

van Oss, C. J. 1994a. *Nature of specific ligand-receptor bonds, in particular the antigen-antibody bond.* New York: Marcel Dekker.

van Oss, C. J. 1994b. *Interfacial forces in aqueous media.* New York: Marcel Dekker.

van Oss, C. J. 1995. Hydrophobicity of biosurfaces: Origin, quantitative determination and interaction energies. *Colloids Surfaces B: Biointerfaces* 5: 91–110.

van Oss, C. J. 2003. Long-range and short-range mechanisms of hydrophobic attraction and hydrophilic repulsion in specific and aspecific interactions. *J Mol Recognit* 16: 177–190.

van Oss, C. J., Chaudhury, M. K., and Good, R. J. 1987. Monopolar surfaces. *Adv Colloid Interface Sci* 28: 35–64.

van Oss, C. J., Chaudhury, M. K., and Good, R. J. 1988. Interfacial Lifshitz-van der Waals and polar interactions in macroscopic systems. *Chem Rev* 88: 884.

van Oss CJ, Good RJ, Chaudhury MK 1986. The role of van der Waals forces and hydrogen bonds in hydrophobic interactions between biopolymers and low energy surfaces. *J Colloid Interface Sci* 111

Wang, Z.-W. 2004. Development of a cell surface hydrophobicity based method to enhance aerobic granulation. Unpublished Interim Report. Nanyang Technological University, Singapore.

Watnick, P. and Kolter, R. 2000. Biofilm, city of microbes. *J Bacteriol* 182: 2675–2679.

Wilschut, J. and Hoekstra, D. 1984. Membrane fusion: From liposome to biological memgrane. *Trends Biochem Sci* 9: 479–483.

Yang, S. F., Tay, J. H., and Liu, Y. 2004. Inhibition of free ammonia to the formation of aerobic granules. *Biochem Eng J* 17: 41–48.

Zheng, Y. M., Yu, H. Q., and Sheng, G. P. 2005. Physical and chemical characteristics of granular activated sludge from a sequencing batch airlift reactor. *Process Biochem* 40: 645–650.

Zita, A. and Hermansson, M. 1997. Effects of bacterial cell surface structures and hydrophobicity on attachment to activated sludge flocs. *Appl Environ Microbiol* 63: 1168–1170.

10 Essential Roles of Extracellular Polymeric Substances in Aerobic Granulation

Yu Liu and Zhi-Wu Wang

CONTENTS

10.1   INTRODUCTION

Extracellular polymeric substances (EPS) are sticky materials secreted by cells; extracellular polysaccharides (PS) are the important component of EPS and are highly involved in the formation of matrix structure and improvement of long-term stability of aerobic granules, as discussed in the preceding chapters. It is thought that EPS act as an effective bioglue to cross-link bacteria into an aerobic granule (figure 10.1), and the EPS matrix of the aerobic granule can protect bacteria from harsh environmental conditions. In view of the importance of PS, this chapter attempts to provide a deeper insight into the functions of EPS in aerobic granulation.

10.2   MAIN COMPOSITION OF EPS IN AEROBIC GRANULES

EPS have been detected in significant quantities in both aerobic and anaerobic granules, and they form a three-dimensional matrix in which bacteria and other particles are embedded. EPS are produced by microorganisms themselves during cultivation, which are advantageous in many respects for their survival in various circumstances. In terms of microbiology, EPS can help stabilize membrane structure and also may serve as a protective barrier (Prescott, Harley, and Klein 1999). EPS produced in biogranules contain variable proportions of proteins, polysaccharides, nucleic acids, humics-like substances, lipids, and heteropolymers-like glycoprotein (Goodwin and

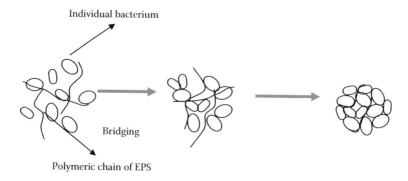

Individual bacterium

Bridging

Polymeric chain of EPS

FIGURE 10.1 A schematic interpretation of extracellular polysaccharides in aerobic granulation.

Forster 1985; Horan and Eccles 1986; Grotenhuis et al. 1991; Urbain, Block, and Manem 1993; Jorand et al. 1995; Frolund et al. 1996). It should be pointed out that polysaccharides are the only component that are synthesized extracellularly for a specific function with neutral and acidic polysaccharides, while proteins, lipids, and nucleic acids can exist in the extracellular polymer network due to the excretion of intracellular polymers or as a result of cell lysis (Durmaz and Sanin 2001; Mahmoud et al. 2003). EPS has been found in almost all forms of microbial aggregates, including bioflocs, biofilm, anaerobic and aerobic granules. However, the content of EPS in aerobic granules was found to be much higher than those in conventional bioflocs and biofilms (Tay, Liu, and Liu 2001c).

In the environmental engineering literature, there are contradictory reports on the composition of EPS in biogranules, especially the ratio of carbohydrates to proteins. Proteins have been reported to be the predominant component of EPS in anaerobic granules (Fukuzaki, Nishio, and Nagai 1995), while other evidence shows that EPS were mainly composed of carbohydrates (Fang 2000). It seems that the quantity and the composition of EPS produced by bacteria depend on a number of factors, such as microbial species, the growth phase of the strain, the type of limiting substrate (carbon, nitrogen, and phosphorous), oxygen limitation, ionic strength, culture temperature, shear force and so on (Nielsen, Jahn, and Palmgren 1997; Tay, Liu, and Liu 2001a; Nichols et al. 2004; Qin et al. 2004). For example, the total amount of EPS and their composition in terms of the ratio of carbohydrates to proteins were influenced by the magnitude of carbon and nitrogen sources (Sheng, Yu, and Yue 2006). This may imply that the composition of extracellular polymers varies and is related to microbial species, the physiological state of bacteria, and operating conditions under which biogranules are developed. In fact, a shift in bacterial species during biogranulation has been reported, and such a microbial shift would affect the production and composition of EPS (Etchebehere et al. 2003; Yi et al. 2003). It appears that the incomparable EPS results reported in the literature result partially from the complexity of a mixed microbial culture under different operation conditions. Another point that needs to be addressed is that the EPS mainly include bound and soluble polymers, the ratio between which may change substantially even within the same species under various growth conditions. The soluble polymers can be transferred to the supernatant after centrifugation, while the bound polymers are still tightly attached to cells and cannot be recovered from the supernatant (Nielsen, Jahn, and Palmgren 1997).

Zhang, Bishop, and Kinkle (1999) compared different EPS extraction methods for a biofilm sample, including the regular centrifugation, ethylenediamine tetraacetic acid (EDTA) extraction, ultracentrifugation, steaming extraction, and regular centrifugation with formaldehyde (RCF), and found that the RCF method gave the highest yield ratio of carbohydrates to proteins of 13.7 for the aerobic/sulfate reducing biofilms and the other methods gave a range of ratios of 1.54 to 2.23. This is indeed consistent with the finding by Azeredo, Lazarova, and Oliveira (1999) that many EPS extraction methods developed for biofilms were not efficient, and somehow could promote leakage of intracellular materials. Sheng, Yu, and Yu (2005a) also compared four extraction methods (EDTA, NaOH, H_2SO_4, heating-centrifugation) of EPS produced from a photosynthetic bacterium, *Rhodopseudomonas acidophila*, and they concluded that the EDTA extraction method was the most effective over the others. A review based on over 200 publications related to EPS in activated sludge reveals that reported composition and quantity of EPS strongly depend on the extraction methods employed (Liu and Fang 2003). Although a number of physical and chemical methods, for example, high-speed centrifugation, boiling treatment in acid or alkali, and utilization of solvent extraction or cation exchange resins are currently available for extracting EPS from biogranules, none of them has been adopted as a standard procedure yet. In addition to the effect of microbial species, the use of nonstandardized procedures causes incomparability of the EPS results available in the present literature.

10.3 MAJOR FACTORS INFLUENCING EPS PRODUCTION IN AEROBIC GRANULES

There is no need for microorganisms to secrete excessive EPS under normal culture conditions. The observed enhanced production of EPS in biogranules is induced by some so-called stressful culture conditions (Nichols et al. 2004; Qin et al. 2004). So far, it has been understood that a number of operating parameters, including reactor type, substrate composition, substrate loading rate, hydraulic retention time, hydrodynamic shear force, settling time in sequencing batch reactors (SBRs), feast-famine regimen in SBRs, culture temperature, etc., may stimulate bacteria to secrete more EPS. The composition of EPS is also related to the characteristics of the feed wastewater, for example EPS in anaerobic granules grown on protein-rich wastewater had high protein and DNA levels, whereas high polysaccharides content was found in anaerobic granules fed by other types of organic wastewaters (Batstone and Keller 2001). In addition, the deficiency of nitrogen was found to favor the production of EPS, which in turn accelerated anaerobic granulation (Punal et al. 2003).

Experimental evidence from aerobic granulation research shows that stressful operating conditions in terms of high hydrodynamic shear force, short settling time/hydraulic retention time, and periodic feast-famine periods would significantly stimulate bacteria to produce more extracellular polysaccharides over proteins in SBRs, as shown in the preceding chapters. The EPS production seems to be positively

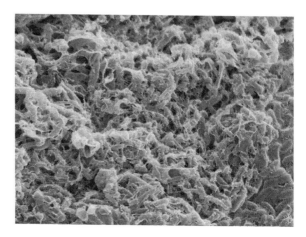

FIGURE 10.2   Extracellular (EPS) matrix structure observed in acetate-fed aerobic granules.

related to the specific oxygen utilization rate (SOUR) of aerobic granules developed in SBRs. In fact, the catabolic activity of microorganisms is directly correlated with the electron transport system activity, which can be roughly described by the SOUR (Trevors 1984; Lopez, Koopman, and Bitton 1986). Moreover, in the aerobic oxidation process the respiratory activity of cells couples to the proton translocation activity and a clear linkage of oxygen reduction to proton translocation has been established (Babcock and Wikstrom 1992). This implies that aggregated bacteria can respond to the stressful culture conditions by regulating their energy metabolism (Liu and Tay 2002).

Chan et al. (2004) thought that the purpose of polymer production was to localize iron oxyhydroxide mineral precipitation in order to enhance metabolic energy generation. In general, the environmental factors influencing EPS production and composition can be attributed to changes in environmental conditions that cause a shift in the microbial community. Subsequently, the distribution of EPS-producing microorganisms varies (increases or decreases) in the whole microbial consortium, and regulation of the metabolic pathway of EPS production occurs in response to changes in the environmental conditions. In a study of EPS production in the presence of toxic substances, addition of toxic substances, such as Cu^{2+}, Cr^{6+}, Cd^{2+}, and 2,4-dichlorophenol (2,4-DCP), was found to stimulate *Rhodopseudomonas acidophila* to secrete more EPS, for example, at respective concentrations of 30 mg L^{-1} Cu^{2+}, 40 mg L^{-1} Cr^{6+}, 5 mg L^{-1} Cd^{2+}, and 100 mg L^{-1} 2,4-DCP, the EPS content in *Rhodopseudomonas acidophila* was increased by 5.5, 2.5, 4.0, and 1.4 times that of the control (Sheng, Yu, and Yue 2005b). These results indicate that EPS can protect cells against toxic effect. In fact, the strong and compact EPS-mediated structure of aerobic granules (figure 10.2) should provide adequate protection against exposure to chemical toxicity.

It has not been demonstrated yet whether the genes for EPS production are expressed before or after bacterial granulation. There are two scenarios for EPS production in time sequence along microbial granulation: (1) microorganisms initially prepare EPS for subsequent self-attachment; (2) microbial attachment comes first, followed by the production of EPS. In scenario 1, EPS production occurs prior to granulation, and the appearance of EPS at the initial site of contact between micro-bial cells may be due to the migration of polymer molecules already existing on the cell surface. In scenario 2, EPS is produced after initial microbial attachment, and bacterial attachment in this case may provide some physiological conditions necessary for EPS excretion.

10.4   THE ROLE OF EPS IN AEROBIC GRANULATION

EPS facilitate cell-to-cell interaction and further strengthen microbial structure through forming a polymeric matrix. In recognition of the importance of EPS quantity in biogranulation process, the contribution of the EPS properties, such as hydrophobicity and charge, also need to be taken into account (Andreadakis 1993; Liao et al. 2001; Wang, Liu, and Tay 2005). Since EPS may accumulate at the cell surface, it could alter cell surface properties, such as cell surface hydrophobicity, surface charge density, binding site, and surface morphology. The surface charge has long been believed to be important in controlling the stability of microbial aggre-gates. It is well known that bacteria carry net negative surface charge when cultivated at physiological pH values (Rouxhet and Mozes 1990). According to the well-known DLVO theory, when the two surfaces have a charge of the same sign, repulsive force between cells will prevent the approach of one cell to another.

Some results showed that EPS could decrease the negative charges of the cell surfaces, and thereby further bridge two neighboring cells physically to each other (Shen, Kosaric, and Blaszczyk 1993; Schmidt and Ahring 1994). Using a colloid titra-tion technique, Morgan, Forster, and Evison (1990) reported that granular sludge was less negatively charged than activated sludge, while Tsuneda et al. (2003) employed a soft particle electrophoresis technique to investigate the influence of EPS on cell surface electrokinetics, and found that EPS could increase the softness of the cell surface and further decrease the negative surface charge density surrounding the cell surface, that is, the EPS layer could hold a lower negative charge compared with those of the native cell surface.

It should be realized that electrostatic interaction between cells is closely associ-ated with the amount of EPS produced as well. Microbial attachment onto a solid surface can be inhibited by electrostatic interaction when the EPS amount is small, whereas cell adhesion is enhanced by polymeric interaction if the EPS amount is large (Tsuneda et al. 2003). Wang et al. (2006) reported that along with aerobic granulation in an SBR, both biomass and EPS concentrations tended to increase, and a decreased sludge volume index (SVI) was found accordingly (figure 10.3), for example, the EPS content in the mature aerobic granules was about 47 mg g^{-1} MLSS (mixed liquor suspended solids) whereas it was only 17 mg g^{-1} MLSS in the seed activated sludge. This seems to imply that aerobic granulation would be associated with an increase in the EPS content.

Cell surface hydrophobicity has been considered as a triggering force of bio-granulation (see chapter 9). Microorganisms with different hydrophobicities were detected in activated sludge (Singh and Vincent 1987; Jorand et al. 1995), and the high cell surface hydrophobicity was usually associated with the presence of fibrillar structure on cell surface and specific cell wall proteins (McNab et al. 1999; Singleton,

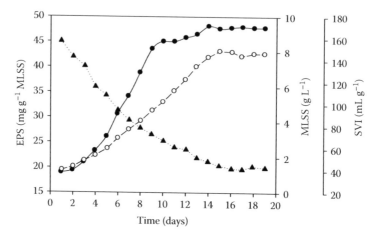

FIGURE 10.3 Changes in EPS (●), MLSS (○), and SVI (▲) in the course of operation. (Data from Wang, Z. P. et al. 2006. *Chemosphere* 63: 1728–1735.)

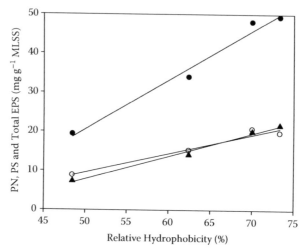

FIGURE 10.4 Correlation of cell surface hydrophobicity to proteins (○) and polysaccharides (▲) and total EPS (●). (Data from Wang, Z. P. et al. 2006. *Chemosphere* 63: 1728–1735.)

Masuoka, and Hazen 2001). In fact, the cell wall of bacteria in anaerobic granules was surrounded by an EPS layer (Forster 1991; de Beer et al. 1996; Veiga et al. 1997). This implies that cell surface hydrophobicity may be related to EPS. Some evidence suggests that proteins and amino acids are the hydrophobic components of the EPS, while polysaccharides are hydrophilic (Dignac et al. 1998). Wang et al. (2006) studied the correlation between EPS and cell surface hydrophobicity of aerobic granules, and a positive correlation of cell surface hydrophobicity to total EPS content was observed, as shown in figure 10.4.

Jorand et al. (1998) studied hydrophobic and hydrophilic properties of extra-cellular polymers produced by activated sludge, and found that a significant proportion of the extracellular polymers was hydrophobic, that is, hydrophobic extracellular polymers are involved in the formation and organization of microbial aggregates. Due to the presence of hydrophobic and hydrophilic groups in EPS, the measured hydrophobicity of EPS indeed reflects an average of the hydrophobicity of its components (Daffonchio, Thaveesri, and Verstraete 1995), whereas the cell surface hydrophobicity and charge is related to the production, composition, and physical characteristics of EPS (Liao et al. 2001), but no specific evidence shows the quantitative contribution of hydrophobic components of EPS to the overall hydrophobicity of biogranules so far. There is evidence that the reduction of surface charge would not be a requirement for the formation of activated sludge, that is, the charge neutralization would not be the main microbial flocculation mechanism (Pavonim, Tenney, and Echelberger 1972; Strand, Varum, and Ostgaard 2003), whereas hydrophobic interaction may be fundamental in biogranulation (see chapter 9). It appears from the study of biofilms that few bacteria were *de facto* in contact with the hydrophilic surface, but more with hydrophobic surfaces of other cells (Ghigo 2003).

So far, different views exist with regard to the role of EPS in aerobic granulation. Di Iaconi et al. (2006) reported that the main component of EPS in aerobic granular biomass in a sequencing batch biofilter reactor was made of proteins, and protein-rich EPS would improve the stability of granular biomass structure. As the main component of proteins is amino acids, which often carry negative charges, they will contribute more than carbohydrates to electrostatic bonds, with consequent increase of biomass structure stability (Di Iaconi et al. 2006). Zheng, Yu, and Sheng (2005) studied the physical and chemical characteristics of aerobic granular sludge from a sequencing batch airlift reactor, and they found that the contents of proteins and carbohydrate in EPS decreased from 126.6 and 15.3 mg g^{-1} volatile suspended solids (VSS) in seed sludge to 51.4 and 5.9 mg g^{-1} VSS in aerobic granular sludge. These results seem to imply that EPS are not involved in aerobic granulation in this special case, but the reason behind this is not yet clear. It may be due to different biodegradation kinetics of carbohydrates and proteins under specific operation conditions.

10.5 EPS-ENHANCED STABILITY OF AEROBIC GRANULES

The accumulation of EPS has been found to correlate with biological aggregation, whereas the metabolic blocking of the synthesis of extracellular polysaccharides prevents microbial aggregation (Cammarota and Sant'Anna 1998; Yang, Tay, and Liu 2004). EPS in granules have been often hypothesized to bridge two neighboring bacterial cells physically to each other as well as with other inert particulate matter, and settle out as aggregates (Ross 1984; Shen, Kosaric, and Blaszczyk 1993; Schmidt and Ahring 1994; Tay, Liu, and Liu 2001b). In the biogranulation process, EPS provide an extensive surface area for bacterial binding. Furthermore, EPS matrixes surrounding aggregated bacteria can provide sites available for attraction of organic and inorganic materials (Yu, Tay, and Fang 2001; Sponza 2002). The total concentrations of electrostatic binding sites on EPS were found to be 20- to 30-fold higher than those reported for bacterial cell surfaces (Liu and Fang 2002). This seems to indicate that EPS of neighboring microbial cells may form a cross-linked network by attraction of organics or inorganics, such as cationic bridging, and further strengthen the structural integrity of a biogranule (Yu, Tay, and Fang 2001).

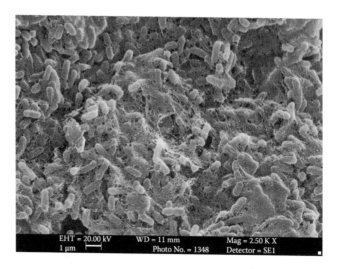

FIGURE 10.5 EPS matrix structure observed in aerobic granule with nitrification capability.

Microscopic observation shows that EPS with a filamentous structure is present within and around the structure of biogranules (Forster 1991; deBeer et al. 1996; Veiga et al. 1997; Tay, Liu, and Liu 2001c), while EPS also can fill in the intercellular spaces in the microcolonies present in aerobic granules (Jiang, Tay and Tay 2002; McSwain et al. 2005). It is most likely that EPS plays an important role in maintaining the structural and functional integrity of aerobic granules (figure 10.5).

Tay, Liu, and Liu (2001b) investigated the role of EPS in the formation and stability of aerobic granules. It was found that the formation of aerobic granules was coupled with a significant increase in EPS expressed as the ratio of polysaccharides (PS) to proteins (PN), while disappearance of aerobic granules, indicated by a significant decrease in bioparticle size, was associated with a simultaneous decrease in extracellular polysaccharides (figure 10.6 and figure 10.7). This indicates that extracellular polysaccharides play a crucial role in the formation as well as in maintaining the stability of aerobic granules.

It is apparent that extracellular polysaccharides excreted by cells can assist in bacterium-to-bacterium self-aggregation by bridging bacterial cells, which induces the formation of initial microbial aggregates. In fact, in the study of anaerobic granulation, Harada et al. (1988) observed that the extracellular polymers excreted by acidogenic bacteria appeared to enhance the strength and structural stability of anaerobic granules. A similar phenomenon was also reported in biofilm systems (Vandevivere and Kirchman 1993). Extracellular polysaccharides indeed play a crucial role in the building architecture of aerobic granules, and subsequently a certain content of extracellular polysaccharides is required in order to maintain the stability of the microbial structure. It has been proposed that extracellular polymers can change the surface negative charge of bacteria, and thereby bridge two neighboring bacterial cells physically to each other as well as other inert particulate matters, and finally settle out as floccus aggregates (Shen, Kosaric, and Blaszczyk 1993; Schmidt and Ahring 1994). It appears from figure 10.6 that the content of cellular

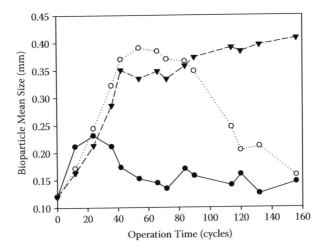

FIGURE 10.6   Change in microbial aggregate size in the course of SBR cycle operation at different specific upflow air velocities of 0.3 (●), 1.2 (○), and 2.4 (▼) cm s^{-1}. (Data from Tay, J. H., Liu, Q. S., and Liu, Y. 2001b. *Lett Appl Microbiol* 33: 222–226.)

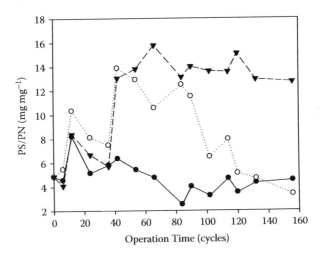

FIGURE 10.7   Change in PS/PN ratio in the course of SBR cycle operation at different specific upflow air velocities of 0.3 (●), 1.2 (○), and 2.4 (▼) cm s^{-1}. (Data from Tay, J. H., Liu, Q. S., and Liu, Y. 2001b. *Lett Appl Microbiol* 33: 222–226.)

polysaccharides is much higher than the content of cellular proteins in both flocci and aerobic granules. A similar phenomenon was also reported in a biofilm system (Vandevivere and Kirchman 1993). It may imply that cellular proteins contribute less to the formation, structure, and stability of granules and biofilms.

So far, not much information is available on the EPS distribution in the earlier stage of biogranulation, whereas the spatial distribution of EPS in mature granules has been reported. By using the calcofluor staining method and fluorescent micro-scopy or confocal scanning electron microscopy, it was revealed that approximately

50% of the total amount of EPS was present in a top 40 μm thick zone from the surface of anaerobic granules, and the rest of the EPS was randomly allocated in the deeper parts of the granules (de Beer et al. 1996). It should be pointed out that the EPS layer on the surface of biogranules was not found in conventional bioflocs.

Study of anaerobic and aerobic granulation in different types of bioreactors under a wide spectrum of operating conditions shows that the total amount of EPS produced is not a decisive factor in the formation and maintaining the stability of biogranules. Instead, the distribution and composition of EPS plays a crucial role in biogranulation (Tay, Liu, and Liu 2001b; Punal et al. 2003; Wang, Liu, and Tay 2005). It has been reported that the content of extracellular polysaccharides in anaerobic granules was almost three times higher than that in anaerobic bioflocs (de Beer et al. 1996), while the formation of aerobic granules was found to be accompanied with a sharp increase of cellular polysaccharides normalized to cellular proteins (Tay, Liu, and Liu 2001c). The ratio of extracellular polysaccharides over proteins by weight in aerobic granules fell into a range of 2 to 16 as discussed in the preceding chapters, which seems to be higher than that reported for the anaerobic granulation process. It seems that the characteristics of biogranules are related to the ratio of polysaccharides to proteins. Previous research showed that anaerobic granules and aerobic bioflocs with a higher proteins to polysaccharides ratio had a lower shear strength and a poorer settleability (Batstone and Keller 2001; Martinez et al. 2004), while Quarmby and Forster (1995) thought that the extracellular polysaccharides could contribute highly to the strength and stability of anaerobic granules. Similar results were also obtained in aerobic granulation, showing that the specific gravity and mechanical strength of aerobic granules increased significantly with increase in the ratio of polysaccharides to proteins (Tay, Liu, and Liu 2001a, 2002).

The formation of biogranules indeed is a microbial evolution instead of a random aggregation of suspended bacteria. It should be a reasonable hypothesis that the spatial distribution of EPS in biogranules would be correlated to microbial evolution and distribution along with granulation. The investigation on the spatial distribution of EPS in heterotrophic biofilms showed that the production yield of EPS tended to decrease with the biofilm depth (Zhang and Bishop 2003). This is probably due to the fact that viable biomass loses its ability to produce EPS at the deeper sections of biofilms because of its lower microbial activity resulting from lower nutrient availability. In addition, the EPS produced by bacteria can be utilized as secondary substrate in the deeper layers or zones of biofilms and biogranules where readily degradable substrates are not available or are limiting. Without doubt, the spatial distribution of EPS in biogranules and biofilms plays an essential role in stabilizing the structure and maintaining the strength of microbial aggregates. This is also supported by the finding that EPS near the edge of granules had a greater effect on shear strength than in the center of the granule (Batstone and Keller 2001). Finally, it should be stressed that literature information on the spatial distribution of EPS in biogranules is very limited, and some questions still remain unanswered, for example, whether the spatial distribution of EPS in biogranules is related to operating conditions or is correlated to microbial distribution and activity in biogranules. Obviously, further research on this aspect is needed.

10.6 CONCLUSIONS

This chapter shows the essential roles of EPS in the formation and maintaining structural stability of aerobic granules. It becomes clear that the composition and the content of EPS in aerobic granules affect the matrix structure and the integrity of aerobic granules. However, it should be realized that there are still a lot of contradictory research reports on the functional roles of EPS in aerobic granulation process because of the complex composition of EPS, different culture conditions, and analytical methods employed.

REFERENCES

Andreadakis, A. D. 1993. Physical and chemical properties of activated sludge floc. *Water Res* 27: 1707–1714.

Azeredo, J., Lazarova, V., and Oliveira, R. 1999. Methods to extract the exopolymeric matrix from biofilms: A comparative study. *Water Sci Technol* 39: 243–250.

Babcock, G. T., and Wikstrom, M. 1992. Oxygen activation and the conservation of energy in cell respiration. *Nature* 356: 301–309.

Batstone, D. J. and Keller, J. 2001. Variation of bulk properties of anaerobic granules with wastewater type. *Water Res* 35: 1723–1729.

Cammarota, M. C. and Sant'Anna, G. L. 1998. Metabolic blocking of exopolysaccharides synthesis: Effects on microbial adhesion and biofilm accumulation. *Biotechnol Lett* 20: 1–4.

Chan, C. S., De Stasio, G., Welch, S. A., Girasole, M., Frazer, B. H., Nesterova, M. V., Fakra, S., and Banfield, J. F. 2004. Microbial polysaccharides template assembly of nanocrystal fibers. *Science* 303: 1656–1658.

Daffonchio, D., Thaveesri, J., Verstraete, W. 1995. Contact-angle measurement and cell hydrophobicity of granular sludge from upflow anaerobic sludge bed reactors. *Appl Environ Microbiol* 61: 3676–3680.

de Beer, D., OFlaharty, V., Thaveesri, J., Lens, P., and Verstraete, W, 1996. Distribution of extracellular polysaccharides and flotation of anaerobic sludge. *Appl Microbiol Biotechnol* 46: 197–201.

Di Iaconi, C., Ramadori, R., Lopez, A., and Passino, R. 2006. Influence of hydrodynamic shear forces on properties of granular biomass in a sequencing batch biofilter reactor. *Biochem Eng J* 30: 152–157.

Dignac, M. F., Urbain, V., Rybacki, D., Bruchet, A., Snidaro, D., and Scribe, P. 1998. Chemical description of extracellular polymers: Implication on activated sludge floc structure. *Water Sci Technol* 38: 45–53.

Durmaz, B. and Sanin, F. D. 2001. Effect of carbon to nitrogen ratio on the composition of microbial extracellular polymers in activated sludge. *Water Sci Technol* 44: 221–229.

Etchebehere, C., Cabezas, A., Dabert, P., and Muxi, L. 2003. Evolution of the bacterial community during granules formation in denitrifying reactors followed by molecular, culture-independent techniques. *Water Sci Technol* 48: 75–79.

Fang, H. H. P. 2000. Microbial distribution in UASB granules and its resulting effects. *Water Sci Technol* 42: 201–208.

Forster, C. F. 1991. Anaerobic upflow sludge blanket reactors: Aspects of their microbiology and their chemistry. *J Biotechnol* 17: 221–231.

Frolund, B., Palmgren, R., Keiding, K., and Nielsen, P. H. 1996. Extraction of extracellular polymers from activated sludge using a cation exchange resin. *Water Res* 30: 1749–1758.

Fukuzaki, S., Nishio, N., and Nagai, S. 1995. High-rate performance and characterization of granular methanogenic sludges in upflow anaerobic sludge blanket reactors fed with various defined substrates. *J Ferment Bioeng* 79: 354–359.

Ghigo, J. M. 2003. Are there biofilm-specific physiological pathways beyond a reasonable doubt? *Res Microbiol* 154: 1–8.

Goodwin, J. A. S. and Forster, C. F. 1985. A further examination into the composition of activated sludge surfaces in relation to their settlement characteristics. *Water Res* 19: 527–533.

Grotenhuis, J. T. C., Smit, M., Van Lammeren, A. A. M., Stams, A. J. M., and Zehnder, A. J. B. 1991. Localization and quantification of extracellular polymers in methanogenic granular sludge. *Appl Microbiol Biotechnol* 36: 115–119.

Harada, H., Endo, G., Tohya, Y., and Momonoi, K. 1988. High rate performance and its related characteristics of granulated sludge in UASB reactors treating various wastewater. *Proceedings of Fifth International Symposium on Anaerobic Digestion,* Bologna, Italy, 521–529.

Horan, N. J. and Eccles, C. R. 1986. Purification and characterization of extracellular polysaccharide from activated sludges. *Water Res* 20: 1427–1432.

Jiang, H. L., Tay, J. H., and Tay, S. T. L. 2002. Aggregation of immobilized activated sludge cells into aerobically grown microbial granules for the aerobic biodegradation of phenol. *Lett Appl Microbiol* 35: 439–445.

Jorand, F., Zartarian, F., Thomas, F., Block, J. C., Bottero, J. Y., Villemin, G., Urbain, V., and Manem, J. 1995. Chemical and structural (2D) linkage between bacteria within activated sludge flocs. *Water Res* 29: 1639–1647.

Jorand, F., Boue-Bigne, F., Block, J. C., and Urbain, V. 1998. Hydrophobic/hydrophilic properties of activated sludge exopolymeric substances. *Water Sci Technol* 37: 307–315.

Liao, B. Q., Allen, D. G., Droppo, I. G., Leppard, G. G., and Liss, S. N. 2001. Surface properties of sludge and their role in bioflocculation and settleability. *Water Res* 35: 339–350.

Liu, H. and Fang, H. H. P. 2002. Characterization of electrostatic binding sites of extracellular polymers by linear programming analysis of titration data. *Biotechnol Bioeng* 80: 806–811.

Liu, Y. and Fang, H. H. P. 2003. Influences of extracellular polymeric substances (EPS) on flocculation, settling, and dewatering of activated sludge. *Crit Rev Environ Sci Technol* 33: 237–273.

Liu, Y. and Tay, J. H. 2002. The essential role of hydrodynamic shear force in the formation of biofilm and granular sludge. *Water Res* 36, 1653–1665.

Lopez, J. M., Koopman, B., and Bitton, G. 1986. INT-dehydrogenase test for activated sludge process control. *Biotechnol Bioeng* 28: 1080–1085.

Mahmoud, N., Zeeman, G., Gijzen, H., and Lettinga, G. 2003. Solids removal in upflow anaerobic reactors: A review. *Bioresource Technol* 90: 1–9.

Martinez, F., Lema, J., Mendez, R., Cuervo-Lopez, F., and Gomez, J. 2004. Role of exopolymeric protein on the settleability of nitrifying sludges. *Bioresource Technol* 94: 43–48.

McNab, R., Forbes, H., Handley, P. S., Loach, D. M., Tannock, G. W., and Jenkinson, H. F. 1999. Cell wall-anchored CshA polypeptide (259 kilodaltons) in *Streptococcus gordonii* forms surface fibrils that confer hydrophobic and adhesive properties. *J Bacteriol* 181: 3087–3095.

McSwain, B. S., Irvine, R. L., Hausner, M., and Wilderer, P. A. 2005. Composition and distribution of extracellular polymeric substances in aerobic flocs and granular sludge. *Appl Environ Microbiol* 71: 1051–1057.

Morgan, J. W., Forster, C. F., and Evison, L. 1990. Comparative study of the nature of biopolymers extracted from anaerobic and activated sludges. *Water Res* 24: 743–750.

Nichols, C. A. M., Garon, S., Bowman, J. P., Raguenes, G., and Guezennec, J. 2004. Production of exopolysaccharides by Antarctic marine bacterial isolates. *J Appl Microbiol* 96: 1057–1066.

Nielsen, P. H., Jahn, A., and Palmgren, R. 1997. Conceptual model for production and composition of exopolymers in biofilms. *Water Sci Technol* 36: 11–19.

Pavonim, J., Tenney, M., and Echelberger, J. 1972. Bacterial extracellular polymers and biological flocculation. *J WPCF* 44: 414–431.

Prescott, L. M., Harley, J. P., and Klein, D. A. 1999. *Microbiology*. Singapore, Indonesia: McGraw-Hill.

Punal, A., Brauchi, S., Reyes, J. G., and Chamy, R. 2003. Dynamics of extracellular polymeric substances in UASB and EGSB reactors treating medium and low concentrated wastewaters. *Water Sci Technol* 48: 41–49.

Qin, L., Liu, Q. S., Yang, S. F., Tay, J. H., and Liu, Y. 2004. Stressful conditions-induced production of extracellular polysaccharides in aerobic granulation process. *Civil Eng Res* 17: 49–51.

Quarmby, J. and Forster, C. F. 1995. An examination of the structure of UASB granules. *Water Res* 29: 2449–2454.

Ross, W. 1984. The phenomenon of sludge pelletization in the treatment of maize processing waste. *Water SA* 10: 197–204.

Rouxhet, P. G. and Mozes, N. 1990. Physical chemistry of the interface between attached micro-organisms and their support. *Water Sci Technol* 22: 1–16.

Schmidt, J. E. and Ahring, B. K. 1994. Extracellular polymers in granular sludge from different upflow anaerobic sludge blanket (UASB) reactors. *Appl Microbiol Biotechnol* 42: 457–462.

Shen, C. F., Kosaric, N., and Blaszczyk, R. 1993. The effect of selected heavy metals (Ni, Co and Fe) on anaerobic granules and their extracellular polymeric substance (EPS). *Water Res* 27: 25–33.

Sheng, G. P., Yu, H. Q., and Yu, Z. 2005a. Extraction of extracellular polymeric substances from the photosynthetic bacterium *Rhodopseudomonas acidophila*. *Appl Microbiol Biotechnol* 67: 125–130.

Sheng, G. P., Yu, H. Q., and Yue, Z. B. 2005b. Production of extracellular polymeric substances from *Rhodopseudomonas acidophila* in the presence of toxic substances. *Appl Microbiol Biotechnol* 69: 216–222.

Sheng, G. P., Yu, H. Q., and Yue, Z. 2006. Factors influencing the production of extracellular polymeric substances by *Rhodopseudomonas acidophila*. *Int Biodeterioration Biodegradation* 58: 89–93.

Singh, K. and Vincent, W. 1987. Clumping characteristics and hydrophobic behavior of an isolated bacterial strain from sewage sludge. *Appl Microbiol Biotechnol* 25: 396–398.

Singleton, D. R., Masuoka, J., and Hazen, K. C. 2001. Cloning and analysis of a *Candida albicans* gene that affects cell surface hydrophobicity. *J Bacteriol* 183: 3582–3588.

Sponza, D. T. 2002. Extracellular polymer substances and physicochemical properties of flocs in steady- and unsteady-state activated sludge systems. *Process Biochem* 37: 983–998.

Strand, S. P., Varum, K. M., and Ostgaard, K. 2003. Interactions between chitosans and bacterial suspensions: adsorption and flocculation. *Colloid Surface B* 27: 71–81.

Tay, J. H., Liu, Q. S., and Liu, Y. 2001a. The effects of shear force on the formation, structure and metabolism of aerobic granules. *Appl Microbiol Biotechnol* 57: 227–233.

Tay, J. H., Liu, Q. S., and Liu, Y. 2001b. The role of cellular polysaccharides in the formation and stability of aerobic granules. *Lett Appl Microbiol* 33: 222–226.

Tay, J. H., Liu, Q. S., and Liu, Y. 2001c. Microscopic observation of aerobic granulation in sequential aerobic sludge blanket reactor. *J Appl Microbiol* 91: 168–175.

Tay, J. H., Liu, Q. S., and Liu, Y. 2002. Characteristics of aerobic granules grown on glucose and acetate in sequential aerobic sludge blanket reactors. *Environ Technol* 23: 931–936.

Trevors, J. T. 1984. The measurement of electron transport system (ETS) activity in freshwater sediment. *Water Res* 18: 581–584.

Tsuneda, S., Aikawa, H., Hayashi, H., Yuasa, A., and Hirata, A. 2003. Extracellular polymeric substances responsible for bacterial adhesion onto solid surface. *FEMS Microbiol Lett* 223: 287–292.

Urbain, V., Block, J. C., and Manem, J. 1993. Bioflocculation in activated sludge: An analytic approach. *Water Res* 27: 829–838.

Vandevivere, P. and Kirchman, D. 1993. Attachment stimulates exopolysaccharide synthesis by a bacterium. *Appl Environ Microbiol* 59: 3280–3286.

Veiga, M. C., Jain, M. K., Wu, W. M., Hollingsworth, R. I., and Zeikus, J. G. 1997. Composition and role of extracellular polymers in methanogenic granules. *Appl Environ Microbiol* 63: 403–407.

Wang, Z. W., Liu, Y., and Tay, J. H. 2005. Distribution of EPS and cell surface hydrophobicity in aerobic granules. *Appl Microbiol Biotechnol* 69, 469–473.

Wang, Z. P., Liu, L., Yao, J., and Cai, W. 2006. Effects of extracellular polymeric substances on aerobic granulation in sequencing batch reactors. *Chemosphere* 63: 1728–1735.

Yang, S. F., Tay, J. H., and Liu, Y. 2004. Inhibition of free ammonia to the formation of aerobic granules. *Biochem Eng J* 17: 41–48.

Yi, S., Tay, J. H., Maszenan, A. M., and Tay, S. T. L. 2003. A culture-independent approach for studying microbial diversity in aerobic granules. *Water Sci Technol* 47: 283–290.

Yu, H. Q., Tay, J. H., and Fang, H. H. P. 2001. The roles of calcium in sludge granulation during UASB reactor start-up. *Water Res* 35: 1052–1060.

Zhang, X. Q. and Bishop, P. L. 2003. Biodegradability of biofilm extracellular polymeric substances. *Chemosphere* 50: 63–69.

Zhang, X. Q., Bishop, P. L., and Kinkle, B. K. 1999. Comparison of extraction methods for quantifying extracellular polymers in biofilms. *Water Sci Technol* 39: 211–218.

Zheng, Y. M., Yu, H. Q., and Sheng, G. P. 2005. Physical and chemical characteristics of granular activated sludge from a sequencing batch airlift reactor. *Process Biochem* 40: 645–650.

11 Internal Structure of Aerobic Granules

Zhi-Wu Wang and Yu Liu

CONTENTS

11.1 INTRODUCTION

The unique features of aerobic granules, as different from biofloc, are their dense and spherical three-dimensional structure. A good perception into the conformation of this granular structure, in comparison with that of bioflocs, will certainly help deepen current understanding of the mechanism of aerobic granulation, as well as its structural stability. As presented in chapter 10, an aerobic granule is mainly build up by microbial cells embedded in their excreted extracellular polysaccharides (PS), that is, PS play a cementing role in connecting individual cells into the three-dimensional structure of an aerobic granule. Moreover, the PS characteristics also influence the surface property of microbial cells (see chapter 9). It seems certain that the structure of an aerobic granule is essentially determined by the distributions and properties of its construction blocks, namely the microbial cells and PS. Thus, this chapter offers up-to-date information about the internal structure of aerobic granules in terms of the distributions of the microbial cells, PS, and cell surface hydrophobicity.

11.2 INTERNAL STRUCTURE OF AEROBIC GRANULES

11.2.1 HETEROGENEOUS STRUCTURE OF AEROBIC GRANULES

An aerobic granule cultivated in an acetate-fed sequencing batch reactor (SBR) was sliced and its internal structure was visualized by imagine analysis technique (Wang,

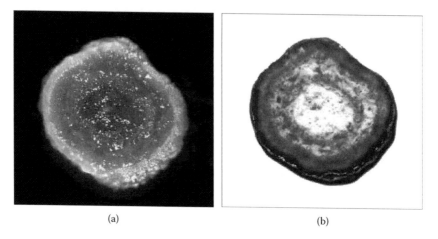

(a)   (b)

FIGURE 11.1 Cross-section view (400 μm thickness) of the aerobic granule in bright field (a) and dark field (b) visualization modes. Scale bar, 500 μm. (From Wang, Z.-W., Liu, Y., and Tay, J.-H. 2005. *Appl Microbiol Biotechnol* 60: 687–695. With permission.)

Liu, and Tay 2005). It was found, as shown in figure 11.1, that the internal structure of an aerobic granule consisted basically of an opaque outer layer and a relatively transparent inner core. The opaque outer layer had a depth of about 800 μm from the granule surface downwards, and the granule center looked transparent.

11.2.2 Porosity of Aerobic Granules

Porosity of biofilm or anaerobic granules can facilitate nutrient transfer (Alphenaar et al. 1992; Zhang and Bishop 1994). J. H. Tay et al. (2003) used 0.1-μm fluorescence beads to study the porosity of aerobic granules, and found that the porosity existed throughout the aerobic granule structure, but it peaked at 150 and 200 μm beneath the surface of aerobic granules with sizes of 0.55 and 1.0 mm, respectively. Nevertheless, the total porous zones decreased with increasing granule diameter, on a unit volume basis (J. H. Tay et al. 2003).

The zigzag pore channel was found to wind through the granule matrix made up by PS, that is, the porosity should be correlated to the richness of PS (Zheng and Yu 2007). A study by size-exclusion chromatography method revealed that the PS content increased, but the porosity decreased with the granule diameter, for example the pore size of an aerobic granule with a size of 0.2 to 0.6 mm was nearly seven times bigger than that of aerobic granules with a larger size of 0.9 to 1.5 mm (figure 11.2). The possible clogging caused by over-produced PS was thus considered to be responsible for the reduced porosity in large-sized aerobic granule. In addition, Chiu et al. (2006) also reported that a large granule would have a high porosity, evidenced by an enhanced oxygen diffusivity, with an increase of granule size, for example, diffusion coefficients of oxygen were measured as 1.24×10^{-9} to 2.28×10^{-9} m^2 s^{-1} as the size of acetate-fed aerobic granules increased from 1.28 to 2.50 mm, and a similar phenomenon was also observed in phenol-fed aerobic granules. Based on these

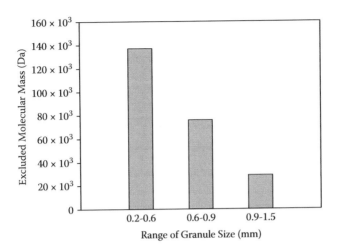

FIGURE 11.2 The penetrable molecular mass for different sized aerobic granules. (Data from Zheng, Y.-M. and Yu, H.-Q. 2007. *Water Res* 41: 39–46.)

controversial findings, it is difficult to conclude that granule porosity is dependent on its particle size.

11.2.3 SIZE-DEPENDENT INTERNAL STRUCTURE OF AEROBIC GRANULES

To investigate the internal structure of aerobic granules with various sizes, mature aerobic granules with a mean diameter of 0.8 to 3.0 mm were sliced and further visualized by image analyzer (Wang, Liu, and Tay 2005). The image analysis revealed that the small aerobic granule with a diameter of 0.8 mm had a nearly homogenous structure, whereas larger aerobic granules with a diameter of 3.0 mm exhibited a layered internal structure in which a clear shell and core could be distinguished (figure 11.3). Furthermore, the granule structure seems to evolve with the growth of the aerobic granule in size, that is, a transition from a homogenous to heterogeneous structure was observed with increase in the granule size (figure 11.3). As can be seen in figure 11.3, this is also evidenced by the gradually brightened transparent space from the granule shell to its center with increased granule size.

As discussed in chapter 8, the occurrence of diffusion limitation is associated with the size of the aerobic granule. The model simulation shows that dissolved oxygen (DO) would become a limiting factor for microbial growth at bulk COD concentration greater than 465 mg L[-1], and the soluble COD can penetrate throughout the aerobic granule with a diameter smaller than 0.8 mm, which exhibited a homogeneous structure (figure 11.1a). On the other hand, severe DO diffusion limitation would be encountered in large-sized aerobic granules of 1.0 to 1.5 mm (chapter 8). These results seem to indicate that the observed layered structure of large-sized aerobic granules would result from diffusion limitation because only those microorganisms living in the shell of the aerobic granule are accessible to DO and substrate.

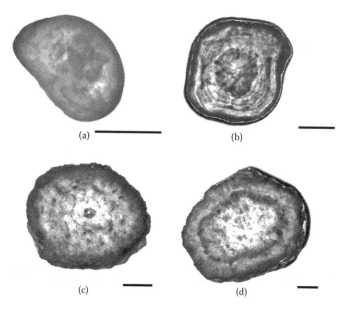

(a)   (b)

(c)   (d)

FIGURE 11.3 Internal structure of sliced aerobic granules with diameters of 0.8 mm (a), 1.3 mm (b), 2.0 mm (c), and 3.0 mm (d); scale bars: 0.5 mm.

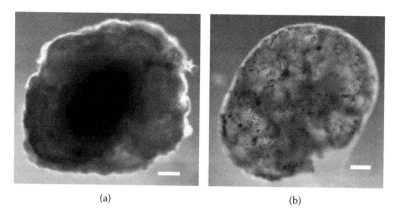

(a)   (b)

FIGURE 11.4 A view of an aerobic granule before (a) and after (b) long-term starvation; scale bar: 300 μm. (From Wang, Z.-W., Liu, Y., and Tay, J.-H. 2005. *Appl Microbiol Biotechnol* 60: 687–695. With permission.)

11.2.4 STRUCTURE CHANGE OF AEROBIC GRANULES DURING STARVATION

Fresh aerobic granules were starved under aerobic condition without addition of carbon and nutrient sources for 20 days. Changes in granule structure before and after the 20-day starvation are shown in figure 11.4. Compared to the fresh aerobic granule (figure 11.4a), the starved granule became more transparent. A transmittance analysis across the intact granule indicates that the opaque core of the fresh aerobic granule had become highly light permeable (figure 11.5), and the sliced, starved granule clearly

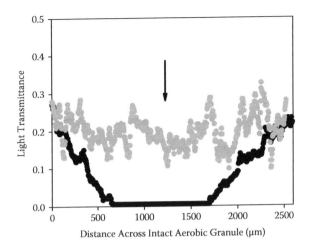

FIGURE 11.5 Light transmittal profiles across intact aerobic granule before (black) and after (gray) long-term starvation (arrow indicates granule center). (From Wang, Z.-W., Liu, Y., and Tay, J.-H. 2005. *Appl Microbiol Biotechnol* 60: 687–695. With permission.)

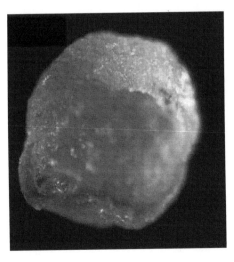

FIGURE 11.6 The hollow structure of the starved aerobic granule. (From Wang, Z.-W., Liu, Y., and Tay, J.-H. 2005. *Appl Microbiol Biotechnol* 60: 687–695. With permission.)

showed a hollow structure even though its outer shell still remained intact (figure 11.6). These observations seem to suggest that the biomass present in the granule shell would not be taken up by bacteria over starvation, while the biomass located in the core of the aerobic granule can be biodegraded under the starvation condition.

11.3 BIOMASS DISTRIBUTION IN AEROBIC GRANULES

The heterogeneous structure of aerobic granules indicates an uneven distribution of biomass. Chen, Lee, and Tay (2007) used fluorescent dyes to visualize the microbial

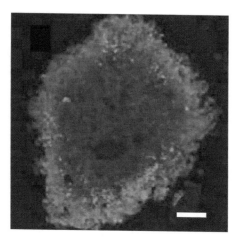

FIGURE 11.7   Cell distribution in aerobic granules, Syto 63 (red)-stained nucleic acid. (From Chen, M. Y., Lee, D. J., and Tay, J. H. 2007. *Appl Microbiol Biotechnol* 73: 1463–1469. With permission.)

cells distribution by means of confocal laser scanning microscopy (CLSM), and found that live cells were concentrated in the granule shell, indicated by a red fluorescence emitted from Syto 63 that stained nucleic acid (figure 11.7). In contrast, the fluorescence from the granule core is rather weak, indicating a limiting number of live bacteria in the core part of the aerobic granule. Similar observation was also reported by Toh et al. (2003) and McSwain et al.(2005). In the study by McSwain et al.(2005), the Syto 63 fluorescence peaked at a depth of 100 μm beneath the granule surface and the granule core part was almost fluorescence free. Detailed distribution of live and dead cells inside the aerobic granule was investigated by J. H. Tay et al. It was demonstrated that most live bacteria, including nitrifiers, only existed in the granule outer shell layer where they were within the reach of mass diffusion, while dead cells and anaerobes were mainly detected at the core of the aerobic granule, indicating an uneven microbial distribution in aerobic granules that should result from diffusion limitation (figure 11.7).

Optical density (OD) has been commonly used to quantify the biomass concentration, that is, a high OD is correlated to a high biomass concentration or density in suspended and biofilm cultures (Gaudy and Gaudy 1980). Figure 11.8 exhibits the OD profile measured across the cross section of an aerobic granule. It was found that the OD in the granule center was close to zero, indicating a very low biomass density or a loose microbial structure at the core. In contrast, the peak OD was observed in the outer layer of the aerobic granule, which would result from a high biomass density or a compact microbial structure (Wang, Liu, and Tay 2005). To confirm these observations, five aerobic granules, namely No. 1 to 5, were sliced and the respective mass density of the outer shell layer and the inner core part was measured. It was found that the mass density of the outer layer of the granule was indeed much higher than that of the core part (figure 11.9). In fact, J. H. Tay et al. (2002) reported a similar biomass distribution in aerobic granules. As discussed earlier, mass diffusion limitation would be responsible for the observed dense surface layer and loose

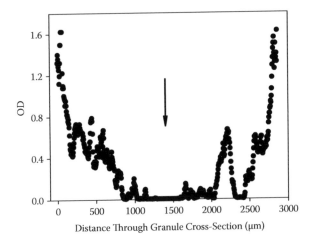

FIGURE 11.8 The OD profile through the granule cross section (arrow indicates granule center). (Data from Wang, Z.-W., Liu, Y., and Tay, J.-H. 2005. *Appl Microbiol Biotechnol* 60: 687–695.)

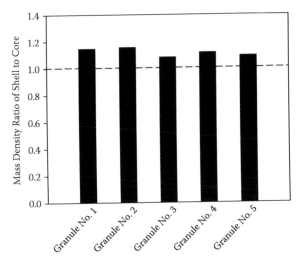

FIGURE 11.9 Biomass density ratios of shell layer to core part of aerobic granules. (Data from Wang, Z.-W., Liu, Y., and Tay, J.-H. 2005. *Appl Microbiol Biotechnol* 60: 687–695.)

inner core of aerobic granules, that is, the unbalanced biomass distribution is due to the diffusion limitation inside the aerobic granule.

11.4 PS DISTRIBUTION IN AEROBIC GRANULES

Calcofluor white is a commonly used fluorescent dye for labeling beta-linked poly-saccharides (PS) (deBeer et al. 1996). The beta-linked polysaccharides are believed to serve as the backbone of the biofilm structure (Sutherland 2001). To localize beta-linked polysaccharides in an aerobic granule, the aerobic granule was sliced and its

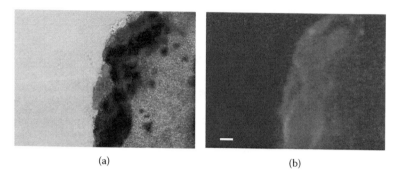

(a)   (b)

FIGURE 11.10 Microscopic view of sectioned aerobic granule cross section before (a) and after (b) calcofluor white staining; scale: 100 µm. (From Wang, Z.-W., Liu, Y., and Tay, J.-H. 2005. *Appl Microbiol Biotechnol* 60: 687–695. With permission.)

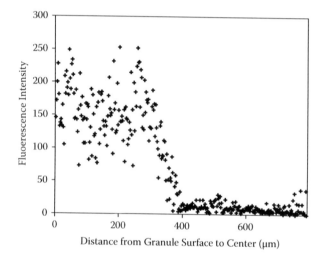

FIGURE 11.11 Profile of the fluorescence intensity from the surface to the center of an aerobic granule. (Data from Wang, Z.-W., Liu, Y., and Tay, J.-H. 2005. *Appl Microbiol Biotechnol* 60: 687–695.)

cross section was stained with calcofluor (Wang, Liu, and Tay 2005). In a fresh granule, the fluorescent dye was attached mainly to the outer shell of the granule, while very weak fluorescence was detected at the center of the aerobic granule (figure 11.10b). The fluorescence intensity profile measured along the direction of the granule radius further showed that most calcofluor white-stained PS was situated in the outer shell of the granule, with a depth of 400 µm below the granule surface (figure 11.11). These findings imply that the beta-linked PS are located mainly in the outer shell of the granule. In fact, a similar distribution of beta-linked PS was also observed in anaerobic granules; the majority of the calcofluor white-stained PS was found in the top 40 µm from the surface of the anaerobic granule (deBeer et al. 1996).

Chen, Lee, and Tay (2007) used three different fluorescence dyes, namely ConA and calcofluor white and FITC, to label the alpha-, beta-linked PS and also protein

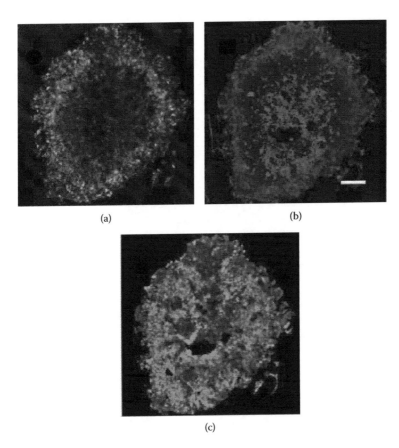

(a)

(b)

(c)

FIGURE 11.12 Fluorescence viewed on granule cross section by staining with ConA for alpha-linked PS (a); calcofluor white for beta-linked PS (b); and FITC for protein (c). Scale bar: 200 µm. (From Chen, M. Y., Lee, D. J., and Tay, J. H. 2007. *Appl Microbiol Biotechnol* 73: 1463–1469. With permission.)

(PN). The observations by CLSM revealed that the alpha-linked PS was distributed mainly on the granule shell, while the granule core was almost alpha-linked PS free (figure 11.12a). A similar distribution of alpha-linked PS was also reported by McSwain et al. (2005). However, the study by Chen, Lee, and Tay (2007) showed a different distribution of the beta-linked PS from what was found in figure 11.10, that is, the beta-linked PS not only appeared on the granule shell, but also was concentrated in the granule core. Moreover, a fluorescent empty layer was found in between the granule shell and core (figure 11.12b). As for PN, a random distribution pattern was found along the granule radium direction (figure 11.12c).

To quantify the PS distribution in the layered aerobic granule, Wang, Liu, and Tay (2005) measured the PS contents in the granule shell as well as in the granule core (figure 11.13). Most PS in the aerobic granule was centralized at the granule core, for example, the PS present in the granule shell only accounts for one-fifth of the PS found in the granule core. Such a finding implies that those gel-like substances observed in the granule center (figure 11.1) could be attributed to PS. As discussed

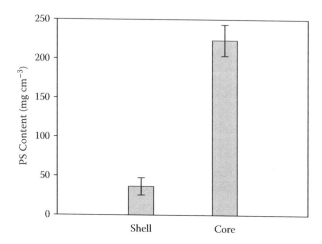

FIGURE 11.13 Distribution of PS in granule shell and core. (Data from Wang, Z.-W., Liu, Y., and Tay, J.-H. 2005. *Appl Microbiol Biotechnol* 60: 687–695.)

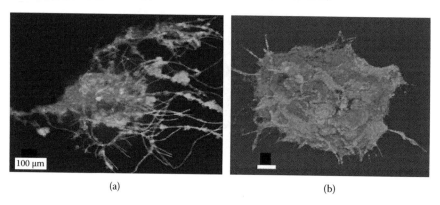

(a)   (b)

FIGURE 11.14 Fluorescence by Syto 63 for cells (bright), ConA for polysaccharides (gray), and FITC for protein (white) in biofloc (a) and aerobic granule (b). (From McSwain, B. S. et al. 2005. *Appl Environ Microbiol* 71: 1051–1057. With permission.)

earlier, the PS present in the granule core is basically biodegradable. In view of the total amount of PS determined in the aerobic granule, it appears that the alpha- or beta-linked PS may not be the dominate constitution of EPS in aerobic granules.

EPS is the extracellular products synthesized by microbial cells. As shown earlier, the cell distribution would be granule size-dependent due to diffusion limitation. Hence, the distribution of PS would also be related to the size of the aerobic granule. McSwain et al. (2005) investigated the PS distribution in small and large aerobic granules with a respective size of 350 μm and 800 μm. The observation by CLSM revealed that in the small bioparticle, both PS and PN were concentrated at the core (figure 11.14a). For the large-sized bioparticle, PS and microbial cells turned out to be only centralized on the granule outer shell, with a random distribution of PN (figure 11.14b). Similar EPS distributions have been reported in anaerobic biofloc and granule, that is, calcofluor-stained PS was mainly distributed on the outer shell of anaerobic granules, but appeared to be concentrated in the center of biofloc (deBeer et al. 1996). It has been recognized that the partial anaerobic condition, like those beneath the granule shell, is able to trigger EPS overproduction (Gamar-Nourani, Blondeau, and Simonet 1998). This provides a plausible explanation for the abundant EPS observed in the granule core (figure 11.13).

11.5 DISTRIBUTION OF CELL SURFACE HYDROPHOBICITY IN AEROBIC GRANULES

Cell surface hydrophobicity has been regarded as a trigger of aerobic granulation (chapter 9). Basically, cell hydrophobicity helps reduce the surface energy of individual cells so as to overcome the dispersive polar force from water, and further promote cell-to-cell co-aggregation. After the formation of the aerobic granule, the primary trigger function of cell surface hydrophobicity may become secondarily important, but cell surface hydrophobicity may continue to play a part in the stability of the aerobic granule structure. Unfortunately, little information is presently available about the distribution of cell surface hydrophobicity in mature aerobic granules.

To give some insights into the cell hydrophobicity distribution in aerobic granules, Wang, Liu, and Tay (2005) separated the granule outer shell from its inner core, and their respective cell surface hydrophobicity was then determined by the method of microbial attachment to solvent (MATS). As shown in figure 11.15, the cell surface hydrophobicity of microorganisms on the granule outer shell was almost twofold higher than that at the granule core. This provides indirect evidence that the gel-like substances found in the granule core (figure 11.1) is relatively hydrophilic as compared to the granule shell.

Solid evidence shows that cell surface hydrophobicity is correlated to the type and properties of EPS (chapter 10). In the three-dimensional structure of aerobic

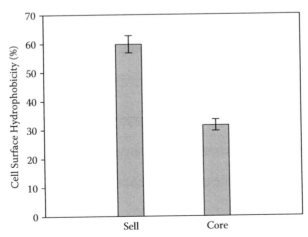

FIGURE 11.15 Cell hydrophobicity distributions in the shell and core parts of an aerobic granule. (Data from Wang, Z.-W., Liu, Y., and Tay, J.-H. 2005. *Appl Microbiol Biotechnol* 60: 687–695.)

granules, only cells in the granule shell have direct contact with the bulk liquid, thus it is reasonable to consider that a hydrophobic granule shell is necessary to keep the integrity of the aerobic granule and further to prevent the granule structure from dissolution into bulk liquid. This would partially illustrate why the granule shell-associated microorganisms favor production of relatively more hydrophobic EPS.

The alpha- and beta-linked EPS have been found on the granule shell layer (figures 11.10, 11.12, and 11.14). In fact, increased production of alpha-linked EPS can improve cell surface hydrophobicity (Lawman and Bleiweis 1991), and insoluble beta-linked EPS was also reported to serve as the backbone of biofilm structure (Sutherland 2001). For the granule core-associated EPS, the hydrophobic property may not be necessary. As shown in figures 11.4 to 11.6, after a 20-day aerobic starvation, most of the granule-core-associated EPS disappeared, that is, those less hydrophobic EPS are highly biodegradable as compared to the shell EPS. In general, only soluble EPS is biodegradable, while insoluble or bound EPS should be nonbiodegradable (Laspidou and Rittmann 2002). In fact, the hydrophilic EPS in the aerobic granule was found to be biodegradable in the course of starvation, but the hydrophobic EPS remained unchanged (Z. H. Li, Kuba, and Kusuda 2006). Furthermore, the tightly bound EPS and loosely bound EPS have been differentiated and the latter was found to decrease with the length of solids retention time (SRT), indicating they could be readily biodegradable (X. Y. Li and Yang 2006).

It is clear that the core readily biodegradable EPS would not play an essential role in maintaining the structural stability of the aerobic granule. Instead, the shell-associated nonbiodegradable and hydrophobic EPS is the key towards the long-term stability of the aerobic granule. In this regard, hydrolysis or disappearance of the core-associated EPS would inevitably result in the disintegration of the aerobic granule. This point is indirectly confirmed by a study of biofilm in which the destruction of the biofilm structure was observed after a key biofilm EPS was hydrolyzed (Skillman, Sutherland, and Jones1999).

11.6 DIFFUSION-RELATED STRUCTURE OF AEROBIC GRANULES

The size-associated structure of aerobic granules was discussed earlier, that is, a small aerobic granule has a homogeneous structure, whereas a heterogeneous structure is found in big aerobic granules (figure 11.3). It appears that mass diffusion is a decisive factor influencing the structure shift of aerobic granules.

In the cycle operation of an aerobic granular sludge SBR, almost all influent COD can be removed in the first hour of aeration, and aerobic granules are thus subjected to substrate starvation in the rest of the cycle time (see chapter 1). As a result, a periodic shift from substrate feast to famine exists in aerobic granular sludge SBRs. Meanwhile, mass diffusion limitation was encountered in aerobic granules with a size bigger than 1 mm (chapter 9). For this reason, the microorganisms beneath the granule shell layer will not only experience DO limitation in the feast phase, but also suffer from substrate deficiency in the subsequent famine phase. This implies that biomass present in the granule core part will have no chance to grow throughout the whole cycle time, thus only facultative or anaerobic microorganisms in the granule core might survive, meanwhile biomass at the granule core might undergo endogenous decay, and their debris in turn would be part of the constituents of the granule EPS. Existence of anaerobes has been found in a layer 800 μm beneath the surface of aerobic granules (S. T. L. Tay et al. 2002). Because of the low growth rate of the granule core-associated microorganisms, they would not be able to out compete the fast-growing microorganisms located at the granule shell. Such an unbalanced growth between the two parts of the aerobic granule naturally results in an uneven biomass distribution, as observed in figures 11.7 to 11.9.

It has been recognized that the partial anaerobic condition can trigger the production of EPS (Gamar-Nourani et al. 1998). As pointed out earlier, aerobic granules with a size larger than 1 mm often have an anaerobic or partially anaerobic core. Under such a circumstance, the overproduction of EPS at the granule core (figure 11.13) would be reasonably explained. On the other hand, the weak fluorescence intensity at the granule core (11.7) and the distribution of live and dead cells in aerobic granules also point to the fact that a substantial portion of microbial cells in the granule core may die of starvation (J. H. Tay et al. 2002). Consequently, it is believed that the observed structure of the aerobic granule is largely related to mass diffusion behaviors of aerobic granular sludge SBRs.

11.7 CONCLUSIONS

The structure of an aerobic granule is related to its size, that is, a small granule has a relatively homogeneous structure, whereas a heterogeneous structure was observed in large aerobic granules. Uneven distributions of granule PS and cell surface hydrophobicity were found. Compared to the granule core, the granule shell had a higher biomass density and was more hydrophobic. It appears that the structure of the aerobic granule is determined by mass diffusive behaviors of the aerobic granular sludge SBR.

REFERENCES

Alphenaar, P. A., Perez, M. C., Vanberkel, W. J. H., and Lettinga, G. 1992. Determination of the permeability and porosity of anaerobic sludge granules by size exclusion chromatography. *Appl Microbiol Biotechnol* 36: 795–799.

Chen, M. Y., Lee, D. J., and Tay, J. H. 2007. Distribution of extracellular polymeric substances in aerobic granules. *Appl Microbiol Biotechnol* 73: 1463–1469.

Chiu, Z. C., Chen, M. Y., Lee, D. J., Tay, S. T. L., Tay, J. H., and Show, K. Y. 2006. Diffusivity of oxygen in aerobic granules. *Biotechnol Bioeng* 94: 505–513.

deBeer, D., OFlaharty, V., Thaveesri, J., Lens, P., and Verstraete, W. 1996. Distribution of extracellular polysaccharides and flotation of anaerobic sludge. *Appl Microbiol Biotechnol* 46: 197–201.

Gamar-Nourani, L., Blondeau, K., and Simonet, J. W. 1998. Influence of culture conditions on exopolysaccharide production by Lactobacillus rhamnosus strain C83. *J Appl Microbiol* 85: 664–672.

Gaudy, A. F. and Gaudy, E. T. 1980. *Microbiology for environmental scientists and engineers.* New York: McGraw-Hill.

Laspidou, C. S. and Rittmann, B. E. 2002. A unified theory for extracellular polymeric substances, soluble microbial products, and active and inert biomass. *Water Res* 36: 2711–2720.

Lawman, P. and Bleiweis, A. S. 1991. Molecular cloning of the extracellular endodextranase of Streptococcus salivarius. *J Bacteriol* 173: 7423–7428.

Li, X. Y. and Yang, S. F. 2006. Influence of loosely bound extracellular polymeric substances (EPS) on the flocculation, sedimentation and dewaterability of activated sludge. *Water Res* 41: 1022–1030.

Li, Z. H., Kuba, T., and Kusuda, T. 2006. The influence of starvation phase on the properties and the development of aerobic granules. *Enzyme Microb Technol* 38: 670–674.

McSwain, B. S., Irvine, R. L., Hausner, M., and Wilderer, P. A. 2005. Composition and distribution of extracellular polymeric substances in aerobic flocs and granular sludge. *Appl Environ Microbiol* 71: 1051–1057.

Skillman, L. C., Sutherland, I. W., and Jones, M. V. 1999. The role of exopolysaccharides in dual species biofilm development. *J Appl Microbiol* 85: 13–18.

Sutherland, I. W. (2001) Biofilm exopolysaccharides: A strong and sticky framework. *Microbiology* 147: 3–9.

Tay, J. H., Ivanov, V., Pan, S., and Tay, S. T. L. 2002. Specific layers in aerobically grown microbial granules. *Lett Appl Microbiol* 34: 254–257.

Tay, J. H., Tay, S. T. L., Ivanov, V., Pan, S., Jiang, H. L., and Liu, Q. S. 2003. Biomass and porosity profiles in microbial granules used for aerobic wastewater treatment. *Lett Appl Microbiol* 36: 297–301.

Tay, S. T. L., Ivanov, V., Yi, S., Zhuang, W. Q., and Tay, J. H. 2002. Presence of anaerobic bacteroides in aerobically grown microbial granules. *Microb Ecol* 44: 278–285.

Toh, S. K., Tay, J. H., Moy, B. Y. P., Ivanov, V., and Tay, S. T. L. 2003. Size-effect on the physical characteristics of the aerobic granule in a SBR. *Appl Microbiol Biotechnol* 60: 687–695.

Wang, Z.-W., Liu, Y., and Tay, J.-H. 2005. Distribution of EPS and cell surface hydrophobicity in aerobic granules. *Appl Microbiol Biotechnol* 69: 469–473.

Zhang, T. C. and Bishop, P. L. 1994. Density, porosity, and pore structure of biofilms. *Water Res* 28: 2267–2277.

Zheng, Y.-M. and Yu, H.-Q. 2007. Determination of the pore size distribution and porosity of aerobic granules using size-exclusion chromatography. *Water Res* 41: 39–46.

12 Biodegradability of Extracellular Polymeric Substances Produced by Aerobic Granules

Zhi-Wu Wang and Yu Liu

CONTENTS

12.1 INTRODUCTION

The essential role of extracellular polymeric substances (EPS) in the formation of biofilm, anaerobic and aerobic granules has been well documented so far (see chapter 9). As EPS has been believed to play an essential role in building and maintaining the spatial structure of immobilized microbial communities, it should not be biodegraded by their own producer, that is, EPS-producing organisms are unable to utilize their own EPS as carbon source (Obayashi and Gaudy 1973; Dudman 1977; Pirog et al. 1977; Sutherland 1999). On the contrary, increasing evidence shows that EPS could be readily biodegradable for their own producers (Patel and Gerson 1974; Boyd and Chakrabarty 1994; Nielsen, Frolund, and Keiding 1996; Ruijssenaars, Stingele, and Hartmans 2000; Zhang and Bishop 2003; Decho, Visscher, and Reid 2005). The EPS distribution in aerobic granules has been presented in chapter 11, showing that a substantial portion of the EPS accumulated at the center of aerobic granules can be utilized over a 20-day starvation period, and the internal structure of aerobic granules becomes hollow compared to the structure of fresh aerobic granules.

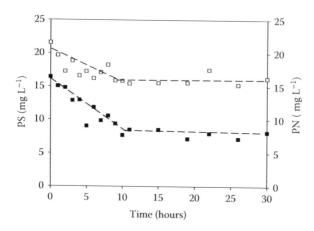

FIGURE 12.1 Biodegradability of PS (■) and PN (□) extracted from fresh aerobic granules. (From Wang, Z.-W., Liu, Y., and Tay, J.-H. 2007. *Appl Microbiol Biotechnol* 74:462–466. With permission.)

In complement to chapter 11, this chapter specifically reviews the biodegradability of EPS produced by aerobic granules as well as its contribution to the stability of aerobic granules.

12.2 BIODEGRADABILITY OF EPS EXTRACTED FROM AEROBIC GRANULES

To investigate the biodegradability of EPS produced by aerobic granules, Wang, Liu, and Tay (2007) extracted EPS from fresh and starved aerobic granules, and the extracted EPS, as the sole carbon source, was then fed to the batch culture of prestarved aerobic granules. Such an experimental design can minimize the interference of EPS stored in fresh aerobic granules.

12.2.1 BIODEGRADABILITY OF EPS EXTRACTED FROM FRESH AEROBIC GRANULES

Figure 12.1 shows the biodegradation profiles of the extracted EPS in terms of polysaccharides (PS) and proteins (PN) observed in the course of the batch culture of prestarved aerobic granules. A rapid biodegradation of both PS and PN was observed in the first 10 hours of the culture until a stationary phase was achieved. It can be seen that nearly 50% of PS was utilized by its own producers as the external carbon source, while only 30% of PN was consumed together with PS. According to figure 12.1, the average linear biodegradation rates of PS and PN were estimated as 15.2 mg PS g^{-1} suspended solids (SS) d^{-1} and 14.8 mg PN g^{-1} SS d^{-1}, respectively, indicating that the biodegradation rates of both PS and PN are highly comparable.

Nielsen, Frolund, and Keiding (1996) investigated the biodegradability of EPS produced from activated sludge during anaerobic storage process. It was found that the sludge EPS content quickly declined as a result of biodegradation, and the biodegradable fraction of EPS accounted for about 40% of the total EPS (figure 12.2). This figure appears to be very close to the fraction of biodegradable EPS produced

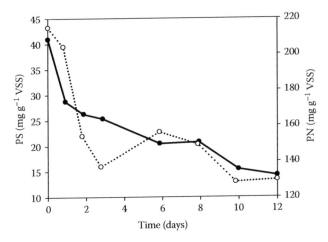

FIGURE 12.2   Change of activated sludge extracellular PS (●) and PN (○) during anaerobic storage. (Data from Nielsen, P. H., Frolund, B., and Keiding, K. 1996. *Appl Microbiol Biotechnol* 44:823–830.)

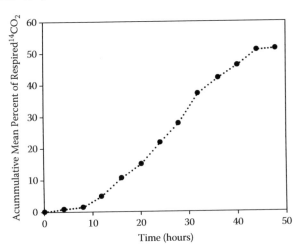

FIGURE 12.3   Biodegradation of ^{14}C-EPS into $^{14}CO_2$ by marine bacteria. (Data from Decho, A. W., Visscher, P. T., and Reid, R. P. 2005. *Palaeogeogr Palaeoclimatol Palaeoecol* 219: 71–86.)

by aerobic granules (figure 12.1). In fact, EPS biodegradation under anaerobic conditions has also been reported previously (Ryssov Nielsen 1975).

    EPS produced by natural bacteria was also found to contain a large biodegradable fraction. Decho, Visscher, and Reid (2005) extracted EPS from marine bacteria that live in a stromatolite, and labeled it with isotope carbon as ^{14}C-EPS. Extraction was fed back to its producer, and the carbon flux was monitored. A clear EPS biodegradation, indicated by the conversion of ^{14}C-EPS to $^{14}CO_2$, was observed in the course of 50 minutes of cultivation (figure 12.3). It can be seen that about 50% of ^{14}C-EPS was finally converted to $^{14}CO_2$. These results are in good agreement with

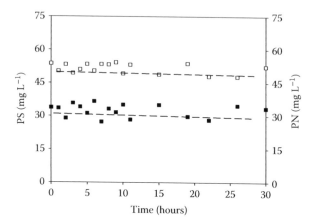

FIGURE 12.4 Biodegradability of PS (■) and PN (□) extracted from starved aerobic granules. (From Wang, Z.-W., Liu, Y., and Tay, J.-H. 2007. *Appl Microbiol Biotechnol* 74:462–466. With permission.)

those found in aerobic granules and activated sludge cultures (figures 12.1 and 12.2). Moreover, it is apparent that the production of biodegradable EPS should be a common phenomenon broadly existing in a wide spectrum of microorganisms, and also the fraction of biodegradable EPS in the total EPS produced is generally around 50% (figure 12.3).

12.2.2 Biodegradability of EPS Extracted from Starved Aerobic Granules

As shown in figure 12.1, nearly 50% of the EPS extracted from fresh aerobic granules were nonbiodegradable in the batch culture of prestarved aerobic granules as users. To confirm such an observation, the EPS extracted from the starved aerobic granules were fed, as the sole carbon source, to the batch culture with the prestarved aerobic granules. In this case, figure 12.4 shows no significant biodegradation of the EPS extracted from the starved granule, indicated by a very low average biodegradation rate of 0.38 mg g^{-1} SS d^{-1} for PS and 0.75 mg g^{-1} SS d^{-1} for PN. These results further confirm that EPS secreted by aerobic granules is basically made up of two major components, that is, biodegradable and nonbiodegradable EPS according to their biodegradability.

12.2.3 Comparison of Biodegradability of Acetate and Extracted EPS

Wang, Liu, and Tay (2007) compared biodegradability of acetate and EPS extracted from the fresh and starved aerobic granules using batch cultures. The initial concentrations of acetate and extracted EPS were kept at 100 mg COD L^{-1}. Figure 12.5 shows that the average biodegradation rates of acetate and EPS extracted from fresh and starved aerobic granules were 5.4 mg COD mg^{-1} SS d^{-1}, 1.1 mg COD mg^{-1} SS h^{-1}, and 0.018 mg COD mg^{-1} SS h^{-1}, respectively. These results point to the fact that bacteria would preferably utilize an external carbon source, such as acetate in this case, whenever it is available. However, after depletion of the external carbon source,

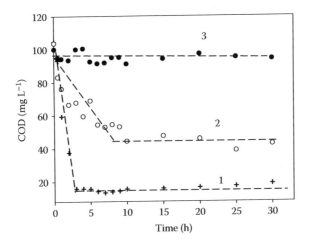

FIGURE 12.5 Biodegradability of acetate and EPS extracted from aerobic granules. 1: Acetate (+); 2: EPS extracted from the fresh granules (O); 3: EPS extracted from the starved granules (●). (From Wang, Z.-W., Liu, Y., and Tay, J.-H. 2007. *Appl Microbiol Biotechnol* 74:462–466. With permission.)

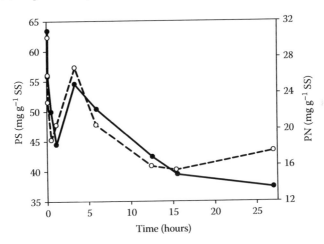

FIGURE 12.6 Biodegradation of PS (●) and PN (O) by biofilm. (Data from Zhang, X. Q. and Bishop, P. L. 2003. *Chemosphere* 50: 63–69.)

the biodegradation of EPS may further provide the minimum energy required for microbial maintenance functions during cell starvation.

Biodegradable EPS has been discovered in many forms of bacteria exploited for environmental engineering. Zhang and Bishop (2003) extracted EPS from biofilm and fed it to their own producer, and found that both PS and PN components in EPS continuously decreased over the culture time (figure 12.6). Once again it can be seen in figure 12.6 that around 50% of the total EPS produced by biofilms is biodegradable, and the respective PS and PN utilization rate of 0.4 mg PS mg^{-1} SS d^{-1} and 0.3 mg PN mg^{-1} SS d^{-1} were obtained. It appears from figure 12.1 (aerobic granules), figure 12.2 (activated sludge), and figure 12.3 (natural bacteria) that the EPS produced by aerobic granules, activated sludge, and biofilms are similar or comparable in the sense of their biodegradability. Evidence thus far from aerobic granules, suspended activated sludge, and biofilms all point to the fact that bacteria are able to vigorously take up EPS as an external food source in the case where an external carbon source is no longer available.

12.3 BIODEGRADATION OF AEROBIC GRANULE-ASSOCIATED EPS DURING STARVATION

To investigate the biodegradation of aerobic granule-associated EPS, fresh aerobic granules taken from a sequencing batch reactor (SBR), without pretreatment, were subjected to aerobic starvation without addition of an external carbon source for 20 days (Wang, Liu, and Tay 2007). The fresh aerobic granules had an initial specific oxygen uptake rate (SOUR) of 62 mg O_2 g^{-1} volatile solids (VS) h^{-1}; a sludge volume index (SVI) of 55 mL g^{-1}, and a mean diameter of 1.6 mm. The content of EPS in aerobic granules was found to decrease with the aerobic starvation, for example about 75% of PS and 78% PN in aerobic granules were removed at the end of the 20-day starvation period (figure 12.7). This seems to indicate that granule microorganisms tended to maximize the use of EPS (75% of EPS degraded) when facing a long term of starvation as compared to what was observed in the short-term batch culture (only 50% of EPS utilized, as shown in figure 12.1).

Wang, Liu, and Tay (2007) used image analysis technique to visualize changes in granule structure before and after aerobic starvation. It was found that the structure of aerobic granules still remained intact even after the 20-day aerobic starvation, but became more transparent compared to the fresh ones (figure 12.8). In the sense of reaction kinetics, it is reasonable to consider that a period of 20 days of starvation would be long enough to reflect the overall response of microorganisms to

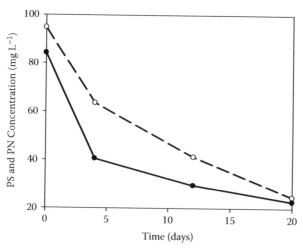

FIGURE 12.7 Change in PS (●) and PN (○) content in the course of aerobic starvation. (Data from Wang, Z.-W., Liu, Y., and Tay, J.-H. 2007. *Appl Microbiol Biotechnol* 74:462–466.)

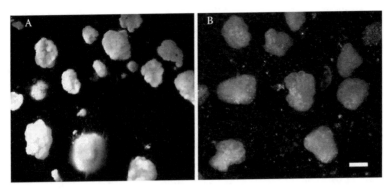

FIGURE 12.8 Morphology of aerobic granules before (a) and after (b) 20 days of starvation; scale bar: 1 mm. (From Wang, Z.-W., Liu, Y., and Tay, J.-H. 2007. *Appl Microbiol Biotechnol* 74:462–466. With permission.)

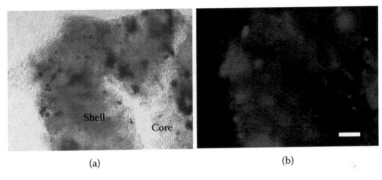

(a)   (b)

FIGURE 12.9 Visualization of beta-linked EPS in aerobic granules after 20 days of starvation by epifluorescence microscopy: (a) unstained granule; (b) stained granule. Bar: 200 µm. (From Wang, Z.-W., Liu, Y., and Tay, J.-H. 2007. *Appl Microbiol Biotechnol* 74:462–466. With permission.)

the imposed starvation. Thus, the portion of EPS left over after the 20-day aerobic starvation would represent the real fraction of nonbiodegradable EPS in the total EPS produced by aerobic granules. If so, about 75% of the granules-associated EPS can be regarded as biodegradable, and the remaining 25% would be not readily biodegradable. It should be realized that this small portion of nonbiodegradable EPS in aerobic granules would be responsible for the structural integrity of aerobic granules (chapter 11).

It becomes clearer now that the EPS produced by aerobic granules can be reasonably classified into two major groups: biodegradable and nonbiodegradable EPS. As discussed in chapter 11, the nonbiodegradable PS would mainly belong to the family of beta-linked PS. To further check the existence and distribution of the beta-linked PS left in starved aerobic granules, sectioned starved aerobic granules were stained with calcofluor white (Wang, Liu, and Tay 2007). Figure 12.9 shows that the most stained PS after the 20-day starvation period were situated at the outer shell of the aerobic granule, while a significant void space was observed in the core part of the starved aerobic granule.

The role of EPS in maintaining the spatial structures of biofilms, aerobic and anaerobic granules has been reported, but with no differentiation of biodegradable and nonbiodegradable EPS (Flemming and Wingender 2001; Liu, Liu, and Tay 2004). It appears from figure 12.9 and chapter 11 that most of the biodegradable PS was located in the core part of aerobic granules, and this part of PS could be utilized after depletion of external carbon source or during long-term starvation. Obviously, disappearance of the soluble PS accumulated at the core of aerobic granules would be responsible for the observed void structure in the starved granules (figure 12.9). Nevertheless, nonbiodegradable EPS situated at the shell of aerobic granules remained nearly unchanged before and after starvation. This reveals the functional role of nonbiodegradable EPS in constructing and maintaining the spatial structure, integrity, and stability of aerobic granules. Without doubt, the large fraction of biodegradable EPS in aerobic granules would serve as a spare energy pool during aerobic starvation.

12.4 EPS BIODEGRADATION IN AN AEROBIC GRANULAR SLUDGE SBR

A large fraction of biodegradable EPS produced by aerobic granules can be utilized by their own producer during starvation (figures 12.1 and 12.7). The cycle operation of aerobic granulation SBR consists of a short substrate feast phase and a relatively long substrate famine phase (figure 12.10). Thus, the EPS production and utilization processes appear to be closely dependent on the availability of external substrate in each SBR cycle. In this regard, Wang et al. (2006) investigated the dynamic change of EPS during an SBR cycle. It was found that the production of PS and PN was coupled to the removal of soluble COD, and this led to a sharp rise of PS and PN contents in the first 2 hours of the SBR cycle (figure 12.10). After the complete depletion of the external COD after 4 hours, the culture came into the starvation phase, and

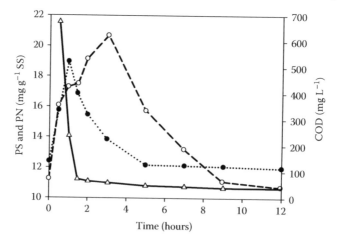

FIGURE 12.10 Profiles of PS (●), PN (○), and COD (△) in one cycle of an aerobic granulation SBR. (Data from Wang, Z. et al. 2006. *Chemosphere* 63: 1728–1735.)

subsequently a significant drop in the PS and PN contents occurred as expected. It is apparent that all biodegradable EPS produced in the feast phase was completely consumed during the subsequent famine phase. Similar to the results presented in figure 12.1, Wang et al. (2006) also showed that the biodegradable EPS accounted for 50% of the total EPS produced by aerobic granules (figure 12.10). These results provide direct experimental evidence that there is a dynamic change of aerobic granules-associated EPS during the cycle operation of an SBR. Such a change, without doubt, will influence the formation and stability of aerobic granules.

So far, it has been believed that EPS would not be an essential cell component under normal culture conditions. Organic carbon flux into EPS production rather than biomass synthesis is indeed energetically unfavorable in normal living conditions. It seems that a certain stressful condition would be responsible for the overproduction of a large amount of biodegradable EPS (e.g. 50%) by aerobic granules, as discussed in chapter 10. It can be seen in figure 12.10 that the famine phase was about 6-fold longer than the feast phase within the cycle operation of an aerobic granular sludge SBR. This in turn indicates that microbial activity of aerobic granules cannot be sustained over a relatively long starvation period without additional energy input. It is reasonable, at least logical, to consider that the EPS produced in the feast phase would be used in the famine phase in order for microorganisms to overcome the energy constraint.

EPS biodegradation in the course of aerobic granulation was also reported by Li, Kuba, and Kusuda (2006). Figure 12.11 shows that a sharp EPS decline in terms of PS and PN contents in sludge along with aerobic granulation, and the EPS biodegradation resulted in a lower cell surface charge. In view of the fact that the cell surface is covered by the EPS, the neutralized cell surface thus can be attributed to the reduction of negatively charged EPS (Li, Kuba, and Kusuda 2006). PN has been regarded as one of the EPS components that can contribute to surface charge (Sponza 2002; Jin, Wilen, and Lant 2003). It is evident that the reduction of PN due to EPS biodegradation is thus able to reduce or neutralize the cell surface charge and in turn facilitates

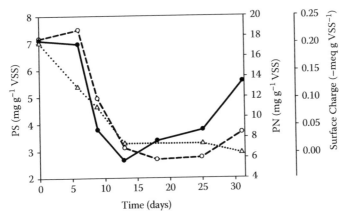

FIGURE 12.11 Biodegradation of PS (●), PN (○), and surface charge (△) in the course of aerobic granulation. (Data from Li, Z. H., Kuba, T., and Kusuda, T. 2006. *Enzyme Microb Technol* 38: 670–674.)

the cell-to-cell co-aggregation process. Similarly, Wang et al. (2006) also found that the cell surface charge dropped from 0.86 to 0.74 meq g^{-1} SS after 14 mg PN g^{-1} SS was biodegraded, and a close correlation between the EPS biodegradation and the reduction in cell surface charge was observed. Furthermore, the EPS biodegradation causes lowered cell surface hydrophobicity, implying that reduction in the biodegradable EPS influences aerobic granulation in SBRs (Wang et al. 2006), while Nielsen, Frolund, and Keiding (1996) also reported a deterioration of the dewaterability with the EPS uptake in the course of anaerobic storage of activated sludge. These results seem to indicate that the utilization of biodegradable EPS would probably make the cell surface more hydrophilic. As shown in chapter 9, more hydrophilic cell surface hydrophobicity would delay or even prevent aerobic granulation.

12.5 ORIGIN OF BIODEGRADABLE AEROBIC GRANULES-ASSOCIATED EPS

EPS present in biomass can be roughly divided into bound EPS (sheaths, capsular polymers, condensed gel, loosely bound polymers, and attached organics) and soluble EPS (soluble macromolecules, colloids, and slimes). Only soluble EPS is biodegradable (Hsieh et al. 1994; Nielsen et al. 1997). Nielsen, Jahn, and Palmgren (1997) developed a conceptual model for the production of EPS in biofilms. This model shows that both bound and soluble EPS are synthesized with the production of new cells, and bound EPS is further hydrolyzed into soluble EPS under appropriate environmental conditions. Meanwhile, cell lysis, decay, and hydrolysis of attached organics also add to the amount of soluble EPS. Eventually, those soluble EPS can be further recycled back to active cells via biodegradation. Analog to the model by Nielsen, Jahn, and Palmgren (1997), the composition of biomass in aerobic granules can be divided into active cells, inert biomass which includes bound EPS, attached organics, biomass residual and inorganic precipitation, and soluble EPS. Assuming that aerobic granules follow the same mechanism for the production of biodegradable EPS, the formation and conversion of biodegradable EPS can thus be illustrated in figure 12.12. It can be seen that there are three possible sources for the production of soluble or biodegradable EPS by aerobic granules: (1) hydrolysis of bound EPS and attached organics; (2) decay of active cells, and (3) direct synthesis from microbial growth. It should be pointed out that there is still a lack of experimental evidence regarding the direct formation of soluble EPS from cell growth-associated substrate utilization even though this has been often hypothesized in model development (Laspidou and Rittmann 2002). In fact, it is also difficult to experimentally distinguish the soluble EPS produced from cell growth-associated substrate utilization from those produced through cells lysis and decay during microbial growth. In view of this uncertainty, soluble EPS formation from the substrate oxidation pathway was plotted with a dashed line (figure 12.12). In addition, since only soluble substrate would be utilized in microbial culture, the contribution of attached organics to the EPS production would be marginal in quantity. Therefore, the most possible pathway of soluble EPS formation in aerobic granules can be attributed to cell lysis, decay, and hydrolysis of bound EPS, which commonly results from insufficiency of electron donor or acceptor present in microbial culture.

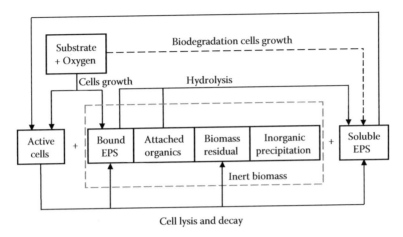

FIGURE 12.12   Diagram of formation and conversion of bound and soluble EPS in relationship to the other biomass components.

It has been shown that diffusion limitation occurs in large-sized aerobic granules (see chapter 8). As discussed in chapter 8, the oxygen concentration would be the rate-limiting factor for the growth of aerobic granules at high bulk COD levels. Consequently, for granules with size larger than 1 mm, oxygen can only penetrate to a depth of less than 300 μm from the surface of aerobic granules. This means that only bacteria situated at the top layer of 300 μm downwards is able to grow aerobically. In this case, severe biomass decay may occur in the core of the aerobic granule, contributing to the production of soluble EPS. As illustrated in chapter 11, the main source of soluble EPS is due to cell decay, including bound EPS hydrolysis. This may explain the extensive accumulation of soluble EPS in the core of the aerobic granule (figure 12.1). In fact, insufficient dissolved oxygen (DO) has been known to cause excessive EPS formation (Kim et al. 2006). To accomplish the conversion of active and inert biomass to soluble EPS, excessive amounts of extracellular proteins in the form of hydrolysis enzyme are needed in the core of the aerobic granule, which has been confirmed by confocal laser scanning micrographs observation (McSwain et al. 2005; Chen, Lee, and Tay 2007).

Figure 12.12 shows that bound EPS is closely associated with the reproduction of active cells. According to chapter 8, the outer shell of the aerobic granule was exposed to sufficient substrate and dissolved oxygen. This provides ideal conditions for new microbial cells to grow as well as the growth-associated production of the bound EPS. As compared to the situation at the core of the aerobic granule, the higher density of active cells can be expected in the surface layer of the aerobic granule (Toh et al. 2003). This in turn partially explains why a high concentration of the bound EPS was detected on the shell of the aerobic granule (chapter 11). In fact, Toh et al. (2003) investigated the viability of microflora residing inside aerobic granules using *in situ* DNA fluorescence staining, in which membrane permeable and nonpermeable DNA stains were used to differentiate live and dead cells (figure 12.13). In their study, about 150 to 200 slices prepared by the cryo-sectioning of a number of aerobic granules from a single size category, were analyzed by CLSM. The following salient

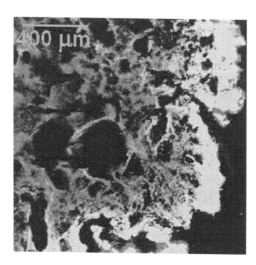

FIGURE 12.13   Cross-section view of an aerobic granule, stained with Syto 9 for the live biomass (bright) and propidium iodide for the dead biomass (gray). (From Toh, S. K. et al. 2003. *Appl Microbiol Biotechnol* 60: 687–695. With permission.)

points can be drawn: (1) regardless of size, the biomass was densest in the outer shell, and the biomass tended to decline along the radius direction downwards to the core of the aerobic granule; (2) active biomass existed mainly within the peripheral zone or void spaces, and non-biomass materials were found at the central part of the aerobic granule. This seems to indicate that the live cells appeared only in the peripheral zone, while dead biomass spread into the inner zone (Toh et al. 2003).

As all the detachment forces are exerted on the granule shell and tend to pull the shell apart, a high amount of bound or nonbiodegradable EPS would be essential to enhance the binding force between cells so as to overcome the external disintegration force (figure 12.9). Therefore, the production of EPS with different properties (e.g. biodegradable and nonbiodegradable) in aerobic granules is a natural need and response of the aerobic granule to counteract environmental stress.

12.6   CONCLUSIONS

The EPS produced by aerobic granules basically comprises biodegradable and nonbiodegradable components. It was shown that at least 50% of the EPS produced by aerobic granules is readily biodegradable by their own producer-aerobic granules, and the remaining part of the EPS seems nonbiodegradable. The readily biodegradable EPS can be taken up by aerobic granules at a rate 5 times slower than the biodegradation rate of acetate, but 50 times faster than that of those not readily biodegradable EPS.

It was further shown that biodegradable and nonbiodegradable EPS have different functions in an aerobic granular sludge SBR, that is, nonbiodegradable EPS were located mainly at the shell layer of the aerobic granule and plays an essential protective role in the granule integrity and stability, while biodegradable EPS accumulated at the central part of the aerobic granule can serve as an additional energy reservoir when external carbon source is no longer available for microbial growth.

REFERENCES

Boyd, A. and Chakrabarty, A. M. 1994. Role of alginate lyase in cell detachment of Pseudomonas aeruginosa. *Appl Environ Microbiol* 60: 2355–2359.

Chen, M. Y., Lee, D. J., and Tay, J. H. 2007. Distribution of extracellular polymeric substances in aerobic granules. *Appl Microbiol Biotechnol* 73: 1463–1469.

Decho, A. W., Visscher, P. T., and Reid, R. P. 2005. Production and cycling of natural microbial exopolymers (EPS) within a marine stromatolite. *Palaeogeogr Palaeoclimatol Palaeoecol* 219: 71–86.

Dudman, W. F. 1977. The role of surface polysaccharides in natural environments. In Sutherland, I. W. ed., *Surface carbohydrates of the prokaryotic cell*. London: Academic Press, 357–414.

Flemming, H. C. and Wingender, J. 2001. Relevance of microbial extracellular polymeric substances (EPSs). I, Structural and ecological aspects. *Water Sci Technol* 43: 1–8.

Hsieh, K. M., Murgel, G. A., Lion, L. W., and Shuler, M. L. 1994. Interactions of microbial biofilms with toxic trace-metals. I. Observation and modeling of cell-growth, attachment, and production of extracellular polymer. *Biotechnol Bioeng* 44: 219–231.

Jin, B., Wilen, B. M., and Lant, P. 2003. A comprehensive insight into floc characteristics and their impact on compressibility and settleability of activated sludge. *Chem Eng J* 95: 221–234.

Kim, H. Y., Yeon, K. M., Lee, C. H., Lee, S., and Swaminathan, T. 2006. Biofilm structure and extracellular polymeric substances in low and high dissolved oxygen membrane bioreactors. *Sep Sci Technol* 41: 1213–1230.

Laspidou, C. S. and Rittmann, B. E. 2002. A unified theory for extracellular polymeric substances, soluble microbial products, and active and inert biomass. *Water Res* 36: 2711–2720.

Li, Z. H., Kuba, T., and Kusuda, T. 2006. The influence of starvation phase on the properties and the development of aerobic granules. *Enzyme Microb Technol* 38: 670–674.

Liu, Y. Q., Liu, Y., and Tay, J. H. 2004. The effects of extracellular polymeric substances on the formation and stability of biogranules. *Appl Microbiol Biotechnol* 65: 143–148.

McSwain, B. S., Irvine, R. L., Hausner, M., and Wilderer, P. A. 2005. Composition and distribution of extracellular polymeric substances in aerobic flocs and granular sludge. *Appl Environ Microbiol* 71: 1051–1057.

Nielsen, P. H., Frolund, B., and Keiding, K. 1996. Changes in the composition of extracellular polymeric substances in activated sludge during anaerobic storage. *Appl Microbiol Biotechnol* 44: 823–830.

Nielsen, P. H., Jahn, A., and Palmgren, R. 1997. Conceptual model for production and composition of exopolymers in biofilms. *Water Sci Technol* 36: 11–19.

Obayashi, A. W. and Gaudy, A. F. 1973. Aerobic digestion of extracellular microbial polysaccharides. *J WPCF* 45: 1584–1594.

Patel, J. J. and Gerson, T. 1974. Formation and utilization of carbon reserves by Rhizobium. *Arch Microbiol* 101: 211–220.

Pirog, T. P., Grinberg, T. A., Malashenko, Y., and Yu, R. 1977. Isolation of microorganism-producers of enzymes degrading exopolysaccharides from *Acinetobacter sp. Mikrobiologiya* 33: 550–555.

Ruijssenaars, H. J., Stingele, F., and Hartmans, S. 2000. Biodegradability of food-associated extracellular polysaccharides. *Curr Microbiol* 40: 194–199.

Ryssov Nielsen, H. 1975. The role of natural extracellular polymers in the bioflocculation and dewatering of sludge. *Vatten* 1: 33–39.

Sponza, D. T. 2002. Extracellular polymer substances and physicochemical properties of flocs in steady- and unsteady-state activated sludge systems. *Process Biochem* 37: 983–998.

Sutherland, I. W. 1999. Polysaccharases for microbial exopolysaccharides. *Carbohydr Polym* 38: 319–328.

Toh, S. K., Tay, J. H., Moy, B. Y. P., Ivanov, V., and Tay, S. T. L. 2003. Size-effect on the physical characteristics of the aerobic granule in a SBR. *Appl Microbiol Biotechnol* 60: 687–695.

Wang, Z.-W., Liu, Y., and Tay, J.-H. 2007. Biodegradability of extracellular polymeric substances produced by aerobic granules. *Appl Microbiol Biotechnol* 74: 462–466.

Wang, Z., Liu, L., Yao, J., and Cai, W. 2006. Effects of extracellular polymeric substances on aerobic granulation in sequencing batch reactors. *Chemosphere* 63: 1728–1735.

Zhang, X. Q. and Bishop, P. L. 2003. Biodegradability of biofilm extracellular polymeric substances. *Chemosphere* 50: 63–69.

13 Calcium Accumulation in Acetate-Fed Aerobic Granules

Zhi-Wu Wang, Yong Li, and Yu Liu

CONTENTS

13.1 INTRODUCTION

A high calcium content has been reported in acetate-fed aerobic granules even though the calcium concentration in the synthetic wastewater was very low (Qin, Liu, and Tay 2004; Wang, Liu, and Tay 2005). Extensive accumulation of calcium was also found in biofilms and anaerobic granules (Batstone et al. 2002; Kemner et al. 2004). Two hypotheses have been put forward to explain the calcium accumulation: (1) calcium links with extracellular polymeric substances (EPS) and forms an EPS-Ca^{2+}-EPS cross-linkage; and (2) calcium is present in the form of $CaCO_3$ (Yu, Tay, and Fang 2001; Wloka et al. 2004). This chapter thus explores the mechanism behind the accumulation, chemical form, and spatial distribution of calcium ion in acetate-fed aerobic granules.

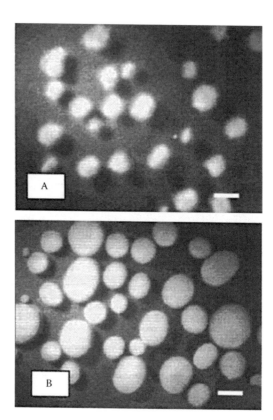

FIGURE 13.1 Aerobic granules cultivated at different calcium concentrations, 0 mg Ca²⁺ L⁻¹ (a) and 100 mg Ca²⁺ L⁻¹ (b). (From Jiang, H. L. et al. 2003. *Biotechnol Lett* 25: 95–99. With permission.)

13.2 EFFECT OF CALCIUM ON AEROBIC GRANULATION

Ca^{2+} has been reported to enhance the formation of anaerobic granules and acidogenic biofilms (Huang and Pinder 1995; Yu, Tay, and Fang 2001). Jiang et al. (2003) studied the effect of calcium on aerobic granulation in sequencing batch reactors (SBRs). For this purpose, two SBRs were operated at the respective Ca^{2+} concentrations of zero and 100 mg L⁻¹. It was found that aerobic granules were formed in both SBRs, and granule sizes were stabilized at around 2 mm and 2.8 mm in the calcium-free and calcium-added SBRs, respectively, after 2 months of operation (figure 13.1). These results indicate that aerobic granulation may not depend on calcium ion, that is, calcium ion is not essential for aerobic granulation in SBRs. Mahoney et al. (1987) investigated anaerobic granulation in two upflow anaerobic sludge blanket (UASB) reactors fed with aero and 100 mg Ca^{2+} L⁻¹, respectively. Similar to the results shown in figure 13.1, successful anaerobic granulation was achieved in both reactors, indicating that calcium is not an essential element for anaerobic granulation either.

Compared to aerobic granules grown on calcium-free medium, aerobic granules cultivated with addition of calcium showed better settleability and higher strength (figure 13.1). It is thought that the Ca^{2+} ion should bind to negatively charged groups

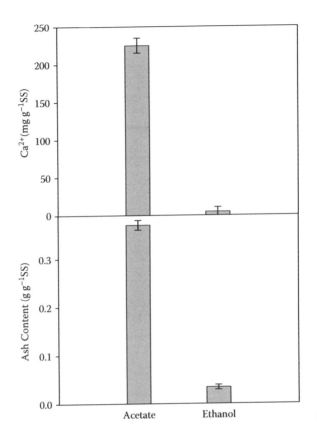

FIGURE 13.2   Calcium and ash contents in ethanol- and acetate-fed aerobic granules. (Data on ethanol from Liu, Yang, and Tay 2003 and on acetate from Wang, Li, and Liu 2007.)

of extracellular polysaccharides present on bacterial surfaces, and act as a bridge to interconnect these components, so as to promote bacterial aggregation and further enhance the structural stability of aerobic granules, anaerobic granules, and biofilms (Costerton et al. 1987; van Loosdrecht et al. 1987; Bruus, Nielsen, and Keiding 1992). It should be pointed out that such a view is still debatable.

13.3   CALCIUM ACCUMULATION IN ACETATE-FED AEROBIC GRANULES

Wang, Li, and Liu (2007) systematically investigated the calcium accumulation in acetate-fed aerobic granules harvested from a column SBR after 2 months of operation, while calcium concentration in influent was as low as 4.65 mg L^{-1}. It was found that acetate-fed aerobic granules had a high calcium content of 225 mg Ca^{2+} mg g^{-1}, contributing to 37% of granule ash content. Compared to acetate-fed aerobic granules, aerobic granules grown on ethanol showed very low calcium and ash contents (figure 13.2). This seems to suggest that calcium accumulation is a phenomenon closely associated with the substrate applied.

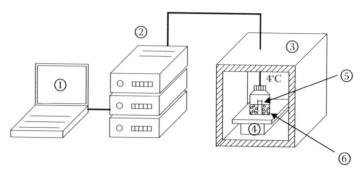

FIGURE 13.3 Respirometer system for analysis of carbonate in the acetate-fed aerobic granule: 1. computer for data collection; 2. respirometer; 3. fridge; 4. shaker; 5. acid containing vial; 6. reaction bottle. (From Wang, Z.-W., Li, Y., and Liu, Y. 2007. *Appl Microbiol Biotechnol* 74: 467–473. With permission.)

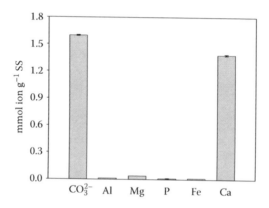

FIGURE 13.4 Ionic composition of acetate-fed aerobic granules. (Data from Wang, Z.-W., Li, Y., and Liu, Y. 2007. *Appl Microbiol Biotechnol* 74: 467–473.)

13.4 CHEMICAL FORM OF CALCIUM IN ACETATE-FED AEROBIC GRANULES

To investigate the chemical form of calcium ion accumulated in acetate-fed aerobic granules, Wang, Li, and Liu (2007) quantified the elemental composition (Ca, Mg, P, Fe, Al) of acetate-fed aerobic granules. The amount of carbonate ion in the acetate-fed aerobic granules was also analyzed. For this purpose, 3 ml of 1 M hydrochloric acid solution was added to 50 ml of 2 g soluble solids (SS) L^{-1} acetate-fed aerobic granules, and the carbon dioxide gas produced was online measured by the carbon dioxide sensor equipped with the respirometer (figure 13.3). Changes in inorganic carbon in the liquid phase were determined by total organic carbon analyzer before and after the experiment (Wang, Li, and Liu 2007). Thus, the content of carbonate in acetate-fed aerobic granules was calculated from the sum of produced carbon dioxide gas and increased inorganic carbon in the liquid phase. Figure 13.4 shows the major inorganic components of acetate-fed aerobic granules. As can be seen, both Ca^{2+} and CO_3^{2-} are dominant over the other inorganic components, such as Mg, P, Fe, and

Al, which are indeed marginal. According to figure 13.4, the molar ratio of granule calcium to carbonate was estimated as 1:1.16, indicating that most calcium ions in aerobic granules exist in the form of calcium carbonate. In terms of chemistry, this also implies that the concentration product of Ca^{2+} and CO_3^{2-} in acetate-fed aerobic granules should be larger than the solubility product constant of calcium carbonate.

13.5 CALCIUM DISTRIBUTION IN ACETATE-FED AEROBIC GRANULES

The calcium distribution in acetate-fed aerobic granules was investigated using a scanning electron microscope (SEM); meanwhile, energy dispersive x-ray spectroscopy (EDX) was also employed for mapping of calcium distribution (Wang, Li, and Liu 2007). The carbonate localization was determined by chemical titration method, that is, 1 M hydrochloric acid solution was dropped on a sliced granule cross section, and the origin of bubbles was visualized by image analysis technique (Wang, Li, and Liu 2007). Fresh acetate-fed aerobic granules with a specific oxygen uptake rate (SOUR) of 64 mg O_2 g^{-1} volatile solids (VS) h^{-1}; sludge volume index (SVI) of 52 mL g^{-1}, and a mean diameter of 1.4 mm were used for the above-mentioned analyses (figure 13.5). Figure 13.5a and b clearly show that calcium was mainly accumulated in the core part of the aerobic granule, while the granule shell was nearly calcium free. The image analysis further showed white deposits localized at 300 μm beneath the granule surface (figure 13.5c). After hydrochloric acid was added to the zone of white deposits (figure 13.5c), gas bubbles were immediately generated (figure 13.5d). The gas phase analysis confirmed that the bubbles generated were carbon dioxide (figure 13.5d). These results clearly indicate that both calcium and carbon ions coexist in the same zone of acetate-fed aerobic granules, that is, calcium exists mainly in the form of $CaCO_3$ in the acetate-fed aerobic granules, which is in good agreement with the stoichiometric analysis (figure 13.4).

The accumulation of calcium was observed in biofilms and anaerobic granules, and Ca^{2+} has been often considered to bridge negatively charged sites on extracellular biopolymers, thus enhancing the matrix stability of attached microbial communities (Bruus, Nielsen, and Keiding 1992; Korstgens et al. 2001; Batstone et al. 2002; Kemner et al. 2004; Wloka et al. 2004). According to such a hypothesis, excessive calcium has often been introduced into the medium for enhanced formation of biofilm and anaerobic granules (Huang and Pinder 1995; Yu, Tay, and Fang 2001). However, it appears from figures 13.4 and 13.5 that calcium detected in acetate-fed aerobic granules was mainly in the form of calcium carbonate rather than in association with extracellular polymeric substances.

13.6 GRANULE SIZE-DEPENDENT CACO$_3$ FORMATION IN ACETATE-FED AEROBIC GRANULES

It should be pointed out that the accumulation of calcium in the form of $CaCO_3$ in acetate-fed aerobic granules was found to be granule size-dependent (Wang, Li, and Liu 2007). As can be seen in figure 13.6, the calcium content of acetate-fed aerobic granules was proportionally related to the granule size, for example, the calcium

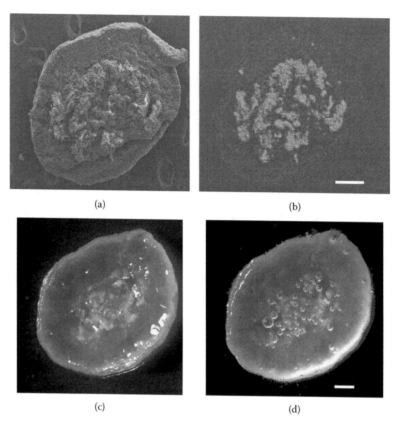

(a) (b)

(c) (d)

FIGURE 13.5 (a) Cross-section view of the acetate-fed aerobic granule by SEM; (b) the EDX mapping for calcium indicated by white color; bar: 100 μm; (c) image analysis cross-section view of the acetate-fed aerobic granule; (d) generation of gas bubbles during the acid-granule reaction; scale bar: 200 μm. (From Wang, Z.-W., Li, Y., and Liu, Y. 2007. *Appl Microbiol Biotechnol* 74: 467–473. With permission.)

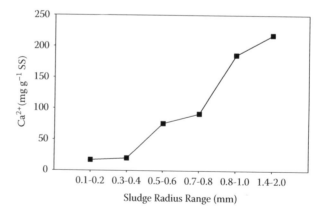

FIGURE 13.6 Calcium contents in aerobic granules with different radius. (Data from Wang, Z.-W., Li, Y., and Liu, Y. 2007. *Appl Microbiol Biotechnol* 74: 467–473.)

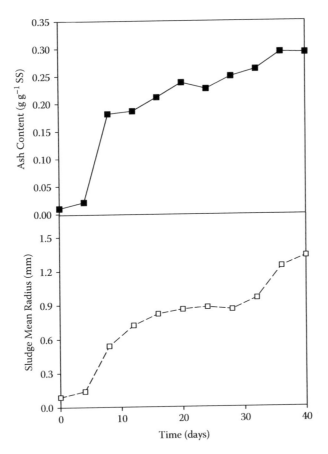

FIGURE 13.7   Ash content and corresponding mean radius of acetate-fed aerobic granules in the course of aerobic granulation. (Data from Wang, Z.-W., Li, Y., and Liu, Y. 2007. *Appl Microbiol Biotechnol* 74: 467–473.)

content in big aerobic granules with radius of 1.4 to 2.0 mm was nearly ten times higher than that in small aerobic granules with radius of 0.1 to 0.2 mm (figure 13.6). In the course of aerobic granulation, it was found that the ash content was very low at the initial stage of aerobic granulation, but it sharply increased on the eighth day in response to a significant increase in granule size, and gradually stabilized at the level of about 0.4 g g^{-1} SS after 40 days of operation (figure 13.7). This implies that the content of $CaCO_3$ or so-called ash content was indeed very low in small aerobic granules, but it tended to increase with the growth in size of acetate-fed aerobic granules.

13.7   MECHANISM OF CALCIUM ACCUMULATION IN ACETATE-FED AEROBIC GRANULES

As discussed earlier, calcium ion is not an essential element necessary for successful aerobic granulation (figure 13.1), and the extensive accumulation of calcium was only found in aerobic granules grown on acetate (figure 13.2). Furthermore, most accumulated calcium actually existed in the form of $CaCO_3$, and it was mainly centralized in the core part of acetate-fed aerobic granules (figures 13.4 and 13.5). One necessary condition for $CaCO_3$ formation at the low calcium ion concentration of 4.65 mg Ca^{2+} L^{-1} is the presence of high CO_3^{2-} concentration at the core of the acetate-fed aerobic granule so that the ionic product of Ca^{2+} and CO_3^{2-} can be higher than the solubility product constant of calcium carbonate.

13.7.1 IONIC EQUILIBRIUM OF CARBONATE ION

In terms of chemistry, the $CaCO_3$ formation is determined by its ionic concentration product:

$$[Ca^{2+}][CO_3^{2-}] = K_{sp,CaCO_3} \tag{13.1}$$

where $K_{sp,CaCO_3}$ is the $CaCO_3$ solubility product constant. Calcium carbonate will form only when the concentration product of calcium and carbonate is greater than $K_{sp,CaCO_3}$. Acetate can be oxidized in a way such that:

$$CH_3COO^- + 2O_2 + H^+ \rightarrow 2CO_2 + H_2O \tag{13.2}$$

Dissolution of carbon dioxide can be expressed as follows:

$$CO_2 + H_2O \leftrightarrow H^+ + HCO_3^- \tag{13.3}$$

and

$$K_{a1} = \frac{[HCO_3^-][H^+]}{[CO_2]} \tag{13.4}$$

$$HCO_3^- \leftrightarrow H^+ + CO_3^{2-} \tag{13.5}$$

and

$$K_{a2} = \frac{[CO_3^{2-}][H^+]}{[HCO_3^-]} \tag{13.6}$$

The overall reaction for carbonate can be expressed as:

$$CO_2 + H_2O \leftrightarrow CO_3^{2-} + 2H^+ \tag{13.7}$$

It should be pointed out that CO_2 produced in equation 13.2 can be dissolved into liquid phase according to Henry's law:

$$P_{CO_2} = K_{h,CO_2}[CO_2] \tag{13.8}$$

where P_{CO_2} is the partial pressure of CO_2 in gas phase, $[CO_2]$ is molar concentration of CO_2 in the liquid phase, and K_{h,CO_2} is the Henry's constant for CO_2.

13.7.2 DIFFUSION KINETICS IN AEROBIC GRANULES

It was assumed in chapter 8 that (1) an aerobic granule is isotopic in physical, chemical, and biological properties; (2) an aerobic granule is ideally spherical; (3) no anaerobic reaction occurs in the process; (4) aerobic granule responses to the change of bulk substrate concentration occur so quickly that the response time can be ignored. As presented in chapter 8, the mass balance equations for a substance between the two layers in granule whose radiuses are, respectively, r and $r + dr$ can be written as:

$$D_s \left(\frac{d^2 s}{dr^2} + \frac{2}{r} \frac{ds}{dr} \right) = R_s \tag{13.9}$$

where D_s and R_s are, respectively, the diffusion coefficient and mass conversion rate of the substance. According to equation 13.9, Wang, Li, and Liu (2007) proposed the following mass diffusion balance equations for O_2, H^+, HCO_3^-, and CO_3^{2-} in aerobic granules:

$$D_{O_2} \left(\frac{d^2 C_{O_2}}{dr^2} + \frac{2}{r} \frac{dC_{O_2}}{dr} \right) = R_{O_2} \tag{13.10}$$

$$D_{H^+} \left(\frac{d^2 C_{H^+}}{dr^2} + \frac{2}{r} \frac{dC_{H^+}}{dr} \right) = R_{H^+} \tag{13.11}$$

$$D_{HCO_3^-} \left(\frac{d^2 C_{HCO_3^-}}{dr^2} + \frac{2}{r} \frac{dC_{HCO_3^-}}{dr} \right) = R_{HCO_3^-} \tag{13.12}$$

$$D_{CO_3^{2-}} \left(\frac{d^2 C_{CO_3^{2-}}}{dr^2} + \frac{2}{r} \frac{dC_{CO_3^{2-}}}{dr} \right) = R_{CO_3^{2-}} \tag{13.13}$$

Li and Liu (2005) showed that dissolved oxygen would be a rate-limiting factor in the growth of aerobic granules, and the oxygen utilization rate can be described by the Monod equation:

$$R_{O_2} = \frac{\rho_x}{Y_{x/O_2}} \mu_{max} \frac{C_{O_2}}{K_{O_2} + C_{O_2}} \tag{13.14}$$

in which ρ_x is biomass density, Y_{x/O_2} is the dissolved oxygen-based growth yield, K_{O_2} is the dissolved oxygen-associated half-rate constant, and μ_{max} is the maximum specific growth rate.

According to equation 13.2, the oxygen utilization rate and the H^+ consumption rate are interrelated by equation 13.15, that is:

$$R_{O_2} = 2R_{H^+ \ (consumption)} \tag{13.15}$$

Similarly, the following relationship can be obtained from equations 13.3 and 13.7 for H^+, HCO_3^-, and CO_3^{2-}:

$$R_{H^+ \ (production)} = R_{HCO_3^-} + 2R_{CO_3^{2-}} \tag{13.16}$$

Thus, the net consumption rate of H^+, namely R_{H^+} in equation 13.11, is given by equation 13.17:

$$R_{H^+} = R_{H^+ \ (consumption)} - R_{H^+ \ (production)} \tag{13.17}$$

The dissolved oxygen (DO) concentration at the granule surface can be reasonably assumed to be equal to its bulk concentration and its rate of change in the granule center would be close to zero in consideration of the DO symmetrical distribution in the granule center (Li and Liu 2005), that is:

$$C_{O_2}\big|_{r=R} = C_{bulk, \ O_2} \tag{13.18}$$

$$\frac{dC_{O_2}}{dr}\bigg|_{r=0} = 0 \tag{13.19}$$

Likewise, C_{H^+} at the granule surface is assumed to be equal to the bulk H^+ concentration, and the derivative of C_{H^+} at the center of the granule is zero (Wang, Li, and Liu 2007):

$$C_{H^+}\big|_{r=R} = C_{bulk, \ H^+} \tag{13.20}$$

$$\frac{dC_{H^+}}{dr}\bigg|_{r=0} = 0 \tag{13.21}$$

Equations 13.10 to 13.13 were solved numerically by Matlab™ 7.0 based on the finite differentiation principle as described in chapter 8.

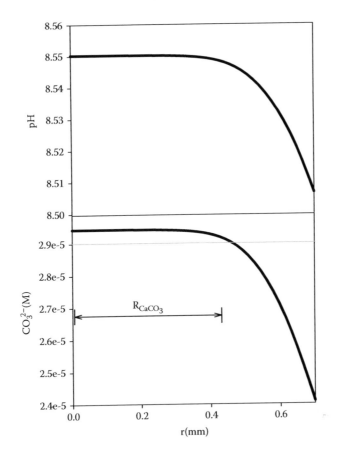

FIGURE 13.8 Simulated profiles of pH and CO_3^{2-} in acetate-fed aerobic granules. (Data from Wang, Z.-W., Li, Y., and Liu, Y. 2007. *Appl Microbiol Biotechnol* 74: 467–473.)

13.7.3 DISTRIBUTION OF pH AND CO_3^{2-} IN ACETATE-FED AEROBIC GRANULES

According to the models presented above, the pH and CO_3^{2-} profiles in aerobic granules were simulated along the granule radius (figure 13.8). For an aerobic granule with a radius of 0.7 mm, the pH and CO_3^{2-} profiles can be divided into two regimens, that is, within a depth of 300 μm from the granule surface, a significant decline in both pH and carbonate concentrations was found, whereas from the depth of 300 μm to the center of the granule, the concentrations of pH and carbonate remain nearly constant at the maximum levels. In fact, a similar pH profile was also observed in the acetate-fed anaerobic granule (de Beer et al. 1992). The maximum calcium concentration in a granule is equal to its bulk concentration, that is, about 1.16×10^{-4} mol $Ca^{2+} L^{-1}$, thus according to equation 13.1, the required minimum CO_3^{2-} concentration for the formation of $CaCO_3$ was estimated as 2.9×10^{-5} 2.9×10^{-5} mol L^{-1}. In view of the CO_3^{2-} profile in figure 13.8, it is reasonable to consider that $CaCO_3$ will be formed mainly in the region with a depth below 300 μm from the granule surface, that is, $r < R_{CaCO_3}$ zone in figure 13.8. Such a theoretical prediction is in good agreement with the experimental observation, as shown in figure 13.5. It can be concluded from figure 13.8 that the calcium accumulation in acetate-fed aerobic granules is mainly due to the fact that alkalinity in the form of hydroxide ion was produced during the biological oxidation of acetate, as illustrated in equation 13.2. In contrast, no hydroxide ion actually can be generated in the biological oxidation of ethanol. This explains why calcium is not accumulated in the ethanol-fed aerobic granules but only in those fed with acetate (figure 13.2).

13.7.4 SIZE-ASSOCIATED FORMATION OF $CaCO_3$ IN ACETATE-FED AEROBIC GRANULES

The volume fraction of the $CaCO_3$ formable region in acetate-fed aerobic granules, namely $(R_{CaCO_3}/R)^3$, was simulated at various granule radiuses in the range of 0 to 2 mm (figure 13.9). It appears that the formation of calcium carbonate is negligible in aerobic granules smaller than 0.5 mm in radius, and calcium carbonate starts to form only in aerobic granules larger than 0.5 mm in radius. This in turn provides a plausible explanation for the observed size-dependent $CaCO_3$ formation in aerobic granules (figures 13.6 and 13.7). Such theoretical estimation is pretty consistent with the experimental results measured as the calcium and ash contents in aerobic granules (figure 13.9). Batstone et al. (2002) has put forward a hypothesis that preformation of a crystal $CaCO_3$ core is a prerequisite of anaerobic granulation. However, both experimental and theoretical evidence in figures 13.6, 13.7, and 13.9 point to the fact that calcium accumulation occurs only after aerobic granulation, and a crystal $CaCO_3$ core is indeed not required for aerobic granulation. In fact, increasing evidence shows that microbial granulation is a cell-to-cell self-immobilization process driven mainly by selection pressures (see chapter 6), that is, calcium may play a very minor role in the microbial granulation process, and aerobic as well as anaerobic granulation is indeed calcium independent (Mahoney et al. 1987; Jiang et al. 2003). This point is also supported by many other previous studies on aerobic and anaerobic granulation (Guiot et al. 1988; Thiele et al. 1990; El-Mamouni et al. 1995; Van Langerak et al. 1998; Jiang et al. 2003).

It appears from microscopic observation and chemical analysis that the accumulation of calcium is in the form of $CaCO_3$, and those accumulations were centralized in the core part of acetate-fed aerobic granules (figures 13.4 and 13.5). The fact that $CaCO_3$ is only formed in aerobic granules with a radius larger than 0.5 mm (figures 13.6 and 13.7) indicates that the formation of calcium carbonate in the acetate-fed aerobic granules is actually granule size dependent, which can be explained by the diffusion limitation described by equations 13.10 to 13.13. This is because the microbial metabolites, such as alkalinity and carbon dioxide produced by acetate oxidation, can react with each other to produce carbonate, and form an ionic equilibrium, as illustrated in equations 13.2, 13.3, and 13.5. As the result of mass diffusion described in equation 13.3, a carbonate gradient along the granule radius is established towards the bulk solution so as to diffuse the carbonate out of the granule. This means that the highest carbonate concentration is found inside the aerobic granule (figure 13.8). It should be realized that this carbonate gradient

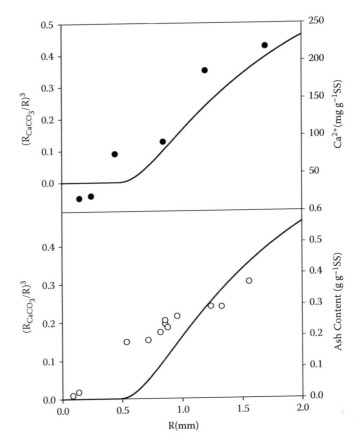

FIGURE 13.9 Comparisons of the simulated and experimentally measured $CaCO_3$ as well as ash contents in the acetate-fed aerobic granules. —: $(R_{CaCO_3}/R)^3$ simulation; ●: calcium content; ○: ash content. (Data from Wang, Z.-W., Li, Y., and Liu, Y. 2007. *Appl Microbiol Biotechnol* 74: 467–473.)

increases with the size of the aerobic granule. Thereby, a large aerobic granule would have a high carbonate concentration in its central part, as shown in figure 13.9.

13.8 CONCLUSIONS

Calcium is indeed not an essential element required for successful aerobic granulation. The excessive calcium accumulation phenomenon appears to be substrate dependent, and it only happens to the acetate-fed aerobic granules. Unlike speculation by some researchers, most calcium excessively accumulated in acetate-fed aerobic granules actually exists in the form of $CaCO_3$ deposit rather than binding with extracellular polysaccharides. It is actually the acetate biological oxidation-associated carbonate ionic equilibrium as well as the mass diffusion limitation that result in the high carbonate concentration centralized in the center of an aerobic granule, and further lead to $CaCO_3$ formation in the granule core part. The volume fraction of this $CaCO_3$ formed space in the granule tends to expand with the granule size increase, as a result of the correspondingly upgraded extent of mass diffusion limitation. $CaCO_3$ is not likely to form in granules smaller than 0.5 mm in radius, providing experimental and theoretical evidence that a crystal $CaCO_3$ core is not required for granulation.

SYMBOLS

D_s	Acetate diffusion coefficient
D_{fo}	Oxygen diffusion coefficient
$\mu(s)$	Growth rate at substrate concentration s
μ_{max}	Maximum growth rate
K_s	Half rate constant
$Y_{x/o}$	Growth yield of oxygen
$Y_{x/s}$	Growth yield of substrate
S_{bulk}	Bulk substrate concentration
ρ_x	Biomass density
R	The radius of a granule
r	One-dimensional coordinate
Δr	The thickness of one layer
$\overline{R}$	Average radius of all granules
s_i	Substrate concentration at the ith layer
s	Substrate concentration at a point of a granule
ΔS_{bulk}	The change of bulk substrate concentration
$S_{bulk\ 0}$	Initial bulk substrate concentration
v	Substrate conversion rate
v_1	Substrate conversion rate of a granule
V	Volume of the reactor
v_{all}	Substrate conversion rate of all granules
X	Biomass concentration
m	Number of aerobic granules in a reactor

REFERENCES

Batstone, D. J., Landelli, J., Saunders, A., Webb, R. I., Blackall, L. L., and Keller, J. 2002. The influence of calcium on granular sludge in a full-scale UASB treating paper mill wastewater. *Water Sci Technol* 45: 187–193.

Bruus, J. H., Nielsen, P. H., and Keiding, K. 1992. On the stability of activated sludge flocs with implications for dewatering. *Water Res* 26: 1597–1604.

Costerton, J. W., Cheng, K. J., Geesey, G. G., Ladd, T. I., Nickel, J. C., Dasgupta, M., and Marrie, T. J. 1987. Bacterial biofilm in nature and disease. *Annu Rev Microbiol* 41: 435–464.

de Beer, D., Huisman, J. W., Van den Heuvel, J. C., and Ottengraf, S. P. P. 1992. Effect of pH profiles in methanogenic aggregates on the kinetics of acetate conversion. *Water Res* 26: 1329–1336.

El-Mamouni, R., Guiot, S. R., Mercier, P., and Safi, B. 1995. Liming impact on granules activity of the multiplate anaerobic reactor (MPAR) treating whey permeate. *Bioprocess Eng* 12: 47.

Guiot, S. R., Gorur, S. S., Bourque, D., and Samson, R. 1988. Metal effect on microbial aggregation during upflow anaerobic sludge bed-filter (UBF) reactor start-up. In *Granular anaerobic sludge, microbiology and technology*. 187–194. Lettinga, G., Zehnder, A. J. B., Grotenhuis, J. T. C., and Hulshoff Pol, L. W. eds. Wageningen, the Netherlands: Purdoc.

Huang, J. and Pinder, K. L. 1995. Effects of calcium on development of anaerobic acidogenic biofilms. *Biotechnol Bioeng* 45: 212–218.

Jiang, H. L., Tay, J. H., Liu, Y., and Tay, S. T. L. 2003. Ca2+ augmentation for enhancement of aerobically grown microbial granules in sludge blanket reactors. *Biotechnol Lett* 25: 95–99.

Kemner, K. M., Kelly, S. D., Lai, B., Maser, J., O'Loughlin, E. J., Sholto-Douglas, D., Cai, Z. H., Schneegurt, M. A., Kulpa, C. F., and Nealson, K. H. 2004. Elemental and redox analysis of single bacterial cells by X-ray microbeam analysis. *Science* 306: 686–687.

Korstgens, V., Flemming, H. C., Wingender, J., and Borchard, W. 2001. Influence of calcium ions on the mechanical properties of a model biofilm of mucoid Pseudomonas aeruginosa. *Water Sci Technol* 43: 49–57.

Li, Y. and Liu, Y. 2005. Diffusion of substrate and oxygen in aerobic granule. *Biochem Eng J* 27: 45–52.

Liu, Y., Yang, S. F., and Tay, J. H. 2003. Elemental compositions and characteristics of aerobic granules cultivated at different substrate N/C ratios. *Appl Microbiol Biotechnol* 61: 556–561.

Mahoney, E. M., Varangu, L. K., Cairns, W. L., Kosaric, N., and Murray, R. G. E. 1987. The effect of calcium on microbial aggregation during UASB reactor start-up. *Water Sci Technol* 19: 249–260.

Qin, L., Liu, Y., and Tay, J. H. 2004. Effect of settling time on aerobic granulation in a sequencing batch reactor. *Biochem Eng J* 21: 47–52.

Thiele, J. H., Wu, W. M., Jain, M. K., and Zeikus, J. G. 1990. Ecoengineering high rate anaerobic digestion systems. Analysis of improved syntrophic biomethanation catalysts. *Biotechnol Bioeng* 35: 990–999.

Van Langerak, E. P. A., Gonzalez-Gil, G., Van Aelst, A., Van Lier, J. B., Hamelers, H. V. M., and Lettinga, G. 1998. Effects of high calcium concentrations on the development of methanogenic sludge in upflow anaerobic sludge bed (UASB) reactors. *Water Res* 32: 1255–1263.

van Loosdrecht, M. C., Lyklema, J., Norde, W., Schraa, G., and Zehnder, A. J. 1987. Electrophoretic mobility and hydrophobicity as a measure to predict the initial steps of bacterial adhesion. *Appl Environ Microbiol* 53: 1898–1901.

Wang, Z.-W., Li, Y., and Liu, Y. 2007. Mechanism of calcium accumulation in acetate-fed aerobic granules. *Appl Microbiol Biotechnol* 74: 467–473.

Wang, Z.-W., Liu, Y., and Tay, J. H. 2005. The role of SBR mixed liquor volume exchange ratio in aerobic granulation. *Chemosphere* 62: 767–771.

Wloka, M., Rehage, H., Flemming, H. C., and Wingender, J. 2004. Rheological properties of viscoelastic biofilm extracellular polymeric substances and comparison to the behavior of calcium alginate gels. *Colloid Polym Sci* 282: 1067–1076.

Yu, H. Q., Tay, J. H., and Fang, H. H. P. 2001. The roles of calcium in sludge granulation during UASB reactor start-up. *Water Res* 35: 1052–1060.

14 Influence of Starvation on Aerobic Granulation

Yu Liu, Zhi-Wu Wang, and Qi-Shan Liu

CONTENTS

14.1 INTRODUCTION

As discussed in the preceding chapters, the unique feature of a sequencing batch reactor (SBR) over the continuous activated sludge process is its cycle operation, which in turn results in a periodic starvation phase during the operation. Such periodical starvation has been thought to be important for aerobic granulation (Tay, Liu, and Liu 2001a). The responses of cells to starvation have been studied intensively. Nevertheless, controversial results have been widely reported in the literature. For instance, starvation has been thought to induce cell surface hydrophobicity, which facilitates microbial adhesion and aggregation (Bossier and Verstraete 1996; Y. Liu et al. 2004); however, the negative effect of starvation on cell surface hydrophobicity was also reported by Castellanos, Ascencio, and Bashan (2000). Moreover, constant cell surface hydrophobicity was observed during carbon starvation as well (Staffan and Malte 1984; Sanin 2003; Sanin, Sanin, and Bryers 2003), and sludge flocculation capacity was found to decrease with prolonged starvation (Rhymes and Smart 1996; Coello Oviedo et al. 2003). In this case, this chapter especially discusses the potential role of starvation in aerobic granulation.

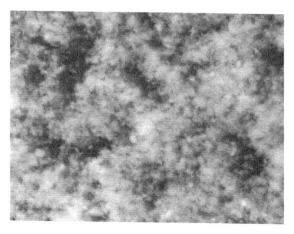

FIGURE 14.1 Morphology of seed sludge. (From Liu, Q.-S. 2003. Ph.D. thesis, Nanyang Technological University, Singapore. With permission.)

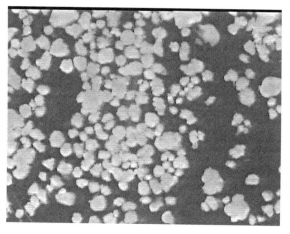

FIGURE 14.2 Morphology of aerobic granules at day 23 in an SBR. (From Liu, Q.-S. 2003. Ph.D. thesis, Nanyang Technological University, Singapore. With permission.)

14.2 POSITIVE EFFECT OF STARVATION ON AEROBIC GRANULATION

Q.-S. Liu (2003) investigated the role of periodic starvation in aerobic granulation, and concluded that it has a positive effect on aerobic granulation in SBRs, as presented below.

14.2.1 Observation of Aerobic Granulation in an SBR

The seed sludge had an average size of 0.07 mm, and exhibited a typical morphology of conventional activated sludge flocs (figure 4.1). On day 23, aerobic granules appeared in the SBR, with a clear, compact physical structure (figure 14.2 and figure 14.3).

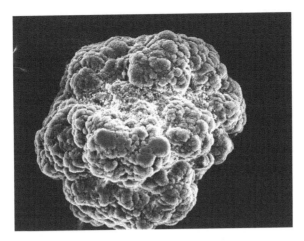

FIGURE 14.3 Morphology of aerobic granules observed by scanning electronic microscopy. (From Liu, Q.-S. 2003. Ph.D. thesis, Nanyang Technological University, Singapore. With permission.)

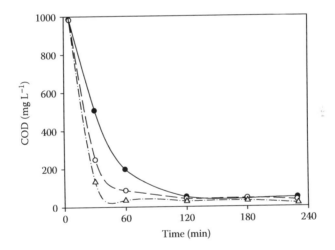

FIGURE 14.4 The COD concentration at the 12th (●), 36th (○), and 60th (△) cycles of an SBR. (From Liu, Q.-S. 2003. Ph.D. thesis, Nanyang Technological University, Singapore. With permission.)

14.2.2 PERIODIC STARVATION IN THE SBR

For aerobic granulation, the SBR is often run in a mode of filling, aeration, settling, and withdrawal of the effluent. Figure 14.4 shows the substrate degradation time at different operation cycles in an SBR. At cycle 12, or the second day after the startup, acetate concentration in terms of COD dropped from 1000 mg L^{-1} to 50 mg L^{-1} within 120 minutes, while the time period for the same COD removal was reduced to 85 minutes at cycle 36, and further to 50 minutes at cycle 60. These results indicate that the degradation time required to reduce substrate concentration to a minimum

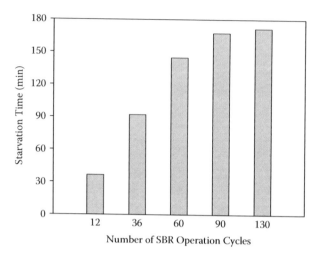

FIGURE 14.5 The observed starvation time versus the number of cycles in the SBR. (Data from Liu, Q.-S. 2003. Ph.D. thesis, Nanyang Technological University, Singapore.)

value was shortened markedly over the operation. Thus, for a given cycle length of 4 hours, a starvation phase would exist even at the beginning of the reactor operation (Q.-S. Liu 2003).

It appears from figure 12.4 that the aeration period can be divided into two consecutive phases, a degradation phase, in which external substrate is depleted to a minimum concentration, followed by an aerobic starvation phase, in which the external substrate is no longer available for microbial growth. It was found that with an increase in the number of cycles in the SBR, the degradation time required to break down the same amount of substrate became shorter (figure 14.4), that is, the starvation time is increased with the number of operation cycles. Buitron, Capdeville, and Horny (1994) studied the relationship of degradation time to SBR operation cycles, and found that after a 10-cycle operation, the degradation time was reduced by 80%, that is, 80% of the aeration period was in a state of aerobic starvation. For a fixed cycle time of 4 hours, the aerobic starvation period was found to be 105 minutes at cycle 12, 140 minutes at cycle 36, and 175 minutes at cycle 60 (Q.-S. Liu 2003). Figure 14.5 further exhibits the direct relationship between the cycle number and starvation time observed in the SBR. Similar results were also reported by Buitron, Capdeville, and Horny (1994). This seems to indicate that there is a periodic aerobic starvation phase in the cycle operation of SBR, but such a periodic starvation pattern does not exist in the continuous activated sludge process.

14.2.3 EFFECT OF PERIODIC STARVATION ON CELL SURFACE HYDROPHOBICITY

Figure 14.6 shows changes in cell surface hydrophobicity and substrate degradation time in the course of the operation of an SBR for aerobic granulation. As can be seen, the seed sludge had a surface hydrophobicity of 49%, while the cell surface hydrophobicity was increased to 70% at cycle 36, and further to 80% at cycle 60, and finally stabilized at 85% after the formation of aerobic granules. The cell surface

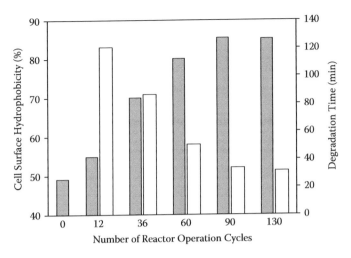

FIGURE 14.6   Changes in cell surface hydrophobicity and substrate degradation.

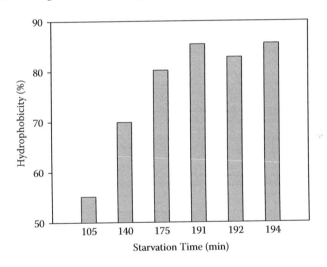

FIGURE 14.7   Cell surface hydrophobicity versus starvation time in an SBR. (Data from Liu, Q.-S. 2003. Ph.D. thesis, Nanyang Technological University, Singapore).

hydrophobicity of aerobic granules was nearly two times higher than that of the seed sludge. As demonstrated in chapter 9, aerobic granulation is associated with an increase in cell surface hydrophobicity.

It can be seen in figure 14.7 that the cell surface hydrophobicity was increased from 55% to 85% when the starvation time was increased from 105 to 190 minutes. Apparently, the periodic starvation in the SBR improves the cell surface hydrophobicity. However, in the continuous activated sludge reactor, no improvement in cell surface hydrophobicity was observed in the course of operation, for example, the cell surface hydrophobicity of sludge cultivated in the continuous reactor was similar to that of the seed sludge over the whole experimental period (Q.-S. Liu 2003).

TABLE 14.1

Biomass Hydrophobicity at Different Filter Bed Depths

Bed Depth (cm)	Biomass Hydrophobicity (%)
0–10	25
10–20	35
20–30	40
30–40	60

Source: Data from Di Iaconi, C. et al. 2006. *Biochem Eng J* 30: 152–157.

The response of bacteria to starvation has been widely reported (Kjelleberg and Hermansson 1984; Hantula and Bamford 1991; Bossier and Verstraete 1996). In a sequencing batch biofilter reactor, Di Iaconi et al. (2006) found that the starvation time increased with the bed height as less and less amount of substrate would reach deeper parts of the filter bed; meanwhile cell surface hydrophobicity tended to increase along the depth of the filter bed (table 14.1). According to such results, Di Iaconi et al. (2006) concluded that starvation could improve cell surface hydrophobicity and the effect of starvation on cell surface hydrophobicity would be more significant than that of hydrodynamic shear force.

Kjelleberg and Hermansson (1984) demonstrated that under starvation conditions, bacteria became more hydrophobic, which in turn facilitated microbial adhesion and aggregation. In fact, aggregation can be regarded as an effective strategy of cells against starvation. A similar phenomenon was also observed by Watanabe, Miyashita, and Harayama (2000), that is, cells showed a higher surface hydrophobicity when they were subject to starvation. It is believed that hydrophobic binding has a prime importance for cell attachment, that is, a higher cell surface hydrophobicity would result in a stronger cell-to-cell interaction and further a dense structure (see chapter 9). This point indeed is confirmed by the results presented in figure 14.8, showing a lower sludge volume index (SVI) at higher cell surface hydrophobicity. In a parallel study, Q.-S. Liu (2003) found that aerobic granulation failed in the continuous activated sludge reactor, and aerobic granules were only developed in the SBR. These findings imply that the periodic starvation-induced hydrophobicity is a governing factor in aerobic granulation in the SBR.

As shown in chapter 9, cell surface hydrophobicity plays a crucial role in the formation of biofilm and biogranules. In a thermodynamic sense, increased cell surface hydrophobicity can result in a lowered surface Gibbs energy, which in turn favors cell-to-cell interaction. In addition, cells in starved colonies were found to form connecting fibrils, which in turn strengthened cell-to-cell interaction and communication (Varon and Choder 2000). Apparently, such starvation-induced changes favor the formation of strong microbial aggregates. Starvation has been proposed to be a trigger in the microbial aggregation process (Bossier and Verstraete 1996). As discussed earlier, in an SBR microorganisms are subject to a periodic aerobic starvation. Tay, Liu, and Liu (2001a) thought that the periodic starvation present in the

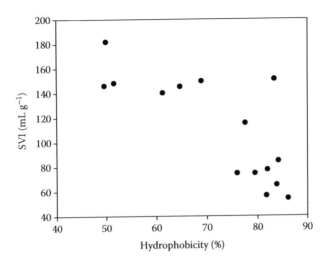

FIGURE 14.8 Sludge volume index (SVI) versus cell surface hydrophobicity in an SBR. (From Liu, Q.-S. 2003. Ph.D. thesis, Nanyang Technological University, Singapore. With permission.)

SBR would be more effective in triggering changes in the cell surface, and further lead to a stronger microbial aggregate.

Based on their study of aerobic granulation in SBRs, Li, Kuba, and Kusuda (2006) thought that "aerobic granulation is initiated by starvation and cooperated by shear force and anaerobic metabolism," and further proposed an EPS-related pathway of aerobic granulation, as illustrated in figure 14.9. According to the interpretation by Li, Kuba, and Kusuda (2006), in the beginning, starvation plays an essential role in aerobic granulation, and subsequently the growth of the aerobic granule provides an anaerobic microenvironment inside the aerobic granule, which favors anaerobic metabolism of facultative microorganisms. Furthermore, both starvation and facultative microorganisms facilitate aerobic granulation. It appears from figure 14.9 that there are two possible pathways leading to aerobic granulation: (1) step 1 → step 3 → step 4 → step 5, and this process is named starvation-driven granulation; (2) step 2 → step 3 → step 4 → step 5, called anaerobic granulation (Li, Kuba, and Kusuda 2006). So far, no solid evidence supports the mechanisms of aerobic granulation, as illustrated in figure 14.9, thus such interpretations of aerobic granulation are subject to further discussion. Consequently, the real role of starvation in aerobic granulation is still debatable and different views exist in the present literature.

14.3 INFLUENCE OF SHORT STARVATION ON AEROBIC GRANULES

To offer in-depth insights into the role of starvation in aerobic granulation, Z.-W. Wang et al. (2006) investigated the influence of short starvation on aerobic granules as presented below.

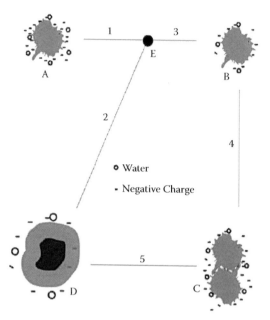

FIGURE 14.9 Effect of EPS on aerobic granulation. (A) Seed sludge with low cell hydrophobicity and high negative charge; (B) flocs or granules with low surface negative charge and high cell hydrophobicity; (C) aggregates of flocs or granules; (D) growth of granule under given shear condition; (E) a reasonable amount of EPS. 1: Starvation-associated EPS consumption; 2: facultative microorganisms-associated EPS consumption and production; 3: EPS-caused modifications of cell surface properties; 4: aggregation of flocs and growth of granules; 5: shear force-enhanced granule structure and detachment. (From Li, Z. H., Kuba, T., and Kusuda, T. 2006. *Enzyme Microb Technol* 38: 670–674. With permission.)

14.3.1 INFLUENCE OF CARBON AND NUTRIENTS STARVATION ON CELL SURFACE PROPERTY

Figure 14.10 shows the effects of carbon, nitrogen, phosphorus, and potassium starvation on cell surface hydrophobicity of aerobic granules. The cell surface hydrophobicity tended to decrease in the course of the N, P, and K starvation cultures, for example, the cell surface hydrophobicity decreased from the initial value of 80% to about 60% after 4 hours of N and P starvation. Meanwhile, no significant change in cell surface hydrophobicity of aerobic granules was found in the K starvation, whereas cell surface hydrophobicity exhibited a slight increase by 7% in the course of C starvation culture. Changes in cell surface zeta potential in the C, N, P, and K starvation batch culture are shown in figure 14.11. The cell surface zeta potential of aerobic granules under the respective C, N, P, and K starvation fluctuated around a certain value, that is, no significant changes can be observed under these starvation conditions.

The fundamental principle of charge interaction shows that oppositely charged objects will exert an attractive influence upon each other, while, in contrast to the attractive force between two objects with opposite charges, two cells that are of like charge will repel each other (figure 14.12). It is obvious that a negatively charged cell

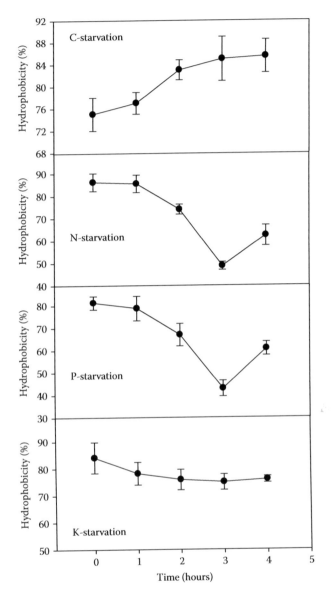

FIGURE 14.10 Changes in cell surface hydrophobicity in the course of the C, N, P, and K starvation batch cultures. (Data from Wang, Z.-W. et al. 2006. *Process Biochem* 41: 2373–2378.)

will exert a repulsive force upon a second negatively charged cell, and this repulsive force will push the two cells apart, and subsequently prevent microbial aggregation.

It is understandable that a weak repulsive force can be expected at a low surface charge density, thus reduced surface charge density has been thought to promote microbial aggregation, which is a key step towards to successful aerobic granulation in SBRs. Furthermore, cell surface hydrophobicity seems to inversely

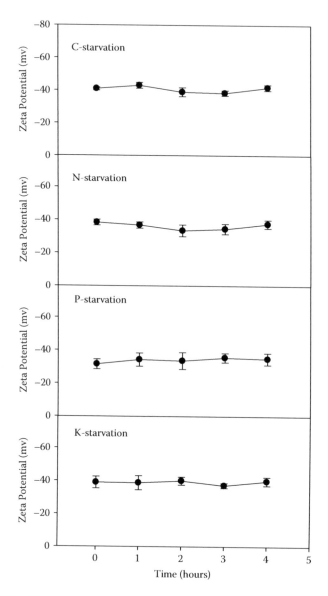

FIGURE 14.11 Changes in cell surface zeta potential in the course of the C, N, P, and K starvation batch cultures. (Data from Wang, Z.-W. et al. 2006. *Process Biochem* 41: 2373–2378.)

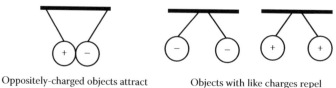

Oppositely-charged objects attract   Objects with like charges repel

FIGURE 14.12 Illustration of charge interaction.

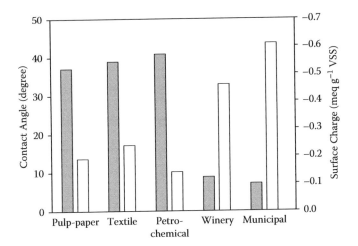

FIGURE 14.13 Biomass surface hydrophobicity (gray) and charge of activated sludge flocs (white) grown on different types of industrial wastewaters. (Data from Sponza, D. T. 2003. *Enzyme Microb Technol* 32: 375–385.)

depend on the surface charge density, which has been reported in activated sludge as well as aerobic granules (Liao et al. 2001; Li, Kuba, and Kusuda 2006). Sponza (2003) investigated physicochemical properties of different activated sludge flocs under steady-state conditions, and correlated cell surface hydrophobicity in terms of contact angle and surface charge density (figure 14.13). The salient points shown in figure 14.13 are as follows:

1. The respective mean contact angle of the winery and municipal activated sludge flocs was 7° and 9°, while contact angles of 37° and 42° were obtained for activated sludge flocs grown on the pulp-paper, textile, and petrochemical wastewaters. Obviously, those tested activated sludge flocs exhibited typical hydrophilic or moderately hydrophobic properties, largely determined by their own culture histories.

2. The activated sludge flocs grown on different types of industrial wastewaters all carried negative charges on their surfaces. A floc surface with lower negative charge exhibits a higher surface hydrophobicity, whereas the more negatively charged floc surface reflect higher hydrophilicity.

It should be recalled here that the contact angle is the angle at which a liquid/vapor interface meets the solid surface. For instance, on extremely hydrophilic surfaces, a water droplet will completely spread with an effective contact angle of 0°. This occurs for surfaces that have a large affinity for water. On the contrary, on many hydrophilic surfaces, water droplets will exhibit contact angles of 10° to 30°, while on highly hydrophobic surfaces, which are incompatible with water, a large contact angle of 70° to 90° can be observed.

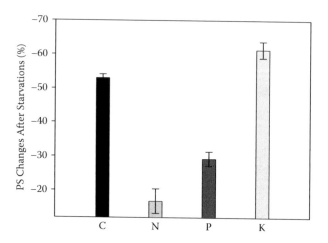

FIGURE 14.14 Changes in PS as a result of C, N, P, and K starvation. (Data from Wang, Z.-W. et al. 2006. *Process Biochem* 41: 2373–2378.)

14.3.2 INFLUENCE OF CARBON AND NUTRIENTS STARVATION ON EPS CONTENT

Extracellular polysaccharides (PS) are sticky materials secreted by cells, and PS are highly involved in adhesion phenomena, formation of matrix structure, microbial physiology, and improvement of long-term stability of granules (see chapter 10). The PS contents in aerobic granules before and after C, N, P, and K starvation were determined (Z. P. Wang et al. 2006). A reduction in the PS content was observed in all four cases (figure 14.4), for example the PS content of aerobic granules was reduced by 15% after 4-hours of N starvation, 30% after 4 hours of P starvation, while the respective PS content was substantially reduced by 53% and 65%, respectively, after 4 hours of C and K starvation. These results imply that the C and K starvation have the most profound negative effect on the PS content in aerobic granules. In addition, Z. P. Wang, ZP et al. (2006) also reported that extracellular polymeric substances (EPS) were produced mainly in the exponential phase, and served as carbon and energy sources in the starvation phase during the granulation process, and they thought that such a function of PS could regulate microbial growth in the interior and exterior of granules and help maintain the integrity of aerobic granules.

14.3.3 INFLUENCE OF CARBON AND NUTRIENTS STARVATION ON MICROBIAL ACTIVITY AND PRODUCTION

Figure 14.15 compares the specific oxygen uptake rate (SOUR) values before and after the 4-hour C, N, P, and K starvation. The SOURs of aerobic granules under the C, N, and P starvation conditions decreased markedly, whereas an increase in the SOUR after K starvation was observed. Furthermore, it appears from figure 14.16 that under the C and K starvation conditions, the growth of aerobic granules was seriously suppressed. This may be due to the fact that carbon and potassium are two essential elements that actively participate in the energy metabolism of cells.

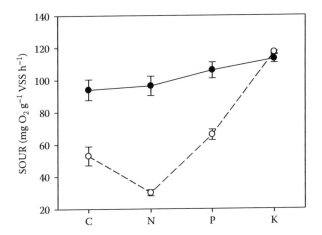

FIGURE 14.15 Changes in SOUR before (●) and after (○) C, N, P, and K starvation. (Data from Wang, Z.-W. et al. 2006. *Process Biochem* 41: 2373–2378.)

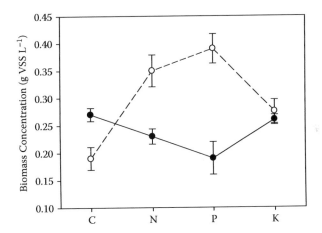

FIGURE 14.16 Changes in biomass concentration before (●) and after (○) C, N, P, and K starvation. (Data from Wang, Z.-W. et al. 2006. *Process Biochem* 41: 2373–2378.)

As discussed earlier, Tay, Liu, and Liu (2001a) thought that periodic starvation observed in SBRs would induce cell surface hydrophobicity which in turn facilitates aerobic granulation in the SBR. It appears from figure 14.10, however, that the cell surface hydrophobicity of aerobic granules subjected to N and P starvation shows a decreasing trend, while change in the cell surface hydrophobicity of the K-starved aerobic granules is insignificant and only the C-starved aerobic granules demonstrate a slight increase in cell surface hydrophobicity. In fact, the N and P starvations have often been found to decrease microbial surface hydrophobicity (Bura et al. 1998; Chen and Strevett 2003; Sanin et al. 2003). Regarding the C starvation, inconsistent results can be found in the literature, for example, Chen and Strevett (2003) reported a positive effect of C starvation on cell surface hydrophobicity, whereas Sanin et al.

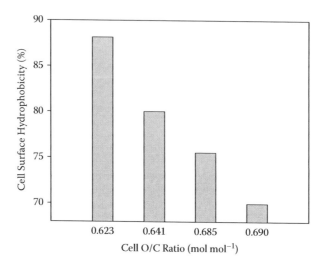

FIGURE 14.17 Effect of cell O/C ratio on cell surface hydrophobicity. (Data from Liu, Y., Yang, S. F., and Tay, J. H. 2003. *Appl Microbiol Biotechnol* 61: 556–561.)

(2003) found no effect of C starvation on cell surface hydrophobicity. Castellanos, Ascencio, and Bashan (2000) even observed the adverse effect of C starvation on cell surface hydrophobicity. Similar to figure 14.14, it has been shown in chapter 10 that a considerable amount of hydrophilic EPS located in the granule core is biodegradable to compensate for the shortage of energy source under the C starvation condition, which subsequently results in a hollow granule structure (Z.-W. Wang, Liu, and Tay 2005). In fact, figure 14.10 indicates that there is no apparent positive effect of the short term C, N, P, and K starvation on cell surface hydrophobicity.

The cell surface zeta potential of microorganisms reflects the charge density of the cell surface. Aerobic granulation was found to be associated with a decrease in the negative charge density of the cell surface (Li, Kuba, and Kusuda 2006). As can be seen in figure 14.11, the fresh and starved aerobic granules both carry negative charge on their surfaces. Compared to the fresh aerobic granules, no significant changes in the surface charges were observed in aerobic granules after the 4-hour starvation. It was thought that cell surface hydrophobicity and surface charge are closely related to the types of proteins located on the outer membrane of the cell, while changes in outer membrane protein profiles have been commonly reported during starvation for nitrogen, phosphorus, and potassium, for example starvation of *Escherichia coli* for potassium and phosphate ions leads to a reversible increase in the rate of protein degradation and an inhibition of ribonucleic acid synthesis (Nystrom, Olsson, and Kjelleberg 1992; Kragelund and Nybroe 1994). By analyzing elemental composition of aerobic granules, Y. Liu, Yang, and Tay (2003) found that cell surface hydrophobicity was closely correlated to cell O/C ratio, that is, a high cell O/C ratio resulted in a low cell surface hydrophobicity (figure 14.17). It seems reasonable to consider that the cell O/C ratio, to a large extent, is determined by membrane surface proteins.

As can be seen in figure 14.14, the content of extracellular polysaccharides decreased slightly in the N- and P-starved aerobic granules, while about 53% and 65% of PS reduction was observed in aerobic granules deprived of carbon and potassium, respectively. These results indicate that under the C- and K-starvation conditions, aerobic granules prefer to metabolize their PS in order to provide energy for maintaining basic functions of the living cells. This phenomenon has been observed in other cultures under C-starvation conditions (Patel and Gerson 1974; Boyd and Chakrabarty 1994; Ruijssenaars, Stingele, and Hartmans 2000; Zhang and Bishop 2003; Z.-W. Wang, Liu, and Tay 2005). It is reasonable to consider that microorganism would by all means make use of any potential carbon source to compensate for the substrate shortage. Moreover, evidence shows that the disintegration of aerobic granules was coupled with a sharp decrease in the PS content in aerobic granules (Tay, Liu, and Liu 2001b), that is, the stability of aerobic granules is tightly associated with the PS of aerobic granules which may serve as matrix materials to strengthen the spatial structure of aerobic granules. One important engineering implication of figure 14.14 is that when aerobic granules are subject to substrate and nutrient starvations, especially C and K starvation, the reduced content of PS weakens the spatial structure and the stability of aerobic granules. This in turn results in instability of aerobic granular sludge. In fact, potassium deficiency is commonly found in industrial wastewater (Murthy and Novak 1998).

Similar to figures 14.15 and 14.16, Hueting, de Lange, and Tempest (1979) also observed a progressive increase in the respiration rate of cells, with a corresponding fall in the biomass production under K limitation. Konings and Veldkamp (1980) suggested that microorganisms had to increase the proton motive force to maintain the internal concentration constant of the limited essential cation, such as K^+, and such a response would lead to an increased rate of dissimilation of the energy source, as shown in figures 14.15 and 14.16. Neijssel and Tempest (1979) further found that there was a linear relationship between the specific rate of oxygen uptake by K-limited cells and the electrochemical potential of the K gradient.

Increased biomass production was observed under the N- and P-starved cultures of aerobic granules (figure 14.16). The increased cell mass during starvation for nitrogen and phosphorus has been thought to be due to the accumulation of storage compounds (Kragelund and Nybroe 1994), while Sidat, Bux, and Kasan (1999) also found that under aerobic conditions, microorganisms can store polyphosphate intracellularly, which serves as phosphorus and energy sources during phosphorus starvation. According to the traditional activated sludge theory (Droste 1997), nitrogen and phosphorus are essentially required for biosynthesis of biomass in a ratio of COD:N:P of 100:5:1. In order to maintain balanced metabolism, cells need to coordinate the assimilation of nitrogen with the assimilation of carbon and other aspects of metabolism. For aerobic granules deprived of nitrogen, hydrolysis of proteins is one way that can provide nitrogen for microbial growth and maintenance. In fact, nitrogen released from hydrolysis of proteins has been reported (St. John and Goldberg 1980; Nystrom, Olsson, and Kjellegerg 1992; Kragelund and Nybroe 1994). As a result of the protein hydrolysis, the SOUR of the N-starved aerobic granules was reduced by 70% as compared to the fresh aerobic granules (figure 14.15). Thus, it seems reasonable to consider that the biomass lost in the C-starvation condition

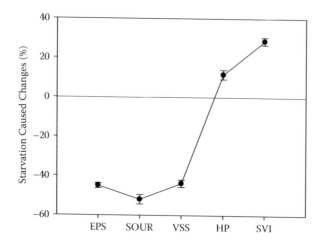

FIGURE 14.18 Changes in EPS, SOUR, VSS, hydrophobicity (HP), and SVI in response to collective substrate and nutrients starvation. (Data from Wang, Z.-W. et al. 2006. *Process Biochem* 41: 2373–2378.)

(figure 14.16) can be attributed to the endogenous decay caused by the electron donor limitation as well as the reduction of PS.

As discussed earlier, C, N, P, and K starvation did not show positive effects on the stability of aerobic granules. A collective substrate and nutrients starvation test was further conducted to look into the collective effects of those limitations together on aerobic granules (Z.-W. Wang et al. 2006). The changes of granule properties in terms of PS, SOUR, volatile solids (VS), hydrophobicity, and SVI as a result of this collective starvation are presented in figure 14.18. It can be seen that PS, SOUR, and VS tended to decrease, while hydrophobicity and SVI seemed to increase in response to the collective starvation of aerobic granules. According to Z.-W. Wang, Liu, and Tay (2005), the PS reduction shown in figure 14.18 should be due mainly to the biodegradation of the hydrophilic PS, which in turn leads to the increased cell surface hydrophobicity. The increased SVI shown in figure 14.18 implies that this collective starvation results in poor settleability and less compact structure of aerobic granules. The markedly reduced SOUR would be the result of the collective limitation of substrate and nutrients.

Bulk liquid turbidity was also measured in the course of the collective starvation (figure 14.19). It was found that the bulk liquid became more and more turbid with the increase in the starvation time. This may indicate a bacterial detachment from aerobic granules in response to the collective starvation. In fact, a similar phenomenon was also found in the biofilm process under substrate- and nutrient-deficient conditions (Sawyer and Hermanowicz 1998; Hunt et al. 2004).

The potential negative effects of starvation on the stability and activity of aerobic granules has been demonstrated. It should be stressed that the genetic regulation of the response to starvation should be understood in the near future, and how starvation influences the cell surface of aerobic granules is still a subject of future discussion.

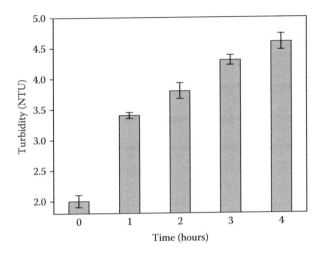

FIGURE 14.19 Changes of bulk liquid turbidity in the course of collective substrate and nutrients starvation. (Data from Wang, Z.-W. et al. 2006. *Process Biochem* 41: 2373–2378.)

14.4 CONCLUSIONS

A periodic aerobic starvation phase exists in SBRs, though different views can be found with regard to the role of starvation in aerobic granulation. Some results showed that starvation is essential in aerobic granulation, and can serve as a trigger for microbial aggregation. However, it appears that aerobic C, P, N, and K starvation can alter the surface properties of aerobic granules and cause negative effects on the stability and activity of aerobic granules. Therefore, discussion on starvation in aerobic granular sludge SBRs will continue and more in-depth research is needed to clarify its role in aerobic granulation.

REFERENCES

Bossier, P. and Verstraete, W. 1996. Triggers for microbial aggregation in activated sludge? *Appl Microbiol Biotechnol* 45: 1–6.

Boyd, A. and Chakrabarty, A. M. 1994. Role of alginate lyase in cell detachment of Pseudo-monas aeruginosa. *Appl Environ Microbiol* 60: 2355–2359.

Buitron, G., Capdeville, B., and Horny, P. 1994. Improvement and control of the microbial activity of a mixed population for degradation of xenobiotic compounds. *Water Sci Technol* 29: 317–326.

Bura, R., Cheung, M., Liao, B., Finlayson, J., Lee, B. C., Droppo, I. G., Leppard, G. G., and Liss, S. N. 1998. Composition of extracellular polymeric substances in the activated sludge floc matrix. *Water Sci Technol* 37: 325–333.

Castellanos, T., Ascencio, F., and Bashan, Y. 2000. Starvation-induced changes in the cell surface of Azospirillum lipoferum. *FEMS Microbiol Ecol* 33: 1–9.

Chen, G. and Strevett, K. A. (2003) Impact of carbon and nitrogen conditions on E. coli surface thermodynamics. *Colloids and Surfaces B: Biointerfaces* 28: 135–146.

Coello Oviedo, M. D., Lopez-Ramirez, J. A., Sales Marquez, D., and Quiroga Alonso, J. M. 2003. Evolution of an activated sludge system under starvation conditions. *Chem Eng J* 94: 139–146.

Di Iaconi, C., Ramadori, R., Lopez, A., and Passino, R. 2006. Influence of hydrodynamic shear forces on properties of granular biomass in a sequencing batch biofilter reactor. *Biochem Eng J* 30: 152–157.

Droste, R. L. 1997. *Theory and practice of water and wastewater treatment.* New York: John Wiley & Sons.

Hantula, J. and Bamford, D. H. 1991. The efficiency of the protein dependent flocculation of Flavobacterium sp. is sensitive to the composition of growth medium. *Appl Microbiol Biotechnol* 36: 100–104.

Hueting, S., de Lange, T., and Tempest, D. W. 1979. Energy requirement for maintenance of the transmembrane potassium gradient in Klebsiella aerogenes NCTC 418: A continuous culture study. *Arch Microbiol* 123: 183–188.

Hunt, S. M., Werner, E. M., Huang, B. C., Hamilton, M. A., and Stewart, P. S. 2004. Hypothesis for the role of nutrient starvation in biofilm detachment. *Appl Environ Microbiol* 70: 7418–7425.

Kjelleberg, S. and Hermansson, M. 1984. Starvation-induced effects on bacterial surface characteristics. *Appl Environ Microbiol* 48: 497–503.

Konings, W. N. and Veldkamp, H. 1980. Phenotypic responses to environmental change. In Ellwood, D. C., Hedger, J. N., Latham, M. J., Lynch, J. M., and Slater, J. M. eds. *Contemporary microbial ecology.* London: Academic Press, 161–191.

Kragelund, L. and Nybroe, O. 1994. Culturability and expression of outer membrane proteins during carbon, nitrogen, or phosphorus starvation of Pseudomonas fluorescens DF57 and Pseudomonas putida DF14. *Appl Environ Microbiol* 60: 2944–2948.

Li, Z. H., Kuba, T., and Kusuda, T. 2006. The influence of starvation phase on the properties and the development of aerobic granules. *Enzyme Microb Technol* 38: 670–674.

Liao, B. Q., Allen, D. G., Droppo, I. G., Leppard, G. G., and Liss, S. N. 2001. Surface properties of sludge and their role in bioflocculation and settleability. *Water Res* 35: 339–350.

Liu, Q.-S. (2003) Aerobic granulation in a sequencing batch reactor. Ph.D. thesis, Nanyang Technological University, Singapore.

Liu, Y., Yang, S. F., and Tay, J. H. 2003. Elemental compositions and characteristics of aerobic granules cultivated at different substrate N/C ratios. *Appl Microbiol Biotechnol* 61: 556–561.

Liu, Y., Yang, S. F., Tay, J. H., Liu, Q.-S., Qin, L., and Li, Y. 2004. Cell hydrophobicity is a triggering force of biogranulation. *Enzyme Microb Technol* 34: 371–379.

Murthy, S. N. and Novak, J. T. 1998. Effects of potassium ion on sludge settling, dewatering and effluent properties. *Water Sci Technol* 37: 317–324.

Neijssel, O. M. and Tempest, D. W. 1979. The physiology of metabolite over-production. *Symp Soc Gen Microbiol* 29: 53–82.

Nystrom, T., Olsson, R., and Kjelleberg, S. 1992. Survival, stress resistance, and alterations in protein expression in the marine Vibrio sp. strain S14 during starvation for different individual nutrients. *Appl Environ Microbiol* 58: 55–65.

Patel, J. J. and Gerson, T. 1974. Formation and utilization of carbon reserves by Rhizobium. *Arch Microbiol* 101: 211–220.

Rhymes, M. R. and Smart, K. A. 1996. Effect of starvation on the flocculation of ale and lager brewing yeasts. *J Am Soc Brew Chem* 54: 50–56.

Ruijssenaars, H. J., Stingele, F., and Hartmans, S. 2000. Biodegradability of food-associated extracellular polysaccharides. *Curr Microbiol* 40: 194–199.

Sanin, S. L. 2003. Effect of starvation on resuscitation and the surface characteristics of bacteria. *J Environ Sci Heal A* 38: 1517–1528.

Sanin, S. L., Sanin, F. D., and Bryers, J. D. 2003. Effect of starvation on the adhesive properties of xenobiotic degrading bacteria. *Process Biochem* 38: 909–914.

Sawyer, L. K. and Hermanowicz, S. W. 1998. Detachment of biofilm bacteria due to variations in nutrient supply. *Water Sci Technol* 37: 211–214.

Sidat, M., Bux, F., and Kasan, H. C. 1999. Polyphosphate accumulation by bacteria isolated from activated sludge. *Water SA* 25: 175–179.

Sponza, D. T. 2003. Investigation of extracellular polymer substances (EPS) and physico-chemical properties of different activated sludge flocs under steady-state conditions. *Enzyme Microb Technol* 32: 375–385.

Staffan, K. and Malte, H. 1984. Starvation-induced effects on bacterial surface characteristics. *Appl Environ Microbiol* 48: 497–503.

St. John, A. C. and Goldberg, A. L. 1980. Effects of starvation for potassium and other inorganic ions on protein degradation and ribonucleic acid synthesis in Escherichia coli. *J Bacteriol* 143: 1223–1233.

Tay, J. H., Liu, Q.-S., and Liu, Y. 2001a. Microscopic observation of aerobic granulation in sequential aerobic sludge blanket reactor. *J Appl Microbiol* 91: 168–175.

Tay, J. H., Liu, Q.-S., and Liu, Y. 2001b. The role of cellular polysaccharides in the formation and stability of aerobic granules. *Lett Appl Microbiol* 33: 222–226.

Varon, M. and Choder, M. 2000. Organization and cell-cell interaction in starved Saccharomyces cerevisiae colonies. *J Bacteriol* 182: 3877–3880.

Wang, Z. P., Liu, L., Yao, J., and Cai, W. 2006. Effects of extracellular polymeric substances on aerobic granulation in sequencing batch reactors. *Chemosphere* 63: 1728–1735.

Wang, Z.-W., Liu, Y., and Tay, J.-H. 2005. Distribution of EPS and cell surface hydrophobicity in aerobic granules. *Appl Microbiol Biotechnol* 69: 469–473.

Wang, Z.-W., Li, Y., Zhou, J. Q., and Liu, Y. 2006. The influence of short-term starvation on aerobic granules. *Process Biochem* 41: 2373–2378.

Watanabe, K., Miyashita, M., and Harayama, S. 2000. Starvation improves survival of bacteria introduced into activated sludge. *Appl Environ Microbiol* 66: 3905–3910.

Zhang, X. Q. and Bishop, P. L. 2003. Biodegradability of biofilm extracellular polymeric substances. *Chemosphere* 50: 63–69.

15 Filamentous Growth in an Aerobic Granular Sludge SBR

Yu Liu and Qi-Shan Liu

CONTENTS

15.1 INTRODUCTION

As shown in the preceding chapters, aerobic granulation is a tailored environmental biotechnology for treating a wide variety of wastewaters. Similar to anaerobic granulation, aerobic granulation is a microbial self-immobilization process that is driven by selection pressures in the sequencing batch reactor (SBR) (see chapter 6). Experimental evidence shows instability of aerobic granules is the major technical problem encountered in operating an aerobic granular sludge SBR, while filamentous growth has been commonly observed in aerobic granular sludge SBRs (Tay, Liu, and Liu 2001; Pan 2003; McSwain, Irvine, and Wilderer 2004; F. Wang et al. 2004; Schwarzenbeck, Borges, and Wilderer 2005). Once filamentous growth dominates the reactor, settleability of aerobic granules becomes poor. This eventually leads to biomass washout and subsequent disappearance of aerobic granules. Thus, filamentous growth, to a great extent, would be responsible for the observed instability of aerobic granules. Instability of aerobic granules is a significant bottleneck in applying this novel wastewater treatment technology. The operating parameters that can encourage filamentous growth and its control are not entirely clear, thus this chapter attempts to discuss the operating conditions that may result in filamentous growth; the major causes of filamentous growth in an aerobic granular sludge SBR; and possible strategies for controlling filamentous growth.

15.2 CAUSES OF FILAMENTOUS GROWTH IN THE ACTIVATED SLUDGE PROCESS

The activated sludge process often suffers from sludge bulking due to overgrowth of filamentous microorganisms in the aeration tank. In order to understand and further control filamentous growth in aerobic granular sludge SBRs, this section examines some key parameters and their combinations that may be responsible for filamentous growth in the activated sludge process.

15.2.1 WASTEWATER COMPOSITION

Carbohydrates have been known to favor the filamentous growth in the activated sludge processes (J. Chudoba 1985; Bitton 1999; Eckenfelder 2000; Richard and Collins 2003). According to Kappeler and Gujer (1994), various wastewater fractions in terms of readily biodegradable substrate, surfactants, hydrophilic and lipophilic slowly biodegradable substrate, and sulfide, can all strongly influence the biocenosis composition. Nevertheless, in the operation of a full-scale activated sludge process, wastewater fractions can hardly be manipulated because of large volume of influent. Adjustment of wastewater composition is not a feasible strategy for controlling filamentous growth.

15.2.2 SUBSTRATE AVAILABILITY

It is thought that filamentous microorganisms grow slowly, that is, they have very low Monod affinity constant (K_s) and maximum specific growth rate (μ_{max}). According to the kinetic selection theory, at low substrate concentration, filamentous organisms achieve a high substrate removal rate compared with that of the floc-forming bacteria that prevail at high substrate concentration (Chiesa and Irvine 1985; J. Chudoba 1985). For example, the growth of *Microthrix parvicella* and the settling problems of the activated sludge resulting from excessive growth of this filamentous species always appear in the municipal wastewater treatment plants with biological oxygen demand (BOD_5) sludge loading rates of less than or equal to 0.1 kg kg^{-1} day^{-1} (Knoop and

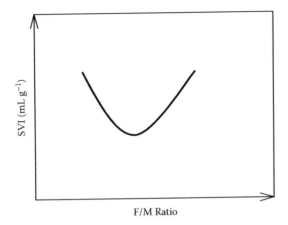

FIGURE 15.1 Effect of ratio of food to microorganisms (F/M) on sludge volume index (SVI). (Adapted from Droste, R. L. 1997. *Theory and practice of water and wastewater treatment.* New York: John Wiley & Sons.)

Kunst 1998). In the continuous activated sludge process, the ratio of food to microorganisms (F/M) is often used to describe the food availability to microorganisms. Figure 15.1 shows the effect of F/M ratio on sludge settleability in terms of the sludge volume index (SVI). The low substrate concentration-associated filamentous growth and formation of pinpoint floc are often referred to as low F/M sludge bulking.

15.2.3 Dissolved Oxygen Concentration

The growth of some filamentous bacteria, such as *Sphaerotilus* and *Haliscomenobacter hydrossis*, is favored by relatively low dissolved oxygen (DO) concentrations (Bitton 1999; Eckenfelder 2000). For example, the growth of *Thiothrix sp.* was favored by low DO concentrations (S. Lee et al. 2003), while other filamentous bacteria, such as *M. parvicella* can grow over a wide range of DO concentrations (Rossetti et al. 2005). T[-1] Benefield, Randall, and King (1975). So far, deficiency of DO is believed to be one of the major causes responsible for most filamentous growth in the activated sludge processes.

15.2.4 Solids Retention Time (SRT)

Because filamentous bacteria are slow growing, a long SRT favors their growth compared to growth of floc-forming microorganisms. For a typical filamentous bacterium, such as *M. parvicella* the maximum specific growth rate is 0.38 to 1.44 day[-1] (Jenkins 1992; Tandoi et al. 1998; Rossetti et al. 2005). However, there are different views with regard to the role of SRT in the growth of filamentous organisms. Droste (1997) thought that in a complete-mix reactor, the activated sludge would tend to become populated with filamentous organisms that exhibit poor settleability and the sludge does not flocculate well; on the other hand, if the complete-mix activated sludge reactor is operated at very long SRT, the sludge would present in the form of pinpoint flocs. It should also be pointed out that SRT and F/M ratio are interrelated in the activated sludge process in a way such that (Droste 1997):

$$\frac{1}{SRT} = Y(F/M) - k_d \qquad (15.1)$$

in which Y is the growth yield of activated sludge and k_d is the decay rate constant.

15.2.5 NUTRIENT DEFICIENCY

Nutrient deficiency can encourage the growth of filamentous organisms. This indeed is in line with the kinetic selection theory for filamentous growth (J. Chudoba, Grau, and Ottava 1973). In general, filamentous organisms have a higher surface-to-volume (A/V) ratio than nonfilamentous bacteria. This high A/V ratio enables them to take up nutrients from culture media containing low levels of nutrients (e.g. nitrogen, phosphorous, and other trace elements). This phenomenon is often observed in the activated sludge process for industrial wastewater treatment. In addition, non-filamentous sludge bulking caused by a nitrogen deficiency in industrial wastewater treatment was also reported, for example, the activated sludge settled properly at an influent BOD/N ratio of 100/4, while filamentous organisms tended to grow excessively at one time during the reaction process when the BOD/N ratio was controlled at 100/3; however, afterwards, the number of filamentous organisms began to reduce. Meanwhile, an excessive growth of viscous Zoogloea was observed and nonfilamentous activated sludge bulking occurred subsequently (Y. Peng et al. 2003).

15.2.6 TEMPERATURE

Temperature affects all biological reactions. The temperature coefficient for floc-forming bacteria is 1.015 for municipal wastewater (Eckenfelder 2000), while the estimated temperature coefficient values for *M. parvicella* strains 4B and RN1 are 1.140 and 1.105, respectively (Rossetti et al. 2002). For example, it was found the high temperature favors the growth of *Nocardia sp.* (S. Lee et al. 2003). In fact, the temperature-dependent filamentous growth can be interpreted well by the kinetic selection theory developed by J. Chudoba, Grau, and Ottava (1973).

15.2 OUTGROWTH OF FILAMENTOUS BACTERIA IN AEROBIC GRANULAR SLUDGE SBRS

As discussed in the preceding chapters, filamentous growth was found to be dominant in glucose-fed aerobic granules, while aerobic granules grown on acetate tended to become populated with nonfilamentous bacteria (figure 15.2). However, figure 15.3 further shows that even in acetate-fed aerobic granules, low levels or moderate levels of filamentous bacteria can still be observed, and they likely serve as a backbone that helps strengthen the spatial structure of aerobic granules. Filamentous bacteria have also been found in phenol-fed aerobic granules (Jiang 2005) and in dairy effluent-fed aerobic granules (Schwarzenbeck, Borges, and Wilderer 2005). These findings imply that filamentous growth in aerobic granules is a very common phenomenon.

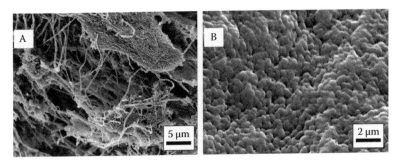

FIGURE 15.2 Microstructures of aerobic granules grown on glucose (a) and acetate (b). (From Tay, J. H., Liu, Q.-S., and Liu, Y. 2002. *Environ Technol* 23: 931–936. With permission.)

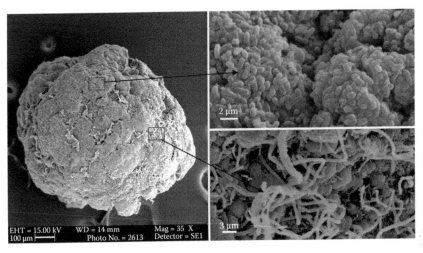

FIGURE 15.3 Coexistence of nonfilamentous and filamentous bacteria in acetate-fed aerobic granules. (From Liu, Y. Q. 2005. Research report, Nanyang Technological University, Singapore. With permission.)

SVI has been commonly used as an excellent indicator of sludge settleability that may indirectly reflect filamentous growth in activated sludge processes. Figure 15.4 shows changes in SVI and biomass concentration observed in a pilot-scale aerobic granular sludge SBR fed with an acetate-based synthetic wastewater (Y. Q. Liu 2005). It is apparent that SVI tended to decline along with the formation of aerobic granules, and such a trend was coupled with an increase in biomass concentration. It was found that aerobic granules were highly stable from day 40 to day 100; afterwards a sharp increase in SVI was observed (figure 15.4), indicating occurrence of filamentous growth in aerobic granules. This point was further confirmed by microscopic observations, as shown in figure 15.5.

In the activated sludge process, the sludge settleability can be classified according to SVI. In general, activated sludge has very good settling characteristics if its SVI value is below 80 mL g^{-1}. Figures 15.2 and 15.5 both show that the excessive growth of filamentous bacteria in or on the aerobic granule causes poor settleability and

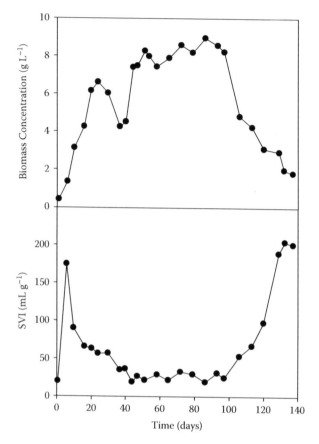

FIGURE 15.4 Changes in biomass concentration and sludge volume index (SVI) in an aerobic granular sludge SBR. (Data from Liu, Y. Q. 2005. Research report, Nanyang Technological University, Singapore.)

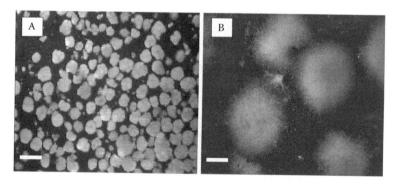

FIGURE 15.5 Morphology of nonfilamentous aerobic granules (a) on day 58 corresponding tofigure 15.4 and filamentous aerobic granules (b) on day 129 corresponding to figure 15.4. Bar: 2 mm. (From Liu, Y. Q. 2005. Research report, Nanyang Technological University, Singapore. With permission.)

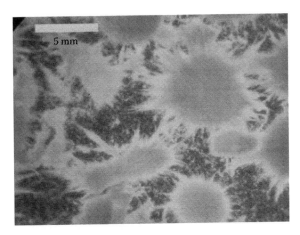

FIGURE 15.6 Outgrowth of filamentous bacteria in aerobic granules grown on dairy effluent. (From Schwarzenbeck, N., Borges, J. M., and Wilderer, P. A. 2005. *Appl Microbiol Biotechnol* 66: 711–718. With permission.)

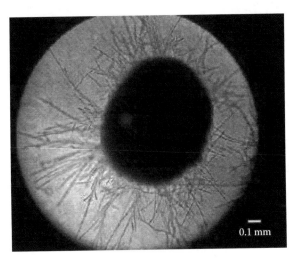

FIGURE 15.7 Filamentous organisms observed on the surface of aerobic granules grown on acetate. (From Wang, F. et al. 2004. IWA Workshop on Aerobic Granular Sludge, Sept. 26–28. Munich, Germany.)

subsequently the washout of granular sludge from the SBR. This in turn can explain the observed significant drop in biomass concentration (figure 15.4). Figure 15.6 shows the outgrowth of filamentous bacteria on aerobic granules grown on dairy effluent, while filamentous growth was also observed in aerobic granules grown on artificial wastewaters (figure 15.7, figure 15.8, and figure 15.9). Similar fluffy aerobic granules were observed in SBR treating low-strength domestic sewage (de Kreuk and van Loosdrecht 2006). It should be emphasized that low levels and moderate levels of filamentous growth do not cause operational problems, and may even stabilize the granule structure (figure 15.3).

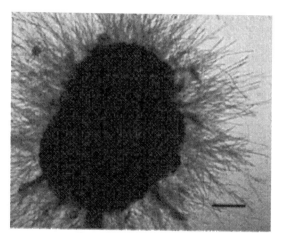

FIGURE 15.8 Filamentous growth in acetate-fed aerobic granules. (From Li, Z. H., Kuba, T., and Kusuda, T. 2006. *Enzym Microb Technol* 39: 976–981. With permission.)

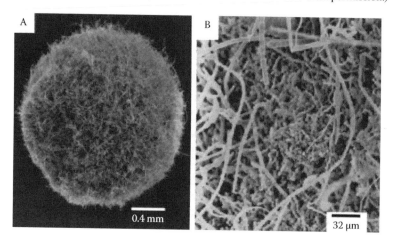

FIGURE 15.9 Filamentous structure observed in glucose-fed aerobic granules: (a) a macro-view and (b) a micro-view. (From Wang, Z. P. et al. 2006. *Chemosphere* 63: 1728–1735. With permission.)

Similar to the situation in the activated sludge process, overgrowth of filamentous bacteria in aerobic granular sludge SBRs is undesirable because it may eventually result in (1) poor settleability of aerobic granules; (2) washout of filamentous granules from the SBR; (3) out competition of filamentous granules over the nonfilamentous granules; (4) increased suspended solids concentration in effluent; and (5) eventual disintegration of aerobic granules. Therefore, the excessive filamentous growth would lead to a failure of the aerobic granular sludge SBR. In fact, occurrence of filamentous growth has been widely reported in aerobic granular sludge SBRs treating different kinds of wastewater (Moy et al. 2002; Pan 2003; McSwain, Irvine, and Wilderer 2004; Tay et al. 2004; F. Wang et al. 2004; Hu et al. 2005; Jiang 2005; Schwarzenbeck, Borges, and Wilderer 2005).

15.4 CAUSES OF FILAMENTOUS GROWTH IN AEROBIC GRANULAR SLUDGE SBRS

It is now clear that many factors can trigger filamentous growth in a biological process. In this section, some possible causes for the overgrowth of filamentous bacteria in aerobic granular sludge SBRs are identified and elaborated on.

15.4.1 TYPE OF SUBSTRATE

Filamentous growth in glucose-fed aerobic granules has been widely observed, while nonfilamentous structures were found in acetate-fed aerobic granules (figure 15.2). Scanning electron microscope (SEM) imaging revealed that glucose-fed granules cultivated at an organic loading rate (OLR) of 6 kg chemical oxygen demand (COD) m^{-3} d^{-1} had a hairy appearance with a loose microbial structure dominated by filamentous bacteria (figure 15.10). These observations clearly show differences in the morphology of both glucose- and acetate-fed granules that arise from the type of substrate. Energy-rich substrates, such as glucose and sucrose, are known to support the proliferation of filamentous bacteria in activated sludge (section 15.2.1) as well as in anaerobic and denitrifying granular sludge (van der Hoek 1988; Thaveesri et al. 1995). Substrate-mediated differences in granule microstructure have also been observed in upflow anaerobic sludge blanket (UASB) reactors treating a wide variety of wastewaters (Fang, Chui, and Li 1995), while the excess of filamentous bacteria caused a delay in anaerobic granulation (D. Zheng, Angenent, and Raskin 2006). It is evident that the substrate may exert a strong selection on filamentous organisms growing in aerobic granules; however, substrate alone may not offer a plausible explanation for the outgrowth of filamentous organisms in aerobic granules, as shown in figures 15.2 and 15.8.

The outgrowth of filamentous bacteria is usually detrimental to the activated sludge processes and can lead to operational disorders, such as sludge bulking and

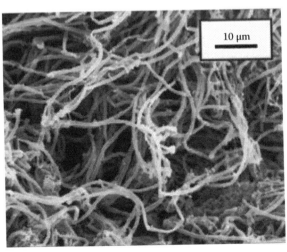

FIGURE 15.10 Microstructure of glucose-fed granules cultivated at 6 kg COD m^{-3} d^{-1}. (From Moy, B. Y. P. et al. 2002. *Lett Appl Microbiol* 34: 407–412. With permission.)

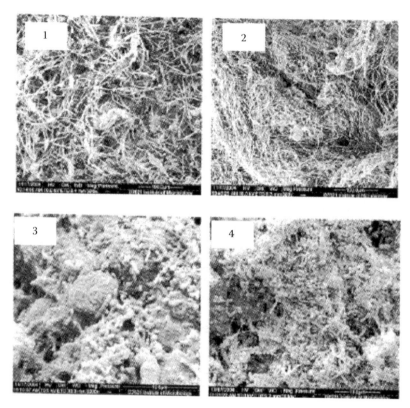

FIGURE 15.11   Aerobic granules grown on different types of carbon source. 1: acetate; 2: glucose; 3: peptone; 4: fecula. (From Sun, F. Y. et al. 2006. *J Environ Sci (China)* 18: 864–871. With permission.)

foaming. Filamentous bacteria create a loose microbial structure in glucose-fed granules with adequate settling and strength characteristics, and such a loose micro-bial structure enables the glucose-fed granules to sustain significantly higher OLRs than the denser and more compact microstructure in the acetate-fed granules, before mass transfer becomes restrictive (Moy et al. 2002).

Sun et al. (2006) investigated the effect of carbon source on the morphology and characteristics of aerobic granules (figure 15.11). Filamentous organisms were devel-oped in aerobic granules fed with acetate and glucose under given culture conditions, while nonfilamentous structures were observed in aerobic granules grown on peptone and fecula. This seems to indicate a close correlation of filamentous growth in aerobic granules to the property of organic carbon source employed. Among all four carbon sources studied, Sun et al. (2006) thought that peptone would be the optimal carbon source for cultivating more stable aerobic granules with excellent settleability.

15.4.2   Long Solids Retention Time in Aerobic Granular Sludge SBRs

SRT represents the average retention time of biomass in a biological system, and is known to be inversely correlated with the specific growth rate of microorganisms:

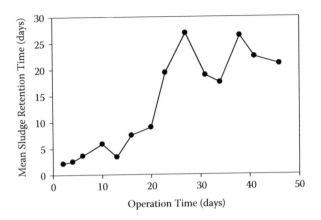

FIGURE 15.12 Fluctuation of sludge retention time in an aerobic granular sludge SBR. (From Liu, Q.-S. 2003. Ph.D. thesis, Nanyang Technological University, Singapore. With permission.)

$$\frac{1}{SRT} = \text{specific growth rate} - \text{specific decay rate} \qquad (15.2)$$

In general, a long SRT means a low specific growth rate. In the field of wastewater biological treatment engineering, the mean SRT can be manipulated according to:

$$SRT = \frac{\text{Total biomass in system}}{\text{Daily desludge rate}} \qquad (15.3)$$

More sludge discharged daily would result in a shorter SRT.

During the formation of aerobic granules, a substantial amount of suspended sludge is discharged out of the SBR in accordance with the preset selection pressures in terms of settling time, volume exchange ratio, and effluent discharge time (see chapter 6). As equation 15.2 shows, such an operation strategy would lead to a low SRT in the period of granulation (figure 15.12). However, along with aerobic granulation, the settleability of the biomass is progressively improved. As a result, the SRT tends to gradually stabilize at about 25 days. In most aerobic granular sludge SBRs, the SRT is not strictly controlled because of the selection pressure-based operation strategy, and it may vary with changes in sludge settleability under given selection pressures. A similar observation to that shown in figure 15.12 was also reported by Pan (2003).

It has been hypothesized that filamentous bacteria have much lower maximum specific growth rates than floc-forming bacteria, as illustrated in figure 15.13 (J. Chudoba 1985). If so, a long SRT would favor filamentous growth. Based on a survey of domestic wastewater treatment plants, it has been concluded that an SRT of longer than 10 days would generally cause serious filamentous growth because of the presence of *M. parvicella* (Richard 1989). Lin (2003) found that microbial granules developed at an SRT of about 10 days were quite stable, with a small granule size and

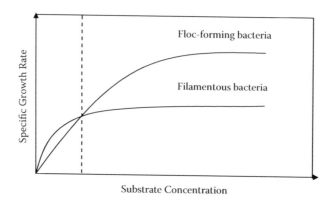

FIGURE 15.13 Specific growth rates of floc-forming and filamentous bacteria versus substrate concentration. (Adapted from Chudoba, J. 1985. *Water Res* 19: 1017–1022.)

absence of a fluffy outer growth, but in SBRs run at the SRT of 70 days, microbial granules turned from nonfilamentous to fluffy or filamentous structure. Consequently, the settleability of granules became poor, and they were eventually washed out of the reactor. Thus, for successful operation of aerobic granular sludge SBRs, SRT should be carefully controlled in order to ensure that it is within a range that is generally acceptable for floc-forming bacteria, as outlined by Metcalf and Eddy (2003).

15.4.3 Substrate Concentration and Concentration Gradients

Generally, aerobic granular sludge SBRs often receive constant influent organics concentration in terms of chemical oxygen demand (COD) (Beun et al. 1999; D. C. Peng et al. 1999; Tay, Liu, and Liu 2001; Moy et al. 2002; Arrojo et al. 2004; L. L. Liu et al. 2005). After the formation of aerobic granules, biomass concentration in the SBR is typically in the range of several to 20 grams per liter, or even higher. For a batch culture, the ratio of initial substrate concentration (S_o) to initial biomass concentration (X_o) can be used to describe the availability of food to microorganisms (P. Chudoba, Capdeville, and Chudoba 1992; Grady, Smets, and Barbeau 1996). An aerobic granular sludge SBR involves a cyclic operation, and the initial biomass concentration (X_o) in each cycle varies with the number of cycles. Figure 15.14 and figure 15.15 exhibit a typical changing trend of the S_o/X_o ratio in an aerobic granular sludge SBR fed with acetate as the sole carbon source. The salient points of these two figures include (1) biomass concentration increases along with aerobic granulation until a stable level is reached; and (2) increased biomass concentration results in a low value of S_o/X_o. This may partially explain why filamentous growth is commonly observed in aerobic granular sludge SBRs under conditions of high biomass concentrations (figure 15.4). As noted by Eckenfelder (2000), with degradable substrates at low concentrations, filamentous growth is favored. As illustrated in figure 15.14, high substrate concentration favors the growth of the floc-forming bacteria over filamentous bacteria so that the floc-formers may dominate the system in this case.

Compared to activated sludge bioflocs, aerobic granules are larger in size, and have a regular shape and compact structure. Y. Li and Liu (2005) have shown that at

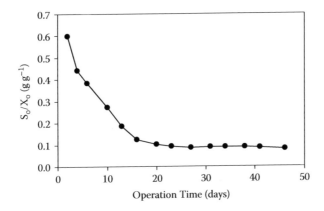

FIGURE 15.14 Change in the S_o/X_o ratio in the operation of aerobic granular sludge SBRs. (Data from Liu, Q.-S. 2003. Ph.D. thesis, Nanyang Technological University, Singapore.)

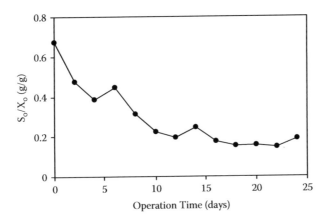

FIGURE 15.15 Change in the S_o/X_o ratio in the operation of an aerobic granular sludge SBR. (Data from Wang, Z.-W. 2007. Ph.D. thesis, Nanyang Technological University, Singapore.)

low bulk substrate concentration, substrate diffusion is a limiting factor in aerobic granular sludge SBRs. An example of the substrate gradient in an aerobic granule is presented in figure 15.16, indicating that the substrate concentration inside the aerobic granule is much lower than that in the bulk solution. Thus, the actual ratio of substrate to biomass in aerobic granules is much smaller than the S_o/X_o values shown in figures 15.14 and 15.15.

At diffusion limitation aerobic granules with porous structure and irregular shape are developed. A similar phenomenon has been reported in the biofilm process. For example, open, filamentous biofilm structures have been observed under low substrate concentration, whereas compact and smooth biofilms arise at high substrate concentrations (van Loosdrecht et al. 1995). In pure culture, the morphology of a microbial colony depends on the micro-gradients of the substrate, and the development of filamentous colonies was observed in low-substrate conditions (Ben-Jacob et al. 1994). Similarly, when compact bioflocs are subject to low substrate conditions,

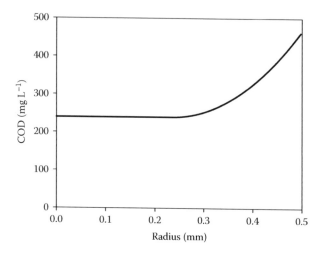

FIGURE 15.16 Concentration profile of substrate within an aerobic granule with a radius of 0.5 mm. (From Li, Y. 2007. Ph.D. thesis, Nanyang Technological University, Singapore. With permission.)

their structures become more open and filamentous (Martins, Heijnen, and van Loosdrecht 2003a). Consequently, substrate concentration can exert a double stress on a microbial community through a low level of substrate in the liquid phase and a steep gradient of substrate concentration within the granule. These factors apparently promote filamentous growth in aerobic granular sludge SBRs.

Y. M. Zheng et al (2006) investigated the instability of aerobic granular sludge cultivated in an SBR operated at a high OLR of 6.0 mg m^{-3} d^{-1}, and found that the compact bacteria-dominated aerobic granules were not stable and gradually transited to large-sized filamentous ones with a diameter of 16 mm after 30 days of operation. As a result, the hydrophobicity and specific gravity of aerobic granules tended to decrease significantly. For large filamentous granules, due to the mass transfer limitation and the possible presence of anaerobes in the core part of the granules, they began to disintegrate and were washed out of the reactor, leading to failure of the reactor. It is obvious that a high OLR is desirable in biological wastewater treatment systems, as it can facilitate the treatment of high-strength wastewaters using compact reactors with small footprints. Moy et al. (2002) reported that glucose-fed aerobic granules were able to sustain the maximum OLR of 15 kg COD m^{-3} d^{-1} (figure 15.17), and these granules initially exhibited a fluffy loose morphology dominated by filamentous bacteria at a low OLR, while the aerobic granules subsequently evolved into smooth irregular shapes characterized by folds, crevices, and depressions at the higher OLRs, and the tight bacteria clusters were observed within an extracellular polymeric matrix.

15.4.4 Dissolved Oxygen Deficiency in Aerobic Granules

An SBR is operated in a repeated cycle mode. Theoretically, the depth of dissolved oxygen (DO) penetration in an aerobic granule is determined by the DO concentration

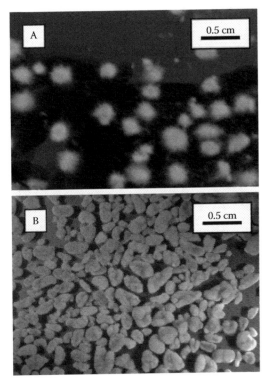

FIGURE 15.17 Morphology of glucose-fed aerobic granules at the OLRs of 6 kg COD m^{-3} d^{-1} (a), and 15 kg COD m^{-3} d^{-1} (b). (From Moy, B. Y. P. et al. 2002. *Lett Appl Microbiol* 34: 407–412. With permission.)

in the bulk solution and the oxygen consumption rate by aerobic granules under given culture conditions. It is well known that DO deficiency favors filamentous growth (Palm, Jenkins, and Parker 1980; Gaval and Pernelle 2003; Martins, Heijnen, and van Loosdrecht et al. 2003b; Rossetti et al. 2005). In aerobic granular sludge SBRs, it has been shown experimentally that a low DO concentration at 40% of the saturation in combination with a pulse feed of easily degradable substrate leads to enhanced filamentous growth and subsequent biomass washout, and as a result, stable granular sludge could not be maintained (Mosquera-Corral et al. 2005). From the point of view of industrial application, how to obtain stable aerobic granules at relatively low dissolved oxygen concentrations remains a challenge and needs to be resolved in the near future.

Figure 15.18 shows the typical DO profile in the liquid phase recorded during one cycle of operation of an aerobic granular sludge SBR, and a valley region of DO was found (Q.-S. Liu 2003). In an SBR the DO profile observed during the aerobic phase should be correlated to the oxidation of external substrate. It was found that the DO profile during one SBR cycle was coupled to the removal of external carbon source in terms of total organic carbon (TOC) (figure 15.19). It is apparent that the reduction of TOC through biological oxidation by aerobic granules results in a significant drop

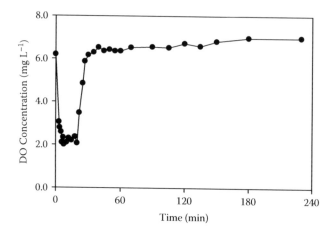

FIGURE 15.18 Dissolved oxygen profile in bulk solution during one cycle of operation of an aerobic granular sludge SBR. (Data from Liu, Q.-S. 2003. Ph.D. thesis, Nanyang Technological University, Singapore.)

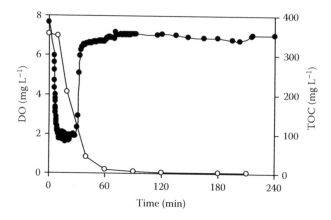

FIGURE 15.19 Profiles of DO (●) and TOC (○) during one cycle of operation of an aerobic granule SBR. (Data from Qin, L. 2006. Ph.D. thesis, Nanyang Technological University, Singapore. With permission.)

of DO in the reactor. After the depletion of external carbon source, the DO returns quickly to the saturation level under given conditions of operation.

The bulk DO concentrations reported in aerobic granular sludge SBRs have varied in the range of 2 mg L^{-1} to the saturation concentration. DO concentration below 1.1 mg L^{-1} has been found to have a negative effect on sludge settleability and eventually leads to growth of filamentous bacteria (Martins, Heijnen, and van Loosdrecht 2003b). A minimum DO concentration of 2 mg L^{-1} has been recommended for preventing the growth of filamentous bacteria, such as *Sphaerotilus natans* (J. Chudoba 1985). The work by Wilen and Balmer (1999) confirms that at low bulk DO concentrations of 0.5 to 2.0 mg L^{-1}, the produced sludge had poor settling property, leading to a high-turbidity effluent compared to those observed at

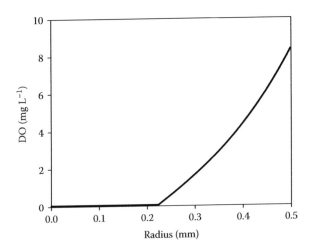

FIGURE 15.20   Concentration profile of DO in an aerobic granule with a radius of 0.5 mm. (From Li, Y. 2007. Ph.D. thesis, Nanyang Technological University, Singapore. With permission.)

high DO concentrations of 2.0 to 5.0 mg L^{-1}. In fact, deteriorated settling properties of sludge have been attributed mainly to excessive growth of filamentous bacteria and the formation of porous flocs (Wilen and Balmer 1999).

The bulk DO concentration in an aerobic granular sludge SBR needs to be sufficiently high in order to prevent filamentous growth. However, compared to conventional bioflocs with a mean size of less than 100 μm and a loose structure, aerobic granules have a large size (0.25 to several millimeters) and a highly compact structure. Therefore, the DO concentration gradient within the aerobic granule can be quite steep, and the bulk concentration of DO does not reasonably reflect the true situation within the aerobic granular sludge in the SBR. Y. Li and Liu (2005) showed that diffusion of organic substrate and DO in the aerobic granule is a dynamic process. In combination with the substrate utilization kinetics and molecular diffusion, a steep gradient of DO has been shown to exist in aerobic granules (figure 15.20 and figure 15.21). Indeed, sludge bulking has been previously hypothesized to originate from the presence of gradients of substrate, DO, and nutrient concentrations in aggregates of microbial sludge (Martins, Heijnen, and van Loosdrecht 2003a; Richard and Collins 2003; Rossetti et al. 2005).

In aerobic granular sludge SBRs, the substrate utilization kinetics can be divided into two regimens: (1) during operation at high bulk substrate concentration, the DO concentration within the aerobic granule is a limiting factor; (2) once the bulk substrate is depleted to low levels, the metabolic activity of the granule is thus determined by the substrate concentration in the granule (Y. Li and Liu 2005). Due to cyclic operation, granules are subjected to repeated limitations of DO and substrate. The repetition of oxygen deficiency or other stresses can be an important inducer of the development of dominant filamentous growth (Gaval and Pernelle 2003). The filamentous growth can be stimulated when DO is reduced in an aerobic granular sludge SBR. This indicates that in aerobic granular sludge SBRs filamentous growth is subject to diffusion-based selection. Overcoming the occurrence of filamentous growth induced by a low DO

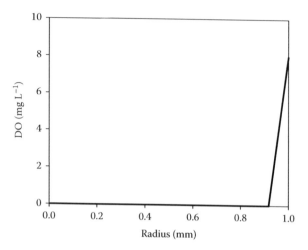

FIGURE 15.21 Concentration profile of DO in an aerobic granule with a radius of 1.0 mm. (From Li, Y. 2007. Ph.D. thesis, Nanyang Technological University, Singapore. With permission.)

level requires increasing the DO concentration in the bulk solution. However, this in turn demands an expanded capacity of the aeration equipment or a reduced load of organic matter. According to Palm, Jenkins, and Parker (1980), the bulk DO concentration (mg L^{-1}) in an activated sludge process that is needed for preventing filamentous growth depends on the specific substrate utilization rate (U, kg COD kg^{-1} MLVSS day^{-1}) and is greater than (U − 0.1)/0.22. It is obvious that aerobic granules are larger and have a more compact structure than the activated sludge flocs. Therefore, the relationship recommended by Palm, Jenkins, and Parker (1980) is not easily translated to the case of an aerobic granular sludge SBR.

15.4.5   NUTRIENT DEFICIENCY IN AEROBIC GRANULES

Nitrogen and phosphorus are two essential elements required for microbial growth. As pointed out earlier, deficiency in the nutrient supply, especially nitrogen, commonly results in bulking of activated sludge (Bitton 1999; Eckenfelder 2000; Metcalf and Eddy 2003; Y. Peng et al. 2003; Richard and Collins 2003; Rossetti et al. 2005). A minimum ammonia concentration of 1.5 mg L^{-1} has been recommended in the effluent to favor the growth of floc-forming microorganisms compared to that of the filamentous microorganisms. In some cases, ammonia concentrations greater than 1.5 mg L^{-1} may be required for effective suppression of filamentous growth (Eckenfelder 2000). Most studies of aerobic granulation have used synthetic organic wastewater with a COD/N ratio of 100:5. Because of potential diffusion limitations in aerobic granules, the situation in a granular sludge SBR is likely more complex than in a conventional activated sludge process.

According to the hypothesis of diffusion-based selection, substrate and nutrient gradients in activated sludge flocs would trigger filamentous growth, that is, a large concentration gradient within the floc favors selection of filamentous bacteria over floc-forming bacteria (Martins et al. 2004). For comparison purpose, diffusion coefficients can be used. The values of the diffusion coefficients are 2.5×10^{-9} m^2 s^{-1} for acetate (Beyenal and Tanyolac 1994), 1.67×10^{-9} m^2 s^{-1} for dissolved oxygen (Chen, Juaw, and Cheng 1991), and 1.01×10^{-9} m^2 s^{-1} for ammonia (Rittmann and Manem 1992). It is apparent that ammonia has the lowest diffusivity. This means that the localized COD:N ratio within the aerobic granules would be much lower than that in the bulk solution. Furthermore, a nitrogen deficiency may be encountered inside the aerobic granule. In the absence of sufficient nitrogen, microorganisms tend to produce a significant amount of extracellular polysaccharides (Aquino and Stuckey 2003; Richard and Collins 2003). This has been confirmed in aerobic granules where a substantial accumulation of polysaccharides in the core part of the granule has been observed (Z.-W. Wang, Liu, and Tay 2005). Therefore, to better understand filamentous growth in aerobic granular sludge SBRs, possible nutrient deficiency within the aerobic granules should be seriously taken into account. For the activated sludge process, the recommended effluent total inorganic nitrogen and phosphate-P concentrations are 1 to 2 mg L^{-1} for ensuring sufficiency of these nutrients (Richard and Collins 2003).

High production of extracellular polymeric substances in aerobic granular sludge has been commonly reported (see chapter 10). According to Richard and Collins (2003), overproduction of extracellular polymeric substances is an important sign of nutrient deficiency in the biological wastewater process. Jelly-like and viscous aerobic granules have been found in aerobic granular sludge SBRs operated under nutrient deficiency conditions (Z.-W. Wang, Liu, and Tay 2005). Similarly, jelly-like activated sludge flocs have been often found in nutrient-deficient cultures in which the floc contains as much as 90% extracellular polymeric substances on a dry weight basis (Bitton 1999; Richard and Collins 2003). Extracellular polymeric substances-rich flocs and aerobic granules have settling and stability problems.

15.4.6 TEMPERATURE SHIFT IN AEROBIC GRANULAR SLUDGE SBRs

As mentioned in chapter 14, section 14.2.6, the influence of the operating temperature on sludge morphology in activated sludge processes can be reasonably explained by the kinetic selection theory (J. Chudoba, Grau, and Ottava 1973). However, there is no published experimental data on the temperature-dependent filamentous growth in aerobic granular sludge SBRs. Preliminary work suggests that in one case at least, aerobic granular sludge SBRs run at 25°C were more susceptible to filamentous growth compared to a similar reactor operated at 17°C (Y. Q. Liu 2005). This observation is consistent with the kinetic selection theory. In a study of temperature-dependent settleability of activated sludge, the SVI of activated sludge was strongly increased with increase in temperature (Krishna and van Loosdrecht 1999). In addition, increased temperature will reduce the DO concentration and this in turn can promote filamentous growth, as discussed earlier.

de Kreuk, Pronk, and van Loosdrecht (2005) investigated the formation of aerobic granules in an aerobic granular sludge SBR operated at moderate and low temperatures. Their results showed that if the SBR was initiated at a low temperature of 8°C, irregular and unstable aerobic granules developed, while in an SBR that was started up at 20°C and subsequently the temperature was lowered to 15°C

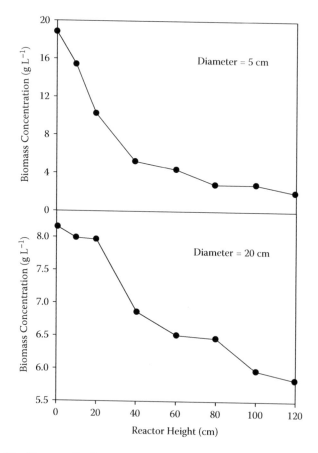

FIGURE 15.22 Biomass distribution in aerobic granular sludge SBR with diameters of 5 and 20 cm. (Data from Tay, J. H. et al. 2004. *J Environ Eng* 130: 1102–1109.)

and further to 8°C, no adverse effects on granule stability and biomass retention were observed. According to de Kreuk, Pronk, and van Loosdrecht (2005), aerobic granular sludge SBRs should be initiated preferentially at high temperature, while lowered temperatures should not pose serious problems for granule stability and the reactor performance in terms of the removal of COD and phosphate after successful development of aerobic granules in the system.

15.4.7 FLOW PATTERNS IN AEROBIC GRANULAR SLUDGE SBRS

The kinetic selection theory shows that a high substrate gradient in the bulk solution can suppress excessive filamentous growth (J. Chudoba, Grau, and Ottava1973). To achieve such a substrate gradient, a plug-flow reactor configuration needs to be employed (J. Chudoba 1985; P. Chudoba 1991; Prendl and Kroiss 1998; Eckenfelder 2000). Tay et al. (2004) examined the granular biomass distribution in two aerobic granular sludge SBRs with respective diameters of 5 cm and 20 cm, operated under identical conditions (figure 15.23). A uniform biomass distribution along the reactor

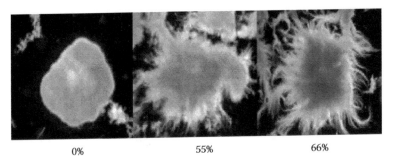

0%   55%   66%

FIGURE 15.23 Morphology of aerobic granules developed in three SBRs with a respective aerated fill of 0%, 55%, and 66%. (From McSwain, B. S., Irvine, R. L., and Wilderer, P. A. 2004. *Water Sci Technol* 49: 19–25. With permission.)

height was found in the 20-cm SBR, while a substantial axial gradient of biomass was observed in the 5-cm SBR. These results suggest a completely mixed flow in the 20-cm SBR and a possible plug-like flow in the 5-cm SBR. The 20-cm aerobic granular sludge SBR was found to favor progressive growth of filamentous bacteria, leading to complete disappearance of aerobic granules over a period of 100 days. In contrast, aerobic granules were stably maintained in the 5-cm SBR, in which no excessive filamentous growth was observed. Y. Liu and Tay (2002) postulated that compared to a completely mixed flow pattern, a circulatory flow in SBRs would facilitate aerobic granulation. Clearly, further study is required on how the hydrodynamics in aerobic granular sludge SBRs affect the development and stability of aerobic granules.

15.4.8 LENGTH OF AEROBIC FEEDING

Aerobic granules are often cultivated in SBRs with a very short fill or feeding time. McSwain, Irvine, and Wilderer (2004) investigated the effect of intermittent feeding on aerobic granule structure. Basically, intermittent feeding in an SBR can create a period of high load, followed by a period of starvation in which external substrate is no long available for microbial growth. It was found that aerobic granules formed in all reactors operated at different aerated fill lengths ranging from 0% to 66% of the total 90-minute fill phase. It can be seen in figure 15.24 that a long aerated fill length seems to promote the development of filamentous granules, while compact and stable aerobic granules were formed in the reactor with dump fill (e.g. 0% aerated fill). It seems that an intermittent feeding strategy may affect the selection and growth of filamentous organisms.

Decreased granule concentration and increased SVI were observed with increase in the length of aerated fill from 0% to 66% (figure 15.25). These results seem to imply that filamentous organisms can compete for organic substrate in the reactors operated under the scheme of long aerated fill because of continuous feeding during aeration, whereas the situation in the reactor run at the aerated fill of 0% is different. As shown in figure 15.14, floc-forming organisms with high substrate utilization kinetics are dominant over filamentous organisms. The following conclusions can be drawn from the study by McSwain, Irvine, and Wilderer (2004): (1) feeding strategy

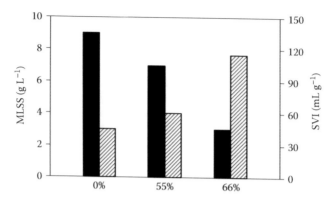

FIGURE 15.24 Average MLSS ■ and SVI ▨ at pseudo-steady state in SBRs operated at the respective aerated fill of 0%, 55%, and 66%. (Data from McSwain, B. S., Irvine, R. L., and Wilderer, P. A. 2004. *Water Sci Technol* 49: 19–25.)

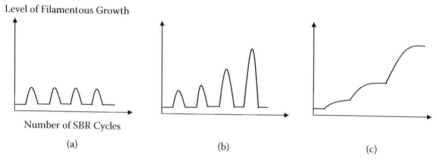

FIGURE 15.25 Propagation patterns of filamentous growth subject to periodic stresses in an aerobic granular sludge SBR. (Adapted from Gaval, G. and Pernelle, J. J. 2003. *Water Res* 37: 1991–2000.)

has a significant effect on the structure of aerobic granules; (2) pulse feeding seems to favor the formation of compact and dense aerobic granules; and (3) the feeding strategy also alters the distribution of floc-forming bacteria over filamentous organisms in aerobic granules.

15.5 PROPAGATION PATTERNS OF FILAMENTOUS GROWTH IN AEROBIC GRANULAR SLUDGE SBRS

Development of filamentous growth can be regarded as a progressive process that is associated with operational stresses, such as DO deficiency, nutrient deficiency, and low substrate availability. Filamentous bacteria are likely to be present in almost all kinds of aerobic granules, but at different levels. Gaval and Pernelle (2003) thought that repetitive stresses would trigger a progressive increase in filamentous bacteria. In aerobic granular sludge SBRs, the stresses are repetitive with the cyclic operation. According to the repetitive stress theory and experimental observations, growth of filamentous bacteria in aerobic granular sludge SBRs subjected to periodic stresses

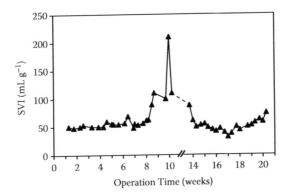

FIGURE 15.26 Change of sludge volume index (SVI) in an aerobic granular sludge SBR. (Data from Schwarzenbeck, N., Borges, J. M., and Wilderer, P. A. 2005. *Appl Microbiol Biotechnol* 66: 711–718.)

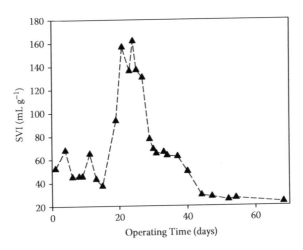

FIGURE 15.27 Change in SVI during aerobic granulation in an SBR fed with a synthetic wastewater containing mainly sucrose, peptone, and beef extract. (Data from Zheng, Y. M., Yu, H. Q., and Sheng, G. P. 2005. *Process Biochem* 40: 645–650.)

can be classified into three types: type 1, low-level or moderate-level filamentous growth (figure 15.26a); type 2, transient filamentous growth (figure 15.26b); and type 3, staircase filamentous growth (figure 15.26c).

In the first response type (figure 15.26a), filamentous growth is negligible or occurs at a low level without significant impact on granule settleability and stability. This type of filamentous growth pattern has been observed experimentally in laboratory-scale aerobic granular sludge SBRs (Q.-S. Liu 2003; Yang, Tay, and Liu 2003; de Kreuk and van Loosdrecht 2004; Tay et al. 2004; L. L. Liu et al. 2005). The second response type (figure 15.26b) has been observed in aerobic granular sludge SBRs treating dairy effluent and a synthetic wastewater, as illustrated in figure 15.26 and figure 15.27 (Schwarzenbeck, Borges, and Wilderer 2005; Y. M. Zheng, Yu, and Sheng 2005). The third response type (figure 15.26c) is similar to that shown in figure 15.4 and it eventually results in operational failure of the aerobic granular sludge SBR.

15.6 CONTROL STRATEGY FOR FILAMENTOUS GROWTH

Filamentous growth in aerobic granular sludge SBRs is a more complicated phenomenon than in conventional activated sludge processes. This is mainly because the diffusion of substrate, DO, and nutrients in aerobic granules can create stresses that result in progressive filamentous growth. In aerobic granular sludge SBRs, multiple stresses can exist simultaneously and filamentous growth may be a consequence of the combined effect of more than one stresses.

In the activated sludge process, almost all control strategies of filamentous growth are based on the kinetic or metabolic selection theory, for example, aerobic, anaerobic, and anoxic selectors have been commonly employed to suppress filamentous growth by creating high substrate gradients in the liquid phase (J. Chudoba, Grau, and Ottava 1973; Wanner et al. 1987; P. Chudoba 1991; Pujol and Canler 1994; Prendl and Kroiss 1998; Eckenfelder 2000; Y. Lee and Oleszkiewicz 2004). Likely, the same principle can be applied to control filamentous growth in aerobic granular sludge SBRs.

In order to control filamentous growth in aerobic granular sludge SBRs, the following points need to be carefully considered:

1. SRT must be controlled through daily granule discharge.
2. To avoid the DO-limiting situation, granule concentration should be controlled within a reasonable range.
3. In view of intraparticle diffusion of substrate, nutrients, and DO, influent COD:N:P ratio and DO level derived from the activated sludge processes should be reexamined.
4. In most aerobic granular sludge SBRs, feeding time is often as short as a few minutes. This diminishes substrate gradients in the liquid phase. Addition of substrate in various aerobic feeding periods could create strong substrate gradients in SBR and, consequently, improve the settleability of sludge (Nowak, Brown, and Yee 1986; Bitton 1999; Martins, Heijnen, and van Loosdrecht 2003a; McSwain, Irvine, and Wilderer 2004). As pointed out by Bitton (1999), intermittent feeding patterns create favorable conditions for the development of nonfilamentous bacteria that have high substrate uptake rates during periods of high substrate concentration and a capacity to store reserve materials during periods of starvation. In fact, increasing evidence shows a positive role of materials storage in control of filamentous growth (Martins et al. 2004).
5. Aerobic granules have a confined structure with different microbial species coexisting. Research shows that selection of slow-growing bacteria (e.g. P-accumulating bacteria and nitrifying bacteria) can help suppress filamentous growth and further improve stability of aerobic granules (Lin, Liu, and Tay 2003; de Kreuk and van Loosdrecht 2004; Y. Liu, Yang, and Tay 2004; Zhu, Liu, and Wilderer 2004).

15.7 CONCLUSIONS

Filamentous growth has been commonly reported in aerobic granular sludge SBRs. This chapter discusses the problem of instability in aerobic granular sludge SBRs from the perspective of filamentous growth. The possible causes of filamentous growth are identified, including long retention times of solids, low substrate concentration in the liquid phase, high substrate gradient within the granules, DO and nutrient deficiency in aerobic granules, temperature shift, and flow patterns. Present experimental evidence shows that intermittent substrate feeding can offer an effective means of creating high substrate gradients in the liquid phase to help control filamentous overgrowth in aerobic granular sludge SBRs.

REFERENCES

Aquino, S. F. and Stuckey, D. C. 2003. Production of soluble microbial products (SMP) in anaerobic chemostats under nutrient deficiency. *J Environ Eng* 129: 1007–1014.

Arrojo, B., Mosquera-Corral, A., Garrido, J. M., and Mendez, R. 2004. Aerobic granulation with industrial wastewater in sequencing batch reactors. *Water Res* 38: 3389–3399.

Ben-Jacob, E., Schochet, O., Tenenbaum, A., Cohen, I., Czirok, A., and Vicsek, T. 1994. Generic modelling of cooperative growth patterns in bacterial colonies. *Nature* 368: 46.

Benefield, L. D., Randall, C. W., and King, P. H. 1975. Stimulation of filamentous microorganisms in activated sludge by high oxygen concentrations. *Water, Air Soil Pollut* 5: 113–123.

Beun, J. J., Hendriks, A., van Loosdrecht, M. C. M., Morgenroth, E., Wilderer, P. A., and Heijnen, J. J. 1999. Aerobic granulation in a sequencing batch reactor. *Water Res* 33: 2283–2290.

Beyenal, H. and Tanyolac, A. 1994. Calculation of simultaneous effective diffusion coefficients of the substrates in a fluidized bed biofilm reactor. *Water Sci Technol* 29: 463–470.

Bitton, G. (1999) *Wastewater microbiology*. New York: Wiley-Liss.

Chen, S. K., Juaw, C. K., and Cheng, S. S. 1991. Nitrification and denitrification of high-strength ammonium and nitrite wastewater with biofilm reactors. *Water Sci Technol* 23: 1417–1425.

Chiesa, S. C. and Irvine, R. L. 1985. Growth and control of filamentous microbes in activated sludge: An integrated hypothesis. *Water Res* 19: 471–479.

Chudoba, J. (1985) Control of activated sludge filamentous bulking. VI. Formulation of basic principles. *Water Res* 19: 1017–1022.

Chudoba, J., Grau, P., and Ottava, V. 1973. Control of activated sludge filamentous bulking. II. Selection of microorganism by means of a selector. *Water Res* 7: 1389–1406.

Chudoba, P. 1991. Etude et intérêt du découplage énergétique dans les processus d'épuration des eaux par voie biologique procédé OSA. Ph.D. thesis, INSA-Toulouse, France.

Chudoba, P., Capdeville, B., and Chudoba, J. 1992. Explanation of biological meaning of the S_O/X_O ratio in batch cultivation. *Water Sci Technol* 26: 743–751.

de Kreuk, M. K. and van Loosdrecht, M. C. M. 2004. Selection of slow growing organisms as a means for improving aerobic granular sludge stability. *Water Sci Technol* 49: 9–17.

de Kreuk, M. K. and van Loosdrecht, M. C. M. 2006. Formation of aerobic granules with domestic sewage. *J Environ Eng ASCE* 132: 694–697.

de Kreuk, M. K., Pronk, M., and van Loosdrecht, M. C. M. 2005. Formation of aerobic granules and conversion processes in an aerobic granular sludge reactor at moderate and low temperatures. *Water Res* 39: 4476–4484.

Droste, R. L. 1997. *Theory and practice of water and wastewater treatment.* New York: John Wiley & Sons.

Eckenfelder, W. W. 2000. *Industrial water pollution control*, 3rd ed. Boston: McGraw-Hill.

Fang, H. H. P., Chui, H. K., and Li, Y. Y. 1995. Effect of degradation kinetics on the microstructure of anaerobic biogranules. *Water Sci Technol* 32: 165–172.

Gaval, G. and Pernelle, J. J. 2003. Impact of the repetition of oxygen deficiencies on the filamentous bacteria proliferation in activated sludge. *Water Res* 37: 1991–2000.

Grady, C. P. L., Smets, B. F., and Barbeau, D. S. 1996. Variability in kinetic parameter estimates: A review of possible causes and a proposed terminology. *Water Res* 30: 742.

Hu, L. L., Wang, J. L., Wen, X. H., and Qian, Y. 2005. The formation and characteristics of aerobic granules in sequencing batch reactor (SBR) by seeding anaerobic granules. *Process Biochem* 40: 5–11.

Jenkins, D. 1992. Towards a comprehensive model of activated-sludge bulking and foaming. *Water Sci Technol* 25: 215–230.

Jiang, H. L. 2005. Development of aerobic granules for enhanced biological wastewater treatment. Ph.D. thesis, Nanyang Technological University, Singapore.

Kappeler, J. and Gujer, W. 1994. Influences of wastewater composition and operating conditions on activated sludge bulking and scum formation. *Water Sci Technol* 30: 181.

Knoop, S. and Kunst, S. 1998. Influence of temperature and sludge loading on activated sludge settling, especially on *Microthrix parvicella*. *Water Sci Technol* 37: 27–35.

Krishna, C. and van Loosdrecht, M. C. M. 1999. Effect of temperature on storage polymers and settleability of activated sludge. *Water Res* 33: 2374–2382.

Lee, S., Basu, S., Tyler, C. W., and Pitt, P. A. 2003. A survey of filamentous organisms at the Deer Island Treatment Plant. *Environ Technol* 24: 855–865.

Lee, Y. and Oleszkiewicz, J. A. 2004. Bench-scale assessment of the effectiveness of an anaerobic selector in controlling filamentous bulking. *Environ Technol* 25: 751–755.

Li, Y. 2007. Metabolic behaviors of aerobic granules developed in sequencing batch reactor. Ph.D. thesis, Nanyang Technological University, Singapore.

Li, Y. and Liu, Y. 2005. Diffusion of substrate and oxygen in aerobic granule. *Biochem Eng J* 27: 45–52.

Li, Z. H., Kuba, T., and Kusuda, T. 2006. Selective force and mature phase affect the stability of aerobic granule: An experimental study by applying different removal methods of sludge. *Enzyme Microb Technol* 39: 976–981.

Lin, Y. M. 2003. Development of P-accumulating microbial granules in SBR. Interim Ph.D. report, Nanyang Technological University, Singapore.

Lin, Y. M., Liu, Y., and Tay, J. H. 2003. Development and characteristics of phosphorus-accumulating microbial granules in sequencing batch reactors. *Appl Microbiol Biotechnol* 62: 430–435.

Liu, L. L., Wang, Z., Yao, J., Sun, X., and Cai, W. 2005. Investigation on the formation and kinetics of glucose-fed aerobic granular sludge. *Enzyme Microb Technol* 36: 712–716.

Liu, Q.-S. 2003. Aerobic granulation in sequencing batch reactor. Ph.D. thesis, Nanyang Technological University, Singapore.

Liu, Y. and Tay, J. H. 2002. The essential role of hydrodynamic shear force in the formation of biofilm and granular sludge. *Water Res* 36: 1653–1665.

Liu, Y., Yang, S. F., and Tay, J. H. 2004. Improved stability of aerobic granules by selecting slow-growing nitrifying bacteria. *J Biotechnol* 108: 161–169

Liu, Y. Q. 2005. Comparative study of aerobic granulation in bubble column and airlift SBRs. Internal research report, Nanyang Technological University, Singapore.

Liu, Y. Q., Liu, Y., and Tay, J. H. 2004. The effects of extracellular polymeric substances on the formation and stability of biogranules. *Appl Microbiol Biotechnol* 65: 143–148.

Martins, A. M. P., Heijnen, J. J., and van Loosdrecht, M. C. M. Effect of feeding pattern and storage on the sludge settleability under aerobic conditions. *Water Res* 37: 2555–2570.

Martins, A. M. P., Heijnen, J. J., and van Loosdrecht, M. C. M. Effect of dissolved oxygen concentration on sludge settleability. *Appl Microbiol Biotechnol* 62: 586–593.

Martins, A. M. P., Pagilla, K., Heijnen, J. J., and van Loosdrecht, M. C. M. Filamentous bulking sludge: A critical review. *Water Res* 38: 793–817.

McSwain, B. S., Irvine, R. L., and Wilderer, P. A. 2004. The effect of intermittent feeding on aerobic granule structure. *Water Sci Technol* 49: 19–25.

Metcalf and Eddy, 2003. *Wastewater engineering: Treatment and reuse*, 4th ed., revised by George Tchobanoglous, Franklin L. Burton, and H. David Stensel. Boston: McGraw-Hill.

Mosquera-Corral, A., de Kreuk, M. K., Heijnen, J. J., and van Loosdrecht, M. C. M. 2005. Effects of oxygen concentration on N-removal in an aerobic granular sludge reactor. *Water Res* 39: 2676–2686.

Moy, B. Y.P., Tay, J. H., Toh, S. K., Liu, Y., and Tay, S. T. L. 2002. High organic loading influences the physical characteristics of aerobic sludge granules. *Lett Appl Microbiol* 34: 407–412.

Nowak, G., Brown, G., and Yee, A. 1986. Effects of feed pattern and dissolved oxygen on growth of filamentous bacteria. *J WPCF* 58: 978–984.

Palm, J. C., Jenkins, D., and Parker, D. S. 1980. Relationship between organic loading, dissolved oxygen concentration, and sludge settleability in the completely-mixed activated sludge process. *J WPCF* 52: 2484–2506.

Pan, S. 2003. Inoculation of microbial granular sludge under aerobic conditions. Ph.D. thesis, Nanyang Technological University, Singapore.

Peng, D. C., Bernet, N., Delgenes, J. P., and Moletta, R. 1999. Aerobic granular sludge: A case report. *Water Res* 33: 890–893.

Peng, Y., Gao, C., Wang, S., Ozaki, M., and Takigawa, A. 2003. Non-filamentous sludge bulking caused by a deficiency of nitrogen in industrial wastewater treatment. *Water Sci Technol* 47: 289–295.

Prendl, L. and Kroiss, H. 1998. Bulking sludge prevention by an aerobic selector. *Water Sci Technol* 38: 19–27.

Pujol, R. and Canler, J. P. 1994. Contact zone: French practice with low F/M bulking control. *Water Sci Technol* 29: 221–228.

Qin, L. 2006. Development of microbial granules under alternating aerobic-anaerobic conditions for carbon and nitrogen removal. Ph.D. thesis, Nanyang Technological University, Singapore.

Richard, M. G. 1989. *Activated sludge microbiology*. Alexandria, VA: Water Environment Federation.

Richard, M. R. and Collins, S. B. F. 2003. Activated sludge microbiology problem and their control. Paper presented at the 20th USEPA National Operator Trainers Conference, June 8, 2003. Buffalo, NY.

Rittmann, B. E. and Manem, J. A. 1992. Development and experimental evaluation of a steady-state, multispecies biofilm model. *Biotechnol Bioeng* 39: 914–922.

Rossetti, S., Tomei, M. C., Levantesi, C., Ramadori, R., and Tandoi, V. 2002. "Microthrix parvicella": A new approach for kinetic and physiological characterization. *Water Sci Technol* 46: 65–72.

Rossetti, S., Tomei, M. C., Nielsen, P. H., and Tandoi, V. 2005. "Microthrix parvicella", a filamentous bacterium causing bulking and foaming in activated sludge systems: A review of current knowledge. *FEMS Microbiol Rev* 29: 49–64.

Schwarzenbeck, N., Borges, J. M., and Wilderer, P. A. 2005. Treatment of dairy effluents in an aerobic granular sludge sequencing batch reactor. *Appl Microbiol Biotechnol* 66: 711–718.

Sun, F. Y., Yang, C. Y., Li, J. Y., and Yang, Y. J. 2006. Influence of different substrates on the formation and characteristics of aerobic granules in sequencing batch reactors. *J Environ Sci (China)* 18: 864–871.

Tandoi, V., Rossetti, S., Blackall, L. L., and Majone, M. 1998. Some physiological properties of an Italian isolate of "Microthrix parvicella". *Water Sci Technol* 37: 1–8.

Tay, J. H., Liu, Q.-S., and Liu, Y. 2001. Microscopic observation of aerobic granulation in sequential aerobic sludge blanket reactor. *J Appl Microbiol* 91: 168–175.

Tay, J. H., Liu, Q.-S., and Liu, Y. 2002. Characteristics of aerobic granules grown on glucose and acetate in sequential aerobic sludge blanket reactors. *Environ Technol* 23: 931–936.

Tay, J. H., Pan, S., He, Y. X., and Tay, S. T. L. 2004. Effect of organic loading rate on aerobic granulation. II. Characteristics of aerobic granules. *J Environ Eng* 130: 1102–1109.

Thaveesri, J., Daffonchio, D., Liessens, B., Vandermeren, P., and Verstraete, W. (1995) Granulation and sludge bed stability in upflow anaerobic sludge bed reactors in relation to surface thermodynamics. *Appl Environ Microbiol* 61: 3681–3686.

van der Hoek, J. P. 1988. Granulation of denitrifying sludge. In Lettinga, G., Zehnder, A. J. B., Grotenhuis, J. T. C., and Hulshoff Pol, L. W. (eds.), *Granular aerobic sludge*, 203–210. Wageningen, the Netherlands, Pudoc.

van Loosdrecht, M. C. M., Eikelboom, D., Gjaltema, A., Mulder, A., Tijhuis, L., and Heijnen, J. J. 1995. Biofilm structures. *Water Sci Technol* 32: 35–43.

Wang, F., Liu, Y. H., Yang, F. L., Zhang, X. W., and Zhang, H. M. 2004. Study on the stability of aerobic granules in SBAR: Effect of superficial upflow air velocity and carbon source. Paper presented at IWA Workshop on Aerobic Granular Sludge, September 26–28. Munich, Germany.

Wang, Z. P., Liu, L., Yao, J., and Cai, W. 2006. Effects of extracellular polymeric substances on aerobic granulation in sequencing batch reactors. *Chemosphere* 63: 1728–1735.

Wang, Z.-W. 2007. Insights into mechanism of aerobic granulation in a sequencing batch reactor. Ph.D. thesis, Nanyang Technological University, Singapore.

Wang, Z.-W., Liu, Y., and Tay, J. H. 2005. Distribution of EPS and cell surface hydrophobicity in aerobic granules. *Appl Microbiol Biotechnol* 69: 469–473.

Wanner, J., Kucman, K., Ottova, V., and Grau, P. 1987. Effect of anaerobic conditions on activated sludge filamentous bulking in laboratory systems. *Water Res* 21: 1541–1546.

Wilen, B. M. and Balmer, P. 1999. The effect of dissolved oxygen concentration on the structure, size and size distribution of activated sludge flocs. *Water Res* 33: 391–400.

Yang, S. F., Tay, J. H., and Liu, Y. 2003. A novel granular sludge sequencing batch reactor for removal of organic and nitrogen from wastewater. *J Biotechnol* 106: 77–86.

Zheng, D., Angenent, L. T., and Raskin, L. 2006. Monitoring granule formation in anaerobic upflow bioreactors using oligonucleotide hybridization probes. *Biotechnol Bioeng* 94: 458–472.

Zheng, Y. M., Yu, H. Q., and Sheng, G. P. 2005. Physical and chemical characteristics of granular activated sludge from a sequencing batch airlift reactor. *Process Biochem* 40: 645–650.

Zheng, Y. M., Yu, H. Q., Liu, S. H., and Liu, X. Z. 2006. Formation and instability of aerobic granules under high organic loading conditions. *Chemosphere* 63: 1791–1800.

Zhu, J. Y., Liu, C. X., and Wilderer, P. A. 2004. Bio-P removal profile of aerobic granule activated sludge (AGAS) from an anaerobic/aerobic SBR system. Paper presented at IWA Workshop on Aerobic Granular Sludge, September 26–28. Munich, Germany.

16 Improved Stability of Aerobic Granules by Selecting Slow-Growing Bacteria

Yu Liu and Zhi-Wu Wang

CONTENTS

16.1 INTRODUCTION

There is evidence showing that the stability of aerobic granules is poorer than that of anaerobic granules developed in upflow anaerobic sludge blanket (UASB) reactors (Morgenroth et al. 1997; Peng et al. 1999; Zhu and Liu 1999). Experimental results from two pilot plants operated as sequencing batch bubble columns demonstrated the feasibility of the aerobic granulation technology in treating real industrial wastewater; however, a big concern remains granule stability, as well as the economic competitiveness (Inizan et al. 2005). Obviously, the poor stability of aerobic granules would limit its application in wastewater treatment practice.

The instability of aerobic granules is probably due to the fact that aerobic bacteria can grow much faster than anaerobic bacteria do. In fact, the stability of biofilm is closely related to the growth rate of bacteria, that is, the higher growth rate of bacteria resulted in a weaker structure of biofilm (Tijhuis, van Loosdrecht, and Heijnen 1995; Y. Liu 1997; Kwok et al. 1998). To date, the question of how to improve the stability of aerobic granules remains unanswered. Therefore, this chapter explores a microbial selection-based strategy for improving the stability of aerobic granules. This would be very useful for the development of a full-scale aerobic granular sludge sequencing batch reactor (SBR) for wastewater treatment.

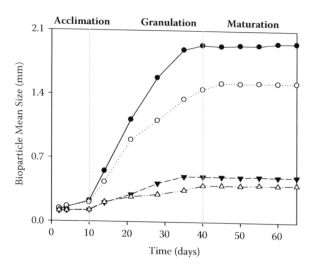

FIGURE 16.1   Changes in size of microbial aggregates. ●: substrate N/COD ratio of 5/100; ○: 10/100; ▼: 20/100; ▲: 30/100. (Data from Liu, Y., Yang, S. F., and Tay, J. H. 2004. *J Biotechnol* 108: 161–169.)

16.2   IMPROVED STABILITY OF AEROBIC GRANULES BY SELECTING SLOW-GROWING NITRIFYING BACTERIA

Under hydrodynamic conditions, the growth of aerobic granules after the initial cell-to-cell attachment is the net result of interaction between bacterial growth and detachment, while the balance between growth and detachment processes in turn leads to an equilibrium or stable granule size (Y. Liu and Tay 2002). Thus, size evolution of the microbial aggregates can be used to describe the growth of granular sludge. Figure 16.1 shows the evolution of microbial aggregates in terms of size observed at different substrate N/COD ratios. It can be seen that the size of microbial aggregates increases gradually and finally stabilizes. According to the granular growth curves shown in figure 16.1, the aerobic granulation process can be categorized in three phases, that is, the acclimation or lag phase, granulation, and maturation, indicated by a stable granule size in the four reactors.

The specific growth rate (μ_d) by size of microbial aggregates can be defined as:

$$\mu_d = \frac{dD/dt}{D} \tag{16.1}$$

in which D is the mean size of the microbial aggregates, and t is operation time. In the granulation phase, as shown in figure 16.1, integrating equation 16.1 gives:

$$\ln D = \mu_d t + \text{constant} \tag{16.2}$$

Hence, the observed size-dependent specific growth rate of microbial aggregate can be determined from the slope of the straight line described by equation 16.2. It should be pointed out that this approach has been successfully employed to estimate the

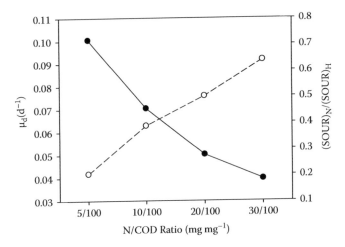

FIGURE 16.2 Effect of substrate N/COD ratio on μ_d (●) and $(SOUR)_N/(SOUR)_H$ (○) of aerobic granules. (Data from Liu, Y., Yang, S. F., and Tay, J. H. 2004. *J Biotechnol* 108: 161–169.)

growth rates of biofilms and anaerobic granules (Y. Liu 1997; Yan and Tay 1997). Figure 16.2 shows the effect of substrate N/COD ratio on μ_d. It is obvious that a higher substrate N/COD ratio results in a lower specific growth rate of aerobic granules.

According to Y. Liu, Yang, and Tay (2004), the overall activity of the heterotrophic population in stable aerobic granules can be quantified by its specific oxygen utilization rate $(SOUR)_H$, while the overall nitrifying activity is represented by the sum of the activities of ammonia oxidizer and nitrite oxidizer, namely $(SOUR)_N$. The relative activity of the nitrifying population over the heterotrophic population in aerobic granules developed at different substrate N/COD ratios is shown in figure 16.2. The $(SOUR)_N/(SOUR)_H$ ratio exhibits an increasing trend with the increase of substrate N/COD ratio. It has been reported that the activity distribution of the nitrifying population over the heterotrophic population in biofilms was proportionally related to the relative abundance of two populations under given conditions (Moreau et al. 1994). Figure 16.3 further indicates that the increased $(SOUR)_N/(SOUR)_H$ ratio would result in a lower observed growth rate of aerobic granules and an improved cell surface hydrophobicity; figure 16.4 and figure 16.5 reveal that aerobic granules with low growth rate have smaller size and more compact structure. As can be seen in figure 16.6, both specific gravity and the sludge volume index (SVI) of aerobic granules are closely correlated to the cell surface hydrophobicity, that is, high cell surface hydrophobicity leads to a compact structure of the aerobic granule.

It appears from figure 16.1 that aerobic granulation is a gradual rather than instant process from dispersed sludge to mature aerobic granules with a stable size. The acclimation phase observed in figure 16.1 implies that a newly inoculated culture does not begin growing immediately, and a period of about 10 days is required for bacteria to adopt to a new environment instead of growth. The observed growth rate by size and mean size at equilibrium of aerobic granules are closely related to the substrate N/COD ratio, that is, higher substrate N/COD ratio results in smaller granules with

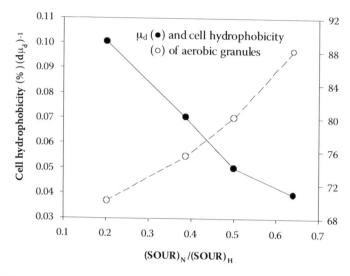

FIGURE 16.3 Effect of $(SOUR)_N/(SOUR)_H$ on μ_d (●) and cell hydrophobicity (○) of aerobic granules. (Data from Liu, Y., Yang, S. F., and Tay, J. H. 2004. *J Biotechnol* 108: 161–169.)

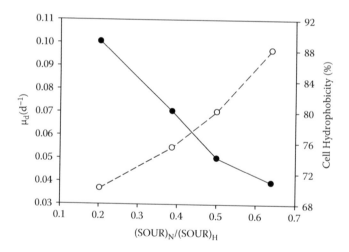

FIGURE 16.4 Effect of μ_d on stable granule size. (Data from Liu, Y., Yang, S. F., and Tay, J. H. 2004. *J Biotechnol* 108: 161–169.)

lower growth rate (figure 16.2). Figure 16.2 also reveals that the nitrifying population in aerobic granules is enriched with the increase of the substrate N/COD ratio. As a result, the heterotrophs in aerobic granules become less and less dominant at high substrate N/COD ratio. It seems that the high substrate N/COD ratio is an important factor that selects nitrifying population. Since the growth of nitrifying bacteria is much slower than heterotrophs (Sharma and Ahlert 1977), aerobic granules may offer a protective matrix for the nitrifying population to grow on without the risk of being washed out of the system.

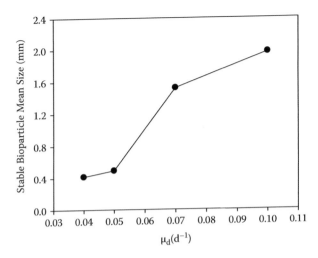

FIGURE 16.5   Effect of μ_d on specific gravity (●) and SVI (○) of aerobic granules. (Data from Liu, Y., Yang, S. F., and Tay, J. H. 2004. *J Biotechnol* 108: 161–169.)

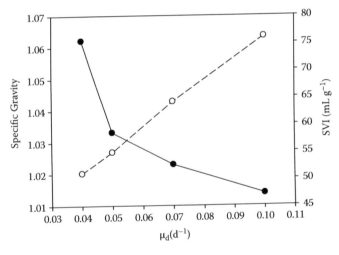

FIGURE 16.6   Relationships of specific gravity (●) and SVI (○) to cell hydrophobicity of aerobic granules. (Data from Liu, Y., Yang, S. F., and Tay, J. H. 2004. *J Biotechnol* 108: 161–169.)

It appears from figure 16.3 that the specific growth rate of aerobic granules is closely related to the distribution of the nitrifying population over the heterotrophic population in aerobic granules. This suggests that the enriched nitrifying population in aerobic granules is mainly responsible for the lowered growth rate of aerobic granules developed at high substrate N/COD ratios. In a study of anaerobic granulation, Yan and Tay (1997) thought that if granulation is purely the result of bacterial aggregation and growth and the granule formed is ideal, a relationship between specific growth rate by size and that by biomass can be derived as follows:

$$\mu_g = \frac{1}{X}\frac{dX}{dt} = \frac{1}{\frac{\pi}{6}\rho D^3}\frac{d\left(\frac{\pi}{6}\rho D^3\right)}{dt} = 3\frac{1}{D}\frac{dD}{dt} = 3\mu_d \qquad (16.3)$$

in which μ_g is specific growth rate by biomass (g biomass g^{-1} biomass d^{-1}), X is biomass concentration of granules, and ρ is density of granules. According to equation 16.3, the specific growth rate by size can be converted to the specific growth rate by biomass. The respective μ_g value of aerobic granules developed at substrate N/COD ratios of 5/100, 10/100, 20/100, and 30/100 is 0.3, 0.21, 0.15, and 0.12 d^{-1} (Y. Liu, Yang, and Tay 2004). The μ_g values of nitrifying population-enriched aerobic granules are comparable with those found in nitrifying biofilms (Oga, Suthersan, and Ganczarczyk 1991).

Aerobic granules have been considered to have relatively low stability (Morgenroth et al. 1997; Zhu and Liu 1999). Obviously, the poor stability of aerobic granules will limit their application in wastewater treatment. The cause behind the poor stability of aerobic granules would be due to the fast growth of heterotrophic bacteria that dominate aerobic granules. The nitrifying population grows much more slowly than heterotrophs, while the physical structure of nitrifying biofilms is much stronger than heterotrophic biofilms (Oga, Suthersan, and Ganczarczyk 1991). Figure 16.3 reveals that the observed growth rate of aerobic granules can be significantly lowered by enrichment of the nitrifying population, and this can be realized through properly controlling the substrate N/COD ratio. As can be seen in figures 16.4 and 16.5, the lowered growth rate in turn results in a smaller size of aerobic granules, but with a higher specific gravity, indicating a compact, strong microbial structure. It further appears from figure 16.7 that large granules have a loose structure. This observation is consistent with those found in biofilms, that is, the compactness of biofilm is reduced with the increase in biofilm thickness (Kwok et al. 1998; Y. Liu and Tay 2002). These all point to the fact that the structural stability of aerobic granules can be significantly improved by selecting slow-growing nitrifying bacteria.

Aerobic granulation is known as a microbial self-immobilization process that should be similar to the growth of biofilm (Y. Liu and Tay 2002). In a study of biofilms, there is evidence that the strength of biofilms is negatively related to the growth rate of microorganisms (Tijhuis, van Loosdrecht, and Heijnen 1995). Kwok et al. (1998) reported that the biofilm density decreased as the growth rate increased, while the density of nitrifying biofilm was found to be higher than that of heterotrophic biofilm (Oga, Suthersan, and Ganczarczyk 1991). This is consistent with the results reported in figures 16.4 and 16.5. Similarly, in the anaerobic granulation process, it was also observed that a high biomass growth rate led to a reduced strength of anaerobic granules, that is, partial loss of structural integrity and disintegration occurs at high biomass growth rates (Morvai, Mihaltz, and Czako 1992; Quarmby and Forster 1995). It becomes clear that the high observed growth rate would encourage the outgrowth of aerobic granules, leading to a rapid increase in the size of the granules, as well as a loose structure with low biomass density.

As discussed earlier, a high substrate N/COD ratio appears to favor the selection of nitrifying bacteria in aerobic granules, thereby one possible operation strategy

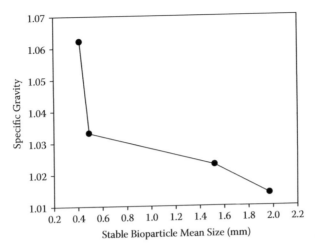

FIGURE 16.7   Relationship between stable mean size and specific gravity of aerobic granules. (Data from Liu, Y., Yang, S. F., and Tay, J. H. 2004. *J Biotechnol* 108: 161–169.)

that can help to improve the stability of aerobic granules is to select slow-growing nitrifying bacteria in aerobic granules by controlling the feed N/COD ratio.

A mushroom-like structure was observed in aerobic granules cultivated at the substrate N/COD ratio of 20/100 (figure 16.8a), and a similar structure was also observed in granules developed at the substrate N/COD ratio of 30/100. However, the aerobic granules developed at the substrate N/COD ratio of 5/100 displayed a nonclustered structure. CLSM (confocal laser scanning microscope) images of FISH (fluorescent *in situ* hybridization) further revealed that the nitrifying population was dominant in the clusters (figure 16.9). Figure 16.8b shows that the top layer mainly consists of cocci-shaped bacteria, while rod-shaped bacteria are dominant subsequently. Tay et al. (2002) also reported that the nitrifying population was mainly located at a depth of 70 to 100 μm from the surface of the granule. In fact, previous research showed that biofilm of mixed bacterial communities formed thick layers consisting of differentiated mushroom-like structures (Costerton et al. 1994), which are very similar to that observed in figure 16.8a. Figure 16.2 shows that the relative abundance of the nitrifying population over the heterotrophic population in the aerobic granules grown at the substrate N/COD ratio of 5/100 is very low as compared to the granules developed at high substrate N/COD ratios. At high substrate N/COD ratio, competition between nitrifying and heterotrophic populations on nutrients is significant.

It has been demonstrated that biofilm can form the mushroom-like structure by simply changing the diffusion rate, that is, the biofilm structure is largely determined by nutrient concentration (Wimpenny and Colasanti 1997). In fact, bacteria may sense and move towards nutrients (Prescott, Harley, and Klein 1999). Because of their slow growth rate, the mushroom-like structure would result from the demand of the nitrifying population on nutrients, and it in turn ensures that the nitrifying population in aerobic granules can maximize access to nutrients. As Watnick and Kolter (2000) noted, in mixed biofilms, bacteria distribute themselves according to who can survive best in the particular microenvironment, and the high complexity

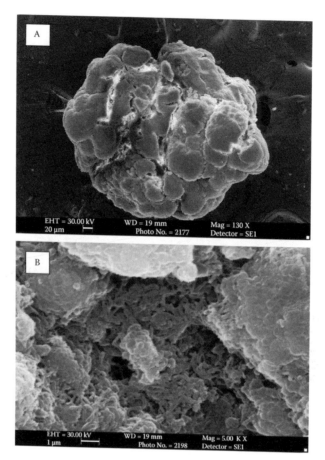

FIGURE 16.8 Mushroom-like structure of an aerobic granule developed at a substrate N/COD ratio of 20/100. (From Liu, Y., Yang, S. F., and Tay, J. H. 2004. *J Biotechnol* 108: 161–169. With permission)

of a microbial community would be beneficial to its stability. These findings seem to indicate that the mushroom-like structure of densely slow-growing nitrifying bacteria would contribute to the stability of aerobic granules developed at high substrate N/COD ratios. In a study of activated sludge floc stability, a similar remark was also made by Wilen, Jin, and Lant (2003). Consequently, the organization of different microbial populations may have an effect on the stability of aerobic granules.

16.3 IMPROVED STABILITY OF AEROBIC GRANULES BY SELECTING SLOW-GROWING P- OR GLYCOGEN-ACCUMULATING ORGANISMS

It is clear that selection of slow growing organisms can improve the density and stability of aerobic granules. de Kreuk and van Loosdrecht (2004) thought that to lower the growth rate of organisms in aerobic granules, easily biodegradable

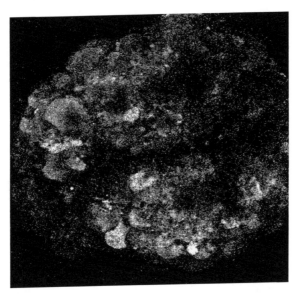

FIGURE 16.9 Distribution of ammonium-oxidizing bacteria (AOB) in an aerobic granule. White color represents AOB. (Courtesy of Dr. V. Ivanov, Nanyang Technological University, Singapore.)

substrate needs to be converted to slowly degradable organics, namely microbial storage polymers. It has been known that phosphate- or glycogen-accumulating organisms can perform such a conversion of external organic carbon to storage polymers. The experimental work by de Kreuk and van Loosdrecht (2004) showed that the selection or enrichment of P-accumulating or glycogen-accumulating organisms in aerobic granules indeed would lead to stable aerobic granules.

Heterotrophic bacteria growing on the slowly biodegradable storage polymers, such as poly-β-hydroxybutyrate (PHB) or glycogen, may have smaller growth rates as compared to those growing on easily biodegradable organic substrates (Carta et al. 2001). For promoting the conversion of an external carbon source to the storage polymers, a long anaerobic feeding period has been often practiced followed by an aerobic reaction phase. By implementing such an operation strategy in an SBR, selection of slow-growing P- or glycogen-accumulating organisms would be expected (de Kreuk and van Loosdrecht 2004). On the contrary, Li, Kuba, and Kusuda (2006) found that when the aerobic filling time was extended from 5 to 30 minutes, the dense and compact aerobic granules were gradually shifted into a light and loose filamentous granular structure, that is, the extension of the aerobic filling time eventually led to instability and the failure of the aerobic granular sludge SBR. It has been reported that when dosage of external phosphate was no longer available, P-accumulating organisms tended to gradually disappear and be replaced by glycogen-accumulating organisms in aerobic granules. Even in this case, the characteristics of aerobic granules seemed not to change significantly, and smooth, dense and stable aerobic granules could be maintained in the SBR (de Kreuk and van Loosdrecht 2004).

So far, evidence shows that a high dissolved oxygen (DO) concentration is necessary for stable aerobic granulation in SBRs (see chapter 8). However, low oxygen concentration is desirable in order to make aerobic granulation technology economically competitive over the conventional activated sludge processes. According to the substrate availability, the operation of an SBR can be roughly divided into two distinct phases or periods, that is, feast and famine periods (Tay, Liu, and Liu 2001; Q.-S. Liu 2003; de Kreuk and van Loosdrecht 2004). Theoretically, the feast period is the period in which the external energy source (e.g. substrate) is available for microbial growth, while after depletion of the external substrate, the culture comes to the famine phase in which only internally stored polymers are available for microbial use.

Y. Q. Liu and Tay (2006) looked into the possibility of variable aeration in an aerobic granular sludge SBR, and they found that after the aeration rate was reduced from 1.66 to 0.55 cm s^{-1} in the famine period, the settleability of aerobic granules in the SBR with reduced aeration was the same as that of aerobic granules in the SBR with constant aeration rate of 1.66 cm s^{-1}. It is apparent from figure 16.10 that reducing the aeration rate during the famine period would not have a significant effect on the stable operation of the aerobic granular sludge reactor, whereas the aeration rate in the feast period is crucial for the stable operation of the aerobic granular sludge. Obviously, by implementing an operation strategy with reduced aeration in the famine phase, a significant reduction in energy consumption would be expected in aerobic granular sludge SBRs.

16.4 IMPROVED STABILITY OF AEROBIC GRANULES BY SELECTING AGED AEROBIC GRANULES

It can be seen in the above discussion that selection of slow-growing bacteria can significantly improve the stability of aerobic granules developed in SBRs. In terms of the process operation, a long solids retention time (SRT) means a low specific microbial growth rate. Based on this basic idea, Li, Kuba, and Kusuda (2006) tried to control the growth rate of aerobic granules by specifically selecting young or aged granules. When young aerobic granules were regularly removed, more and more aged granules would accumulate in the system, leading to a reduced biodiversity of those remaining aerobic granules. It has been thought that the reduced biodiversity due to enriched aged aerobic granules would help to select slow-growing bacteria and thus increase the stability of aerobic granules (Li, Kuba, and Kusuda 2006). Along with the takeout of young aerobic granules, granules remaining in the SBR would become more aged, and subsequently a remarkable increase in the granule ash content was observed (Li, Kuba, and Kusuda 2006). If the aged aerobic granules were removed from SBR, Li, Kuba, and Kusuda (2006) found that large, loose aerobic granules appeared and dominated the system. This may be due to the fact that filamentous microorganisms grew excessively in the system, eventually leading to instability of aerobic granules.

16.5 CONCLUSIONS

The stability of aerobic granules is key to long-term and stable operation of aerobic granular sludge bioreactors. In this respect, the selection and enrichment of slow-growing organisms, such as nitrifying bacteria, P-accumulation and glycogen-accumulating organisms, appears to be the most feasible engineering strategy.

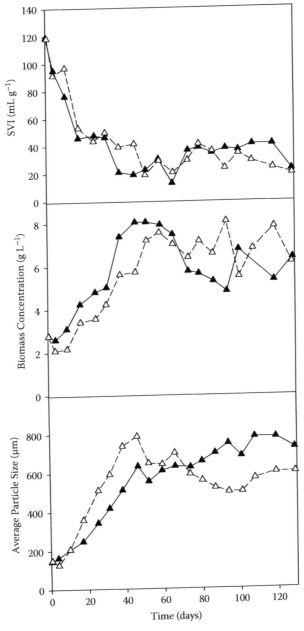

FIGURE 16.10 SVI, biomass concentration, and average particle size of aerobic granular sludge in two SBRs operated at (▲) reduced aeration rate of 0.55 cm s⁻¹ and (△) constant aeration rate of 1.66 cm s⁻¹. (Data from Liu, Y. Q. and Tay, J. H. 2006. *J Biotechnol* 124: 338–346.)

REFERENCES

Carta, F., Beun, J. J., van Loosdrecht, M. C. M., and Heijnen, J. J. 2001. Simultaneous storage and degradation of PHB and glycogen in activated sludge cultures. *Water Res* 35: 2693–2701.

Costerton, J. W., Lewandowski, Z., Debeer, D., Caldwell, D., Korber, D., and James, G. 1994. Biofilms, the customized microniche. *J Bacteriol* 176: 2137–2142.

de Kreuk, M. K. and van Loosdrecht, M. C. M. 2004. Selection of slow growing organisms as a means for improving aerobic granular sludge stability. *Water Sci Technol* 49: 9–17.

Inizan, M., Freval, A., Cigana, J., and Meinhold, J. 2005. Aerobic granulation in a sequencing batch reactor (SBR) for industrial wastewater treatment. *Water Sci Technol* 52: 335–343.

Kwok, W. K., Picioreanu, C., Ong, S. L., van Loosdrecht, M. C. M., Ng, W. J., and Heijnen, J. J. 1998. Influence of biomass production and detachment forces on biofilm structures in a biofilm airlift suspension reactor. *Biotechnol Bioeng* 58: 400–407.

Li, Z. H., Kuba, T., and Kusuda, T. 2006. Selective force and mature phase affect the stability of aerobic granule: An experimental study by applying different removal methods of sludge. *Enzyme Microb Technol* 39: 976–981.

Liu, Q.-S. 2003. Aerobic granulation in a sequencing batch reactor. Ph.D. thesis, Nanyang Technological University, Singapore.

Liu, Y. (1997) Estimating minimum fixed biomass concentration and active thickness of nitrifying biofilm. *J Environ Eng* 123: 198–202.

Liu, Y. and Tay, J. H. 2002. The essential role of hydrodynamic shear force in the formation of biofilm and granular sludge. *Water Res* 36: 1653–1665.

Liu, Y., Yang, S. F., and Tay, J. H. 2004. Improved stability of aerobic granules by selecting slow-growing nitrifying bacteria. *J Biotechnol* 108: 161–169.

Liu, Y. Q. and Tay, J. H. 2006. Variable aeration in sequencing batch reactor with aerobic granular sludge. *J Biotechnol* 124: 338–346.

Moreau, M., Liu, Y., Capdeville, B., Audic, J. M., and Calvez, L. 1994. Kinetic behavior of heterotrophic and autotrophic biofilms in wastewater treatment processes. *Water Sci Technol* 29: 385–391.

Morgenroth, E., Sherden, T., van Loosdrecht, M. C. M., Heijnen, J. J., and Wilderer, P. A. 1997. Aerobic granular sludge in a sequencing batch reactor. *Water Res* 31: 3191–3194.

Morvai, L., Mihaltz, P., and Czako, L. (1992) Kinetic basis of a new start-up method to ensure the rapid granulation of anaerobic sludge. *Water Sci Technol* 25: 113–122.

Oga, T., Suthersan, S., and Ganczarczyk, J. J. 1991. Some properties of aerobic biofilms. *Environ Technol* 12: 431–440.

Peng, D. C., Bernet, N., Delgenes, J. P., and Moletta, R. 1999. Aerobic granular sludge: A case report. *Water Res* 33: 890–893.

Prescott, L., Harley, J., and Klein, D. 1999. *Microbiology*. Boston: McGraw-Hill.

Quarmby, J. and Forster, C. F. 1995. An examination of the structure of UASB granules. *Water Res* 29: 2449–2454.

Sharma, B. and Ahlert, R. C. 1977. Nitrification and nitrogen removal. *Water Res* 11: 897–925.

Tay, J. H., Liu, Q.-S., and Liu, Y. 2001. The effects of shear force on the formation, structure and metabolism of aerobic granules. *Appl Microbiol Biotechnol* 57: 227–233.

Tay, J. H., Ivanov, V., Pan, S., and Tay, S. T. L. 2002. Specific layers in aerobically grown microbial granules. *Lett Appl Microbiol* 34: 254–257.

Tijhuis, L., van Loosdrecht, M. C. M., and Heijnen, J. J. 1995. Dynamics of biofilm detachment in biofilm airlift suspension reactors. *Biotechnol Bioeng* 45: 481–487.

Watnick, P. and Kolter, R. 2000. Biofilm, city of microbes. *J Bacteriol* 182: 2675–2679.

Wilen, B. M., Jin, B., and Lant, P. 2003. The influence of key chemical constituents in activated sludge on surface and flocculating properties. *Water Res* 37: 2127–2139.

Wimpenny, J. W. T. and Colasanti, R. 1997. A unifying hypothesis for the structure of microbial biofilms based on cellular automaton models. *FEMS Microbiol Ecol* 22: 1–16.

Yan, Y. G. and Tay, J. H. 1997. Characterisation of the granulation process during UASB start-up. *Water Res* 31: 1573–1580.

Zhu, J. and Liu, C. 1999. Cultivation and physico-chemical characteristics of granular activated sludge in alternating anaerobic/aerobic process. *Chin J Environ Sci* 20: 38–41.

17 Pilot Study of Aerobic Granulation for Wastewater Treatment

Qi-Shan Liu and Yu Liu

CONTENTS

17.1 INTRODUCTION

Aerobic granulation technology has been applied for the high-efficiency treatment of a wide variety of wastewater including toxic wastewater, and it has been demonstrated in pilot-scale plants (de Bruin et al. 2004; de Kreuk, de Bruin, and van Loosdrecht 2004; Liu et al. 2005), while its full scale application has not yet been reported. In industrial practice, the fast and easy startup of upflow anaerobic sludge blanket (UASB) reactors can be realized by seeding anaerobic granules directly into the reactor. This will significantly reduce the time required for anaerobic granulation which usually takes 2 to 8 months. A similar startup strategy is also applicable in initiating aerobic granular sludge sequencing batch reactors (SBRs). Existing evidence shows that aerobic granules can be stored over a period of 7 weeks, and its activity quickly recovered (Zhu and Wilderer 2003), while J. H. Tay, Liu, and Liu (2002) also found that aerobic granules can be stably stored for 4 months at 4°C. These findings suggest that use of the stored aerobic granules as seed would be feasible in full-scale operation of aerobic granular sludge SBRs.

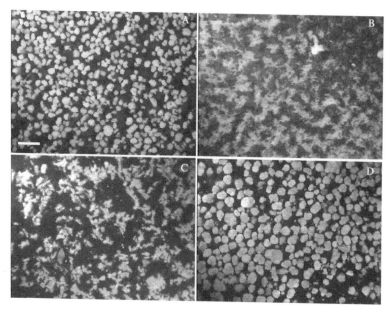

FIGURE 17.1   Morphology of bioparticles in the pilot-scale SBR at day 1 (A), day 5 (B), day 20 (C), and day 65 (D). Scale bar: 4 mm. (From Tay, J. H. et al. 2004. *Proceedings of Workshop on Aerobic Granular Sludge, Munich, Germany.* With permission.)

17.2   STARTUP OF PILOT-SCALE AEROBIC GRANULAR SLUDGE SBRS

J. H. Tay et al. (2004) investigated aerobic granulation in a pilot-scale SBR. The pilot-scale aerobic granular sludge SBR was initiated by seeding mature aerobic granules harvested from a laboratory-scale SBR.

17.2.1   COMPARISON OF PILOT- AND LABORATORY-SCALE SBRs

The seed aerobic granules used in the pilot study had a mean diameter of 0.83 mm (figure 17.1A). It was found that aerobic granules tended to disintegrate shortly after seeding into the pilot-scale SBR, and loose flocs became dominant in the reactor on day 5 (figure 17.1B). As a result, the mean diameter of bioparticles decreased to 0.19 mm and the sludge volume index (SVI) increased from 19 to 175 mL g^{-1} (figure 17.2A). However, compact aggregates were gradually re-formed on day 20, indicated by a mean diameter of 0.4 mm and an SVI of 63 mL g^{-1} (figure 17.1C). The granule size continued to increase up to a peak value of 1.4 mm on day 50, and finally stabilized at this level with an SVI of around 26 mL g^{-1} (figure 17.2A). It can be seen that the steady-state granules in the pilot-scale SBR had a compact structure similar to the seed granules, but they were larger in size (figure 17.1A and D). Unlike the evolution of aerobic granules in the pilot-scale SBR, aerobic granules in the laboratory-scale SBR remained stable throughout the whole study period, indicated by relatively constant granule size and SVI, and there was an increase in size during the first 2 weeks (figure 17.2B).

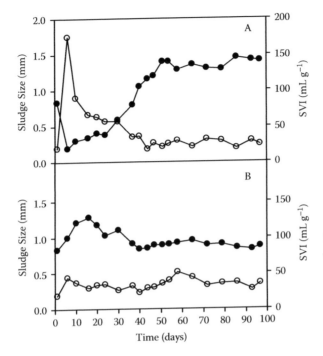

FIGURE 17.2   Changes in mean diameter (●) and SVI (○) of bioparticles in the course of operation of a pilot-scale SBR (A) and a laboratory-scale SBR (B). (Data from Tay, J. H. et al. 2004. *Proceedings of Workshop on Aerobic Granular Sludge, Munich, Germany.*)

The initial increase in granule size in the laboratory-scale SBR was probably due to the relatively low sludge concentration when the reactor was started up (figure 17.3B). Low biomass concentration would lead to fewer collisions among granules and weaker detachment from individual granule. Difference in granule size observed in the pilot- and laboratory-scale SBRs at steady state can likely be attributed to different shear and detachment forces in the two reactors. Figure 17.4 shows that more biomass was retained in the bottom half of the laboratory-scale SBR, whereas an even biomass distribution was observed in the pilot-scale SBR. The high accumulation of granular sludge at the bottom of the laboratory-scale SBR would certainly increase the collision and detachment rate among the granule particles. Consequently, smaller granules were developed in the laboratory-scale SBR. As discussed in chapter 2, the size of aerobic granules is inversely correlated to the shear force generated by the air bubbles and collisions among the sludge particles (J. H. Tay, Liu, and Liu 2004). The size of the granules developed in the stable laboratory-scale SBR was similar to the seed granules (figure 17.2). The similarity in size is not unexpected because the reactor configuration and operating conditions were similar in the laboratory-scale SBR and the reactor used for precultivation of seed granules.

The biomass concentration in both pilot- and laboratory-scale SBRs was the same at the level of 0.4 g L^{-1} at the reactor startup (figure 17.3). The biomass concentration tended to gradually increase to 6.5 g L^{-1} in the first 3 weeks of operation in both reactors. A drop in biomass concentration was observed in the period of day 30

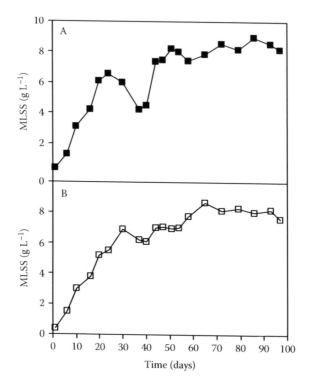

FIGURE 17.3 Sludge concentration versus time in pilot-scale (A) and laboratory-scale (B) SBRs. (From Tay, J. H. et al. 2004. *Proceedings of Workshop on Aerobic Granular Sludge, Munich, Germany.*)

to 40, which resulted from a reduced settling time in the SBR from 5 to 2 min. Obviously, this would cause the washout of slow-settling sludge, leading to a temporary reduction in biomass concentration. On the other hand, the shorter settling time also exerted a stronger selection pressure on biomass, which in turn encourages the retention of biomass with excellent settleability, as discussed in chapter 6. Consequently, the biomass concentration gradually increased over time and finally stabilized at 8.0 g L^{-1} in both reactors. These results seem to indicate that the seed granules in the laboratory-scale SBR can be successfully maintained, and new granules can grow immediately after the reactor startup, while granules can be lost but re-formed shortly from disintegrated granules in the pilot-scale SBR.

The distribution of biomass concentration along the reactor height was different in the pilot- and laboratory-scale SBRs (figure 17.4). The biomass was distributed rather evenly along the reactor height in the pilot-scale SBR, whereas more biomass was accumulated in the lower half of the laboratory-scale SBR. This may be due to the difference of hydrodynamic conditions in the two reactors. It is believed that the initial disappearance of aerobic granules and dominant growth of bioflocs in the pilot-scale SBR was likely linked to the prevailing hydrodynamic conditions due to different reactor diameters. Moreover, the size and location of air diffusers in the column SBR would also affect the hydrodynamic flow pattern. However, the cycle

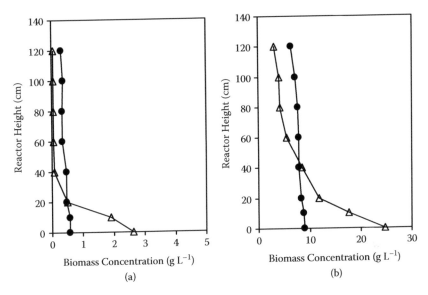

FIGURE 17.4 Sludge distribution along the reactor height during the first day (A) and at steady state (B). ●: pilot-scale SBR; △: laboratory-scale SBR. (From Tay, J. H. et al. 2004. *Proceedings of Workshop on Aerobic Granular Sludge, Munich, Germany.*)

operation of SBR provides a selective process that allows for the gradual redevelopment of granular biomass with good settling characteristics. For instance, a short settling time promotes the selection of fast-settling bioparticles.

17.2.2 Characteristics of Aerobic Granules Developed in Pilot- and Laboratory-Scale SBRs

17.2.2.1 Granule Size and Morphology

The mean diameter of aerobic granules developed in the pilot-scale SBR was 1.37 mm (table 17.1), which was larger than that of the seed granules with a typical size of 0.83 mm. Aerobic granules cultivated in the laboratory-scale SBR had a similar size as the seed granules. These two kinds of aerobic granules exhibited a similar morphology in terms of aspect ratio and roundness. In fact, a very wide size range of aerobic granules has been reported, from 0.2 mm up to 16 mm (Morgenroth et al. 1997; J. H. Tay, Liu, and Liu 2001; Zheng et al. 2006).

17.2.2.2 Settling Property

Aerobic granules in the pilot-scale SBR had an SVI as low as 26.5 mL g⁻¹ and a high specific gravity of 1.017, while the SVI was 34.4 mL g⁻¹ and specific gravity was 1.015 for those granules cultivated in the laboratory-scale SBR. The volatile solids content of granules in the pilot-scale SBR (62.4%) was lower than that of granules cultivated in the laboratory-scale SBR (74.9%). This indicates a significant accumulation of inorganic materials in the granules developed in the pilot-scale SBR. The higher inorganic content was partially responsible for the observed low SVI.

TABLE 17.1

Characteristics of Aerobic Granules Cultivated in Pilot-Scale and Laboratory-Scale SBRs.

Items	Pilot-Scale SBR	Laboratory-Scale SBR
Mean diameter (mm)	1.37 ($\pm$ 0.09)	0.89 ($\pm$ 0.07)
Aspect ratio	0.67 ($\pm$ 0.16)	0.69 ($\pm$ 0.15)
Roundness	0.69 ($\pm$ 0.15)	0.69 ($\pm$ 0.16)
SVI (mL g^{-1})	26.5 ($\pm$ 5.9)	34.4 ($\pm$ 6.9)
Specific gravity	1.017 ($\pm$ 0.0005)	1.015 ($\pm$ 0.0005)
VSS/SS (%)	62.4 ($\pm$ 2.8)	74.9 ($\pm$ 3.6)
Integrity coefficient (%)	96.0 ($\pm$ 2.0)	96.9 ($\pm$ 2.5)
SOUR (mg O$_2$ g^{-1} VSS h^{-1})	74.1 ($\pm$ 12.4)	80.6 ($\pm$ 18.2)

17.2.2.3   Physical Strength

The physical strength of aerobic granules, expressed as integrity coefficient, was 96.0% in the pilot-scale SBR and 96.9% in the laboratory-scale SBR, that is, aerobic granules developed in the laboratory-scale SBR were comparable with those in the pilot-scale SBR.

17.2.2.4   Microbial activity

The specific oxygen uptake rate (SOUR) as an indicator of microbial activity was 74.1 mg O$_2$ g^{-1} volatile suspended solids (VSS) h^{-1} for granules in the pilot-scale SBR and 80.6 mg O$_2$ g^{-1} VSS h^{-1} for the granules in the laboratory-scale SBR (table 17.1). The slightly low microbial activity of aerobic granules in the pilot-scale SBR is thought to be size-related. In fact, the limitation of mass transport and diffusion is generally more pronounced for larger granules, which would result in low microbial activity, as discussed in chapter 8. It is apparent that use of fresh aerobic granules as seed is feasible to quickly start up an aerobic granular sludge SBR.

17.3   STARTUP OF A PILOT-SCALE SBR USING STORED GRANULES AS SEED

Liu et al. (2005) used aerobic granules that had been stored for 4 months to initiate a pilot-scale SBR, and found that the seed granules were maintained stably, and new granules could be successfully formed thereafter. The size of granules gradually increased from 1.28 to 1.7 mm within 1 week (figure 17.5), and then decreased to a size similar to the seed granules. Similar to figure 17.2B, the initial increase in granule size is due to the fewer collisions among granules and subsequent weak detachment, because of low biomass concentration in the reactor in the initial period. New granules began to form after day 5, and biomass concentration gradually increased accordingly (figure 17.6).

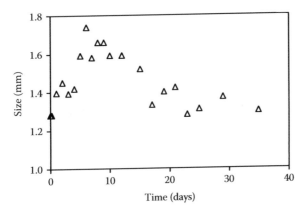

FIGURE 17.5 Sludge particle size versus operation time in the pilot-scale SBR. (From Liu, Q.-S. et al. 2005. *Environ Technol* 26: 1363–1369. With permission.)

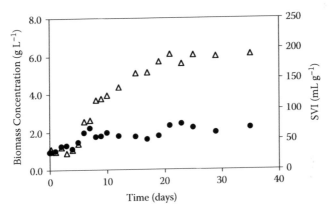

FIGURE 17.6 Biomass concentration (△) and SVI (●) versus operation time in the pilot-scale SBR. (From Liu, Q.-S. et al. 2005. *Environ Technol* 26: 1363–1369. With permission.)

A biomass concentration of the stored aerobic granules of 1.03 g L^{-1} was initially seeded into the pilot-scale SBR, and remained unchanged in the first 4 days. Afterwards, it gradually increased to a stable level of 6.0 g L^{-1}. It should be pointed out that the pilot-scale SBR was initiated with an initial biomass concentration of 1.03 g L^{-1} and low influent COD of 400 mg L^{-1}. This resulted in a granule surface loading rate of 8.7 g COD m^{-2} d^{-1} at the beginning of the study, as shown in figure 17.7. The granule surface loading rate then fluctuated from 6.5 to 11.0 g COD m^{-2} d^{-1} till day 6, depending upon the biomass concentration in the reactor and the organic loading rate applied. At steady state, the surface loading rate dropped to 1.4 g COD m^{-2} d^{-1} because of the high biomass concentration in the reactor. It is mostly likely that a high granule surface loading rate would promote the growth of suspended bacterial cells instead of granules. Thus, a low granule surface loading rate might be applied for the reactor startup in order to prevent the outgrowth of sludge flocs, particularly during the initial period.

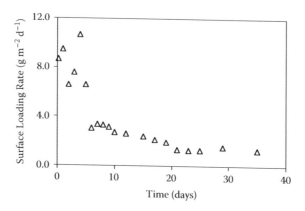

FIGURE 17.7   Granule surface loading rate versus operation time. (From Liu, Q.-S. et al. 2005. *Environ Technol* 26: 1363–1369. With permission.)

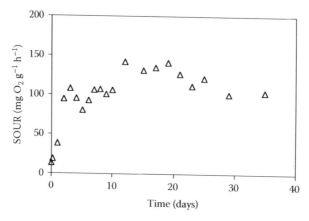

FIGURE 17.8   Activity recovery of stored aerobic granules during the operation of the pilot-scale SBR. (From Liu, Q.-S. et al. 2005. *Environ Technol* 26: 1363–1369. With permission.)

The seed aerobic granules after 4 months of storage had a SOUR of 13.4 mg O_2 g^{-1} VSS h^{-1} (figure 17.8). After 2 days of cultivation in the pilot-scale SBR, the SOUR increased to 94.5 mg O_2 g^{-1} VSS h^{-1}, which is comparable to that of fresh aerobic granules. These results clearly showed that the stored aerobic granules can be revived with a full recovery of microbial activity within 2 days. The short recovery time of the microbial activity of stored granules would be very much advantageous for its application in industrial practice. Seed granules had a light grey color with a black core (figure 17.9A), which is suspected to be due to the sulfide generated by sulfate-reducing bacteria during storage, while fresh aerobic granules often have brownish-yellow color. However, after 2 days of reviving, the apparent color of the stored aerobic granules turned to that of fresh granules (figure 17.9B).

Figure 17.10 shows the reactor performance of the pilot-scale SBR in terms of influent and effluent COD. The reactor was initiated by supplying an influent COD of 400 mg L^{-1}, and after the first SBR cycle, the effluent COD was 173 mg L^{-1}. With

FIGURE 17.9   Apparent colors of stored aerobic granules (A) and those after 2 days of reviving (B). (From Liu, Q.-S. et al. 2005. *Environ Technol* 26: 1363–1369. With permission.)

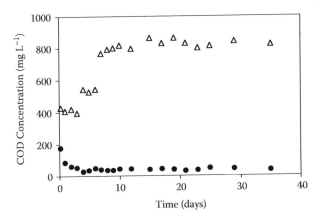

FIGURE 17.10   COD concentration profiles observed in the pilot-scale SBR seeded with stored aerobic granules. △: influent; ●: effluent. (From Liu, Q.-S. et al. 2005. *Environ Technol* 26: 1363–1369. With permission.)

the gradual recovery of granule microbial activity, the effluent COD decreased to 82 mg L^{-1} after 1 day of SBR operation, and further to 60 mg L^{-1} at the end of the second day. The influent COD was increased to 550 mg L^{-1} on day 4 and further to 800 mg L^{-1} on day 7 because aerobic granule activity had been fully recovered and new granule development was also observed. It appears from figure 17.10 that the increase in the influent COD had little impact on the removal efficiency, and stable effluent COD concentration of 37 mg L^{-1} was recorded, corresponding to a COD removal efficiency of 96%.

    Successful startup of the pilot-scale aerobic granular sludge SBR by seeding stored granules was demonstrated to be feasible. The microbial activity of stored granules can fully recovered within 2 days. In fact, the granules cultivated from benign substrates, such as acetate, can be used as the microbial seeds to produce granules to degrade toxic substrates, such as phenol (S. T. L. Tay et al. 2005). This

FIGURE 17.11 Aerobic granular sludge pilot plant installed in the Netherlands. (From De Kreuk, M. K., De Bruin, L. M. M., and van Loosdrecht, M. C. M., 2004. Paper presented at IWA Workshop on Aerobic Granular Sludge, Munich, Germany.)

further extends the application of seed granules to other types of wastewater or toxic wastewater treatment.

17.4 STARTUP OF A PILOT-SCALE SBR USING ACTIVATED SLUDGE AS SEED

Aerobic granulation directly from activated sludge flocs with municipal wastewater was successfully demonstrated in a pilot-scale plant in the Netherlands (De Bruin et al. 2004). Two column SBRs 6 m in height and 0.6 m in diameter were operated in parallel treating wastewater at a flow rate of 5.0 m³ h⁻¹ (figure 17.11). Formation of aerobic granules with an SVI of 55 mL g⁻¹ could take place in a few weeks. Granular sludge also had a good capability for the removal of nitrogen and phosphate present in the municipal wastewater. It was found that pretreatment to remove suspended solid particles in order to improve granulation and the post-treatment might be needed so as to satisfy the stringent discharge limits. It appears that there will be no problem for aerobic granulation from bioflocs in pilot-scale SBRs.

17.5   CONCLUSIONS

Aerobic granulation in pilot-scale SBRs is an important step for its full application in the industrial-scale plant. Aerobic granular sludge SBRs can be initiated by seeding either fresh or stored aerobic granules. The microbial activity of stored aerobic granules can be quickly recovered within a short time of 2 days. Meanwhile, aerobic granulation directly from bioflocs as seed has also been demonstrated in the pilot plant. Results from the pilot studies show that aerobic granulation is a promising technology for wastewater treatment due to its compact footprint, lower sludge production, and simultaneous organic and nutrient removal.

REFERENCES

de Bruin, L. M. M., van Der Roest, H. F., De Kreuk, M. K., and van Loosdrecht, M. C. M. 2004. Promising result pilot plant research aerobic granular sludge technology at WWTP. Paper presented at IWA Workshop on Aerobic Granular Sludge, September 27–28. Munich, Germany.

de Kreuk, M. K., De Bruin, L. M. M., and van Loosdrecht, M. C. M. 2004. Aerobic granular sludge: From idea to pilot plant. Paper presented at IWA Workshop on Aerobic Granular Sludge, September 27–28. Munich, Germany.

Liu, Q.-S., Liu, Y., Tay, S. T. L., Show, K. Y., Ivanov, V., Benjamin, M., and Tay, J. H. 2005. Startup of pilot-scale aerobic granular sludge reactor by stored granules. *Environ Technol* 26: 1363–1369.

Morgenroth, E., Sherden, T., van Loosdrecht, M. C. M., Heijnen, J. J., and Wilderer, P. A. 1997. Aerobic granular sludge in a sequencing batch reactor. *Water Res* 31: 3191–3194.

Tay, J. H., Liu, Q.-S., and Liu, Y. 2001. Microscopic observation of aerobic granulation in sequential aerobic sludge blanket reactor. *J Appl Microbiol* 91: 168–175.

Tay, J. H., Liu, Q.-S., and Liu, Y. 2002. Characteristics of aerobic granules grown on glucose and acetate in sequential aerobic sludge blanket reactors. *Environ Technol* 23: 931–936.

Tay, J. H., Liu, Q.-S., and Liu, Y. 2004. The effect of upflow air velocity on the structure of aerobic granules cultivated in a sequencing batch reactor. *Water Sci Technol* 49: 35–40.

Tay, J. H., Liu, Q.-S., Liu, Y., Show, K. Y., Ivanov, V., and Tay, S. T. L. 2004. A comparative study of aerobic granulation in pilot- and laboratory-scale SBRs. Proceedings of Workshop on Aerobic Granular Sludge, September 27–28. Munich, Germany.

Tay, S. T. L., Moy, B. Y. P., Jiang, H. L., and Tay, J. H. 2005. Rapid cultivation of stable aerobic phenol-degrading granules using acetate-fed granules as microbial seed. *J Biotechnol* 115: 387–395.

Zheng, Y. M., Yu, H. Q., Liu, S. H., and Liu, X. Z. 2006. Formation and instability of aerobic granules under high organic loading conditions. *Chemosphere* 63: 1791–1800.

Zhu, J. R. and Wilderer, P. A. 2003. Effect of extended idle conditions on structure and activity of granular activated sludge. *Water Res* 37: 2013–2018.

Index

F

G